THE ROMAN IMPERIAL ARMY

Plate I
Tombstone of the centurion Facilis, at Colchester (p. 132).

THE ROMAN IMPERIAL ARMY

OF THE FIRST AND SECOND CENTURIES A.D.

GRAHAM WEBSTER

M.A., PH.D., F.S.A., A.M.A.

ADAM & CHARLES BLACK

LONDON

FIRST PUBLISHED 1969
REPRINTED WITH CORRECTIONS 1969 AND 1974
BY A. AND C. BLACK LIMITED
4, 5 AND 6 SOHO SQUARE LONDON W.1

ISBN 0 7136 0934 6

DEDICATION

To Professor Eric Birley
A modest recognition of his
life-long work on the Army

PRINTED AND BOUND
IN GREAT BRITAIN BY
REDWOOD BURN LIMITED
TROWBRIDGE AND ESHER

CONTENTS

ILLUSTRATIONS

PHOTOGRAPHS

ILLUSTRATIONS

DRAWINGS

ACKNOWLEDGEMENTS FOR ILLUSTRATIONS

The author and publishers wish to thank the following for the illustrations appearing in this book; Mr. Frank Bayley for his help with some of the photographic illustrations; Lady Richmond for permission to reproduce illustrations drawn by the late Sir Ian Richmond, Figs. 26, 34, 39, 41 and 42, and also for allowing the author to redraw Figs. 47 and 50; Mr. David T.-D. Clarke and the Colchester and Essex Museum for Pl. I; Professor Dr. H. von Petrikovits and the Rheinisches Landesmuseum, Bonn, for Pl. VI; The Römisch-Germanisches Zentralmuseum, Mainz, for Pl. VII; Mr. D. R. Petch and the Grosvenor Museum, Chester, for Pls. XI and XXVII; The Museum of Fine Arts, Boston, for Pl. IX(b) (Amos Cummings Bequest Fund, 59.336); Dr. David J. Smith and the Museum of Antiquities, Newcastle-upon-Tyne, for Pl. XVI; Professor Dr. A. Klasens and the Rijksmuseum van Oudheden, Leiden, for Pl. XVIII; The Yale University Press for Pl. XX; The Trustees of the British Mueum for Pl. XXII(a); C. H. Beck'sche of Munich for Pl. XXII(b); Mr. Herbert Clegg and the Cumberland and Westmorland Antiquarian and Archaeological Society for Figs. 26 and 39; Major C. S. Higgins and the Society of Antiquaries of Scotland for the loan of the block for Pl. XXV(a); Professor Dr. W. Glasbergen of the Instituut voor Prae-en Protohistorie, Amsterdam for Fig. 49 and Pl. XXVI; Mr. F. Austin Child, Mr. C. M. Daniels and the Society of Antiquaries of Newcastle-upon-Tyne for Fig. 32; Mr. F. A. Lepper and the Society for the Promotion of Roman Studies for the loan of the blocks for Figs. 34, 41 and 42; Professor Dr. H. Schönberger and the Römisch-Germanische Kommission for Fig. 38; Mr. G. C. Boon and the National Museum of Wales for Figs. 35, 36 and 37, and also for allowing the author to redraw Fig. 33; Mr. A. H. A. Hogg and H.M.S.O. for Fig. 48; Mr. P. K. Baillie Reynolds for Fig. 45; Dr. A. Baatz and the Saalburgmuseum for Pl. XXIX(b); Mr. H. Russell Robinson for drawing Figs. 15, 16 and 19; Dr. Edith Wightman for Pl. XIII; Mr. J. E. Hedley of Hexham for Pl. XIV(b); and also the Loeb Classical Library for permission to quote passages.

ABBREVIATIONS

A.A.	*Archaeologia Aeliana*
Ann. Épig.	*Année Épigraphique*
Ant. J.	*Antiquaries Journal*
Arch.	*Archaeologia*
Arch. Camb.	*Archaeologia Cambrensis*
Arch. Cant.	*Archaeologia Cantiana*
Arch. Ért.	*Archaeologiai Értesitö*
Arch. J.	*Archaeological Journal*
Bell. A.	Caesar, *de Bello Alexandrino*
Bell. Af.	Caesar, *de Bello Africo*
Bell. C.	Caesar, *de Bello Civili*
Bell. G.	Caesar, *de Bello Gallico*
Bell. H.	Caesar, *de Bello Hispaniensi*
Bell. Jug.	Sallust, *Bellum Jugurthinum*
B.J.	*Bonner Jahrbücher*
B.M.	British Museum
Bull.	*Bulletin*
C.A.H.	*Cambridge Ancient History*
Cheesman, *Auxilia*	G. L. Cheesman, *The Auxilia of the Roman Imperial Army*, 1914
Cichorius	C. Cichorius, *Die Reliefs der Traianssäule*, 1896 and 1900
C.I.L.	*Corpus Inscriptionum Latinarum*
Class. J.	*Classical Journal*
Class. P.	*Classical Philology*
Class. Q.	*Classical Quarterly*
Class. R.	*Classical Review*
Durden Coll.	J. W. Brailsford, *Antiquities from Hod Hill in the Durden Collection*, 1962
E.E.	*Ephemeris Epigraphica*
Ep.	Pliny the Younger, *Epistulae*
Frere	S.S. Frere, *Britannia*, 1967
Geog. J.	*Geographical Journal*
Geog. Rev.	*Geographic Review*
G. pro Vindonissa	*Gesellschaft pro Vindonissa*
Hist.	Tacitus, *Histories*

Hofheim	E. Ritterling, 'Das Frührömische Lager bei Hofheim im Taunus', *Annalen des vereins für Nassauische Altertumskunde und Geschichtsforschung*, 40, Wiesbaden 1913
I.L.S.	H. Dessau, *Inscriptiones Latinae Selectae*
J.	*Journal* or *Jahrbuch*
J. Brit. Arch. Ass.	*Journal of the British Archaeological Association*
J. Öest.	*Jahreshefte des Österreichischen Archaölogischen Instituts*
Josephus	Josephus, *War of the Jews*
J.R.G.Z.M.	*Jahrbuch des Römisch-Germanischen Zentralmuseum Mainz*
J.R.S.	*Journal of Roman Studies*
Lewis and Reinhold	N. Lewis and M. Reinhold, *Roman Civilisation*, Vol. 2, The Empire, New York 1955
M.Z.	*Mainzer Zeitschrift*
Nat. Hist.	Pliny the Elder, *Naturalis Historiae*
Newstead	J. Curle, *A frontier post and its people, the fort of Newstead in the Parish of Melrose*, Glasgow 1911
O.R.L.	*Der Obergermanisch-Raetische Limes des Römerreiches*
P.	*Proceedings*
Pap.	*Papers*
Parker	H. M. D. Parker, *The Roman Legions*, 1928
P. Brit. Acad.	*Proceedings of the British Academy*
P. Soc. Ant. Scot.	*Proceedings of the Society of Antiquaries of Scotland*
P.-W.	Pauly-Wissowa-Kroll, *Realencyclopädie der Klassischen Altertumswissenschaft*
Rang.	A. von Domaszewski, *Die Rangordnung des Römischen Heeres*, ed., by Brian Dobson, Köln 1967
R.C.H.M.	*The Royal Commission on Historical Monuments*
R.I.B.	R. G. Collingwood and R. P. Wright, *The Roman Inscriptions of Britain*, Vol. 1, 1965
R.I.C.	*Roman Imperial Coinage*
R.L.Ö.	*Der römische Limes in Österreich*
Saal.-J.	*Saalburg-Jahrbuch*
S.H.A.	*Scriptores Historiae Augustae*
Starr	Chester G. Starr, *The Roman Imperial Navy 31 B.C.–A.D.* 324, 2nd ed., 1960
Strat.	Frontinus, *Strategemata*
T.	*Transactions*
T. Cumb. and West.	*Transactions of the Cumberland and Westmorland Antiquarian and Archaeological Society*

Val. Max.	Valerius Maximus
Westd. Z.	*Westdeutsche Zeitschrift*
Wright and Richmond, 1955	R. P. Wright and I. A. Richmond, *The Roman Inscribed and Sculptured Stones in the Grosvenor Museum, Chester*, 1955
Epig. Stud.	*Epigraphische Studien*

FOREWORD

The booklet on the Roman Army produced in 1956 for the Grosvenor Museum, Chester, has been well received and this has encouraged me to attempt a more extensive version. This has been a difficult task since it has brought me into areas of knowledge with which I have no familiarity. This present work may be considered as little more than a compilation, but even as such, it should be of use to students in classics and archaeology venturing into this specialist field. I hope also that the general reader will find much to interest and stimulate him. I have been very fortunate in being able to lean heavily for assistance and advice on many colleagues and I am very pleased to be able to record my gratitude for their help so freely and willingly given, help which has undoubtedly saved me many times from grievous error. The late Sir Ian Richmond was always extremely kind and encouraging with my early drafts and Professor D. R. Dudley and Mr. B. R. Hartley have also read the proofs and made many helpful suggestions. Professor Eric Birley, to whom we owe so much on Roman military studies, has been exceptionally helpful and kindly read the proofs. Dr. John Mann, Dr. John Wilkes and Mr. R. W. Davies have kindly read and commented on chapters. Professor Hans Schönberger has been very helpful with the selective bibliography, and Dr. B. Dobson, Dr. A. R. Birley, Mr. C. M. Daniels, Mr. G. Boon and Mr. G. R. Watson are among those who have assisted with helpful discussions on specific points. I am also grateful to Mr. Brian Hobley for drawing the frontier maps, to Miss Valerie Singer for reading the manuscript and proofs and to my secretary Mrs. Muriel Stanley for her constant help, and in particular for compiling the index.

A third printing has enabled the author to make some corrections and to insert where possible more recent references, but publications since 1969 have made some sections of the book in need of more serious revision which must wait a new edition.

Chapter I

INTRODUCTORY CHAPTER

Rome began in the eighth century B.C. as a group of small Iron Age peasant communities of differing origins[1] and some of their huts and burials have been excavated on the Palatine Hill and in other parts of the city where it has been possible to uncover remains deeply buried below later structures. The importance of Rome lay in its geographic setting at a crossing of several important land routes, the bridgehead across the Tiber and the connection with the sea. But it is doubtful if the site would have developed had it not been taken over in the sixth century by the Etruscans, who initiated large-scale drainage schemes and public works which converted a group of villages into an urban community. Although the Roman historians looked back to the expulsion of the kings in 509 B.C., the traditional date, as the great turning-point in their history, Rome may have remained under Etruscan influence until at least 475 B.C.[2] The Etruscans had much wealth and prosperity and together with many other cities, Rome suffered a cultural recession as the brilliant civilization dimmed. But although the people of Rome had been pawns in an Etruscan power complex, ironically this had implanted the ideas from which grew Rome's constitution shaped by her peculiar practical genius. In order to strengthen their position in the declining years, the Etruscan monarchs had favoured the plebs against the landed aristocracy and instituted the joint magistracies. Some of the most revered political and religious institutions saw their origins in this period at the beginning of the fifth century B.C. The Romans had absorbed much from the Etruscans and not least in military matters,[3] a debt they were later reluctant to acknowledge. Only in

---

[1] For a valuable summary of this evidence see Raymond Bloch, *The Origins of Rome*, English Edition, 1960.

[2] A. Momigliano, *J.R.S.*, 56 (1966), p. 21.

[3] McCartney, 'Military indebtedness of early Rome to Etruria', *Memoirs of the American Academy in Rome*, 1915, p. 121.

matters concerning divination did they openly accept Etruscan superiority and many times called upon their priests to expound difficult omens. The Etruscans borrowed much from the Greeks, with whom they maintained close trading contacts. Our knowledge of Etruscan civilization is derived from the remarkable series of paintings and objects found in their tombs. Portrayals of warriors in paintings and bronze statuettes with finds of actual helmets and weapons give an indication of their appearance. It is important to realize that the wealth of the Etruscans was due mainly to their skill as metallurgists and workers in iron and bronze. Thus their military equipment would have been of excellent quality.

The stout bronze helmets are normally close-fitting and offer some protection to the neck. Whether the elaborate crests were worn in battle or adorned only the effigies of their war-gods is not clear.[1] Body armour consisted of an elaborate system of rectangular bronze plates held together, presumably, by thongs.[2] The main effect of this cuirass was, like that of Japanese armour, to deflect the sword-blade; against the point of a heavy spear thrust towards the body it could not have been so useful, but shields were carried for such contingencies. This type of cuirass survived, but in other patterns, into much later centuries. The shoulder-pieces appear in precisely the same shape in the equipment of the early imperial centurion (Pl. I) and the bronze plates are still seen on the officers' 'kilts'.

Representations of Etruscan warriors and their equipment often appear as frescoes in tombs. They clearly show that the foot-soldiers were essentially spearmen and carried two spears with large foliate heads, and small round

---

[1] A fine example of an early bronze helmet of the eighth century B.C. used as a lid to a cinerary urn was found at Tarquinia (M. Pallottino, *The Etruscans*, Pelican edition, 1955, Pl. 3.A; *Notizie degli Scavi*, 1907, p. 63). A sturdy round helmet like a bowler hat without a brim was found in 1817 and is now in the British Museum. It has a dedicatory inscription indicating that it was offered as a gift to Zeus at Olympia in 474 B.C., by Hieron of Syracuse; this booty had been collected after the great naval defeat of the Etruscans off Cumae. The small bronze figurines and frescoes of gods and heroes often show very large crests attached to the helmet (see also the Aules Feluskes stele, M. Pallotino, *Etruscologia*, 1963, Pl. 21).

[2] One of the best examples is seen on the fine bronze figure of Mars found at Todi and now in the Vatican Museum (R. Bloch, *L'Art Étrusque*, 1956, Pl. 40). In the Villa Giulia there is a remarkable handle to the lid of a bronze cist consisting of three soldiers, two carrying a wounded or dead comrade, showing precisely the same details, although less well defined (*ibid.*, Pl. 77). A fine example from the wall paintings is in the Tomb of Orcus, where the demon Geryón is shown in this type of body armour (A. Stenico, *Roman and Etruscan Painting*, 1963, Pl. 51).

shields.[1] The sword appears to be a subsidiary weapon and where it is illustrated is short and leaf-shaped. As McCartney[2] has indicated, it was the loss of the spear which was the greatest indignity a Roman soldier of the early period could suffer and, after the disaster at the Caudine Forks, the Roman had to pass under a yoke of spears, not swords.[3] As one might suppose, the Etruscans were quite capable of forging fine sword-blades, so their dependence on the spear would seem to be the result of long-established custom. The amount of horse gear and representations of men and horses with foot-soldiers may merely indicate that those who could afford it provided themselves with this means of transport, but fought on foot with their comrades.

There is no reason to suppose that the Roman warriors at this period differed much in appearance from their opponents and they probably used similar tactics in battle. At this time the massive Greek phalanx was the mode and wars were won on the clash of these tight-packed masses of armed warriors. The Etruscans were probably more at home on the sea than on the land, but their naval power was weakened by an alliance with Carthage against the Greeks,[4] which resulted in their loss of independence.

The Etruscan method of recruitment was, according to Livy (ix, xxxix, 5), by a law (*lex sacrata*) which required each man to enrol a comrade. In Rome, however, a more precise arrangement had been established based on citizenship and voting rights. Roman authorities attributed this to Servius Tullius in the sixth century B.C.[5] The assembly of the Roman people, the *comitia centuriata* seems to have had a military basis and the summons to the then meeting-place, the Field of Mars, was by a trumpet blast and the hoisting of

---

[1] As on the silver bowl from the Regolini-Galassi tomb, D. Randall-MacIver, *Villanovans and early Etruscans*, 1924, Pl. 38, 1. The soldiers are here seen carrying two spears, as much later legionaries had two *pila*.

[2] *Loc. cit. supra*, p. 123; this point has also been made by Clark Hopkins in *Class. J.*, 60 (1965), pp. 214–19.

[3] Livy, ix, 6; see also for the significance of the spear, A. Alföldi, '*Hasta summa imperii*', *American J. Arch.*, 63 (1959), pp. 1–27.

[4] Herodotus (i, 166, 2) describes one of these battles against the Phocaeans in which the Etruscans and Carthaginians combined.

[5] Servius Tullius, whose Etruscan name was Mastarna, appears on the frescoes on the François tomb at Vulci. He was a warrior from Vulci who with the help of the Vibenna brothers succeeded Tarquin the Elder by force. His reforms could be interpreted as efforts to win the support of the rising merchant class of Rome against the Etruscans. He built the first defences of Rome but fell to Tarquinius Superbus, who had enlisted the help of the noble families.

red flags.[1] Tullius instituted the census and graded citizens into five classes according to their wealth.[2] Of the most wealthy, the first class, he formed eighty centuries, forty seniors and forty juniors, the former to guard the city, the latter to form an expeditionary force for war elsewhere. These men were equipped like the Greek hoplite with helmet, round shield, greaves and breastplate, all of bronze, and carried a spear and sword. They had the services of two centuries of artisans from the proletariat, not to bear arms, but to make and maintain siege engines, and, presumably, similar equipment. The second class were enrolled in twenty centuries and they carried an oblong shield and similar gear to the first class, with the exception of a breastplate. In the third class, also in twenty centuries, the equipment was the same except for the omission of greaves, and the fourth class had no armour at all, but carried spear and javelin. The fifth class had thirty centuries and carried only slings with stone missiles; in the same grade were the horn-blowers (*cornicines*) and trumpeters (*tubicines*) forming two centuries. Of the rest of the population, those owning less than about a tenth of the property required as the minimum qualification of the first class were exempt from military service and formed only a single century in the *comitia*. This arrangement allowed the wealthy citizens to control the voting in the assembly, since, with the second class, they had more centuries than all the others; the number of men in each century must have varied considerably, the largest being the proletariat. The advantage of the system was in placing the whole of the propertied citizens on a war footing and it was evidently felt that the only reliable soldiers were those who had something for which to fight. The Army officers were drawn from the leading citizens, who were enrolled as knights in twelve centuries, but their horses were provided by the State with the help of a tax on spinsters.

There was, however, little warfare on a massive scale in these early days, mainly because the cities of the Etruscan League refused to combine against a common enemy and this failure led to their ultimate downfall. Mostly the clashes were due to small raiding parties and private feuds. One of the first serious campaigns Rome undertook was against her rival and neighbour, Veii, at the end of the fifth century B.C. This struggle had economic causes: the need for Rome to expand her growing population and the need for Veii to maintain her trade route down the Tiber. But the Etruscan city, like many others, was built on a hill-top in a good defensible position and could not be

[1] This was the signal for the commencement of hostilities, but this practice may have stemmed from its use with the *comitia*.

[2] Livy, i, 42–43.

taken by storm. The account given by Livy is one which has been reshaped by tradition, myth mingling with reality. A siege lasting several years would include winter conditions and Rome was obliged to devise ways of maintaining her troops in the field for long periods, introducing pay[1] and protection against the weather. It is worthy of note that the city was only captured by prodigious feats of engineering. In siege-works the Roman Army was to develop in logical and practical ways the answers to most situations and to reach a stage which they regarded as perfection.[2]

About 390 B.C. the Romans received their greatest humiliation. Tribes of Celts had crossed the Alps from the Danube and swept down Italy, tossing aside a Roman army sent out to protect the city, which was ravaged and burnt to the ground. Only the Capitol held out, but its defenders capitulated after a seven-month siege.[3] The Celts departed with their booty as suddenly and swiftly as they had appeared. If Rome was to re-establish her authority over the peoples of Central Italy, and be prepared to meet any similar disasters of the future, some drastic reorganization was needed. Although this was traditionally held to have been the work of the great hero Furius Camillus, it seems more likely that the reforms were introduced over a longer period during the second half of the fourth century.

Undoubtedly the most important change was the abandonment of the tactic of the Greek phalanx. This solid mass of heavily armoured hoplites with their large spears had in its day been superb and invincible as Xenophon so ably demonstrates; but against lightly armed, fast-moving barbarians it was like a tortoise in a rat-race. In changing the mode of fighting so drastically the Romans showed their genius for adaptability. Livy describes in detail the new organization of the legion,[4] but there are difficulties in reaching a full understanding. There were now three lines of soldiers, the *hastati* in the front, and behind them the *principes* and the *triarii* in the rear. The first two ranks comprised the main fighting force,[5] each having fifteen *manipuli.*

---

[1] Cavalry received three times the pay of the infantry. The introduction of pay was of importance in the evolution of the citizen-militia.

[2] *Strat.*, iii, introduction.

[3] The evidence for this is, however, confusing and the Capitol may have been taken, O. Skutsch, 'The Fall of the Capitol', *J.R.S.*, 43 (1953), pp. 77–78.

[4] viii, 8.

[5] They were known at this stage as the *antepilani*, which, as Parker has indicated (p. 13, fn. 1), has nothing to do with the *pilum*, but is derived from *pilus*—a closed rank; it is equivalent to the later term *antesignani.*

It was this subdivision into smaller tactical units which allowed greater freedom of movement and flexibility over the solid phalanx. The first line, the *hastati*, contained the flower of youth and had two kinds of troops; there were in each maniple twenty lightly armed men each carrying a spear and javelins.[1] The rest of the unit carried body armour and a rectangular shield, the *scutum*,[2] which at this time replaced the smaller round *clipeus* and became the distinctive equipment of the legionary. In the second line, the *principes*, were the picked men of experience and maturity. They were the best equipped with *scutum* and armour. Behind these two foremost ranks and the standards came the rear rank divided into fifteen companies. It may be significant that Livy does not use the term maniple but *ordo* to describe this division. Each company consisted of three sections (*tres partes*), also referred to as *vexilla*,[3] each composed of sixty men, two centurions, and a *vexillarius* or standard-bearer. The total number in the *ordo* is then given by Livy as 186.[4] The three *vexilla* in each *ordo* contained different types of men, the first the *triarii*, who were veterans, the second the *rorarii*, younger men, and the third the *accensi*, the least dependable and thus relegated to the rearmost position. When seen in diagrammatic form (Fig. 1) the legion is not as clumsy an organization as might appear from the description. The tactics Livy describes as follows: First the *hastati* would engage the enemy and if they were not successful they retired between the maniples of the second line, leaving the battle to the best troops, while the front rank of the third line, the *triarii*, knelt down with their spears thrust forward making a bristling hedge, but not visible to the enemy in front. If the second line failed or was thrust back, the third rank closed in and attacked. As Livy indicates, the effect of these new troops suddenly rising and moving into battle would have been a shock for the enemy. Finally Livy tells us there were 300 cavalry to each legion, which totalled 5,000 infantry. One piece of information which Livy omits is the

---

[1] The words used by Livy are *hasta* and *gaesum*, this latter being a heavy type of javelin, thought to have been derived from the Gauls (Servius *ad Aen.*, vii, 664).

[2] This change is confirmed by Plutarch and Dionysius of Halicarnassus (see below p. 24), but is not reflected on the monuments. Couissin (pp. 240–7) suggests that a large oval shield similar to that of the Gauls continued in use until the time of Caesar. There is much dispute over the origin of the *scutum*. Diodorus thought it was a Roman invention (xxiii, 2), others that it came from the Samnites, but Livy describes their distinctive shield as kite-shaped (ix, 40).

[3] A term also used for the cloth standard or flag which is considered below, p. 139.

[4] This would be correct if the *vexillarius* was included in the main body of sixty, which seems reasonable, as he would not be regarded as an officer of special rank.

size of the maniple. The *ordo* of the third rank which comprised three *vexilla* was 186 strong and fifteen of these would amount to 2,790 men, more than half the strength of the legion. This means that the strength of the maniples of the first and second rank must have been sixty to seventy each, i.e. the same as a *vexillum* of the third rank.[1]

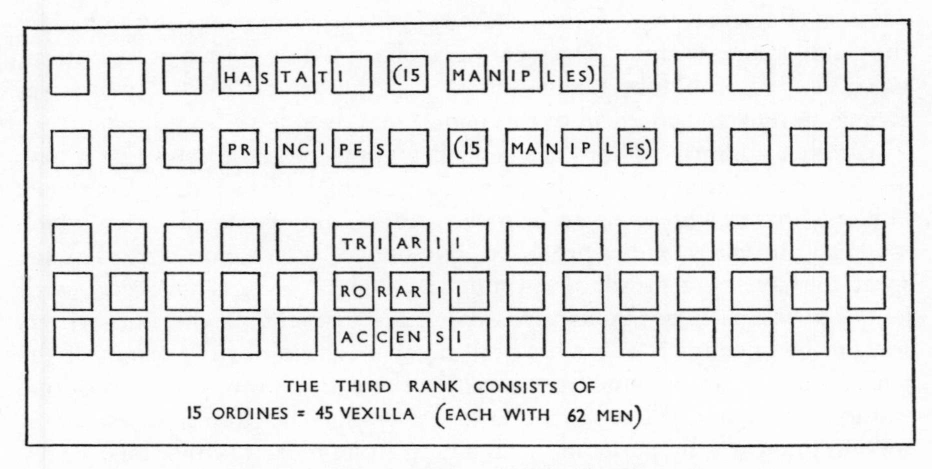

FIG. 1 *The Legion described by Livy*

Plutarch[2] adds to Livy's account of the innovations of Camillus some details which have an authentic ring; one can imagine this experienced and wily soldier weighing up the barbarians with their effective use of their long heavy swords, wielded perhaps with more vigour than science. The bronze helmets of the Romans proved to be an inadequate protection from these swords, so he ordered the issue of helmets made of iron with a polished surface to cause the swords to be deflected. The pattern of the shield was also changed; as Livy indicates, the smaller round shield was replaced by the *scutum*, giving much better body protection.[3] Plutarch adds the interesting

---

[1] In an attempt to make better sense of Livy's figures, Conway (*Class. Q.*, 12 (1918), pp. 9–14) assumed that the Roman historian had been in error and that the number of maniples and *ordines* in each rank had been ten, the figure given by Polybius for the legion of the third century; if, however, at this stage the sizes of the maniple and *vexillum* were the same, but three *vexilla* = one *ordo*, Livy cannot be faulted.

[2] *Camillus*, 40.

[3] Parker, p. 11.

detail that the edges were now made of bronze and presumably this was a binding which covered the complete edge instead of a mere border. He also credits Camillus with a new javelin tactic, that of thrusting the javelin under the enemy sword. But the purpose of this, to take the impact of the sword blow, makes little sense, since the wooden shaft of the weapon would be easily sheared. What seems more likely is that the Romans were attempting to come to close quarters with their short swords and were being outreached by the Gauls' longer weapon. The spear or javelin could be used to greater effect, especially if it could be thrust into the enemy armpit as the sword was swung. A new weapon drill of this nature may well have been needed.

Another authority, Dionysius of Halicarnassus (xiv, 9), states that a new weapon was introduced, 'Better arms than the barbarians possess have been fashioned for us—breastplates, helmets, greaves, mighty shields, with which we keep our entire bodies protected, two-edged swords, and instead of the spear the javelin, a missile that cannot be dodged . . .' The Greek word is *hyssos*[1] which is not conclusive, and the statements of the two writers seem to be in conflict. There has been much discussion of the origin of the *pilum*, some saying it came from Iberia,[2] others that it was a weapon of one of the Roman allies;[3] but *pilum* also means pestle. The primitive method of milling grain was by pounding it in a deep stone mortar with a long heavy wooden pestle which could be hollowed in the centre to facilitate handling. Such an implement sharpened at the end and its point hardened by fire would be an admirable spear. Most of the early Roman weapons had to be made of wood, as the Etruscans at first denied the Romans iron, except for purely agricultural purposes. Only when this metal became available in quantity could it be used to tip spears and arrows. The *pilum* is thus likely to have undergone considerable changes over a long period of time and if this is indeed the weapon attributed to Camillus, he may have taught the Roman legionaries to use it as a thrusting spear to outrange the long Gallic sword. To achieve the best effect it would have needed a small iron point. Another

---

[1] E. Egger, 'Une note sur le mot 'ΥΣΣΟΣ' (*Mémoires de la Société des Antiquaires de France*, ix, 1866, p. 288; the whole question is fully discussed by P. Couissin, *Les Armes Romaines*, 1926, pp. 20–23; 129–38, 181–5, 279–94.

[2] von Groller, *Der Ursprung des Pilums*, 1911 (Rheinland Museum); *Numantia*, iii (1927), 249; it is here associated with the Spanish *phalarica*.

[3] Mayer inferred from Diodorus (xxiii, 3) that it came from the Samnites; Pliny suggests an Etruscan origin (*Nat. Hist.*, vii, 56, 201)

(a) a sacrifice (*suovetaurilia*), (p. 267)

Plate II
Scenes from Trajan's Column

(b) legionaries working without their body armour and using *dolabrae* (p. 123).

Plate III
Cutting and handling turves, a scene on Trajan's Column (p. 177).

possibility, preferred by Couissin,[1] is that the *pilum* originated from the *verutum*, a long thin thrusting spear used by the *velites*.[2]

There is little doubt that the Romans always showed a readiness to adopt and to borrow ideas from other peoples, a practice they were not always ready to admit.[3]

There are two important descriptions of the Roman Army of the early Republic by the two great historians, Livy and Polybius. Livy was writing of the Army as it was constituted in the second half of the fourth century B.C. and Polybius chose the period of Cannae (216 B.C.). Before considering the Polybian account, a few general comments on the background history in these centuries is advisable, since an army cannot be judged merely on its organization or equipment. The historical aspects as seen in the tasks which confronted Rome, her resources of man-power, war materials and finances are of paramount importance.

The fourth century B.C. was devoted to the establishment of control over Central Italy. This was not merely a question of gaining political ascendancy. The extension of Roman citizenship and Latin rights with the planting of colonies assisted in an integration of peoples. This led to political solidarity and above all extensions of the recruiting areas for the Army. As the franchise was granted to more and more peoples so the Army grew in strength and the wisdom of linking this responsibility with rights of citizenship became apparent.[4] An aspect overlooked by some historians is the growing control the Romans must have exercised over the rich mineral deposits of the hills of Central Italy. It was this wealth which had been responsible for the rise of Etruria and which became an inheritance of the growing power of Rome and supplied the munitions of war.[5]

---

[1] *Les Armes Romaines*, pp. 129–38, 181–5.

[2] Livy, I, 43, 6; as its name implies, it was developed from the long roasting spit and may have had a long iron shank at the outset; see also Vegetius, ii, 15.

[3] Sallust puts into the mouth of Caesar a saying that the Ancient Fathers took both offensive and defensive weapons from the Samnites (*Catiline*, li, 38), whose armour made a particular impression on the Romans (Livy, ix, 40); see also Schuchhardt, *Sitzungsbericht der Preussichen Akademie der Wissenschaften, Phil. Hist. Klasse*, 23 (1931), pp. 608–34.

[4] The responsibility for raising and maintaining troops was an important part of the treaty arrangements made with the allies, who were yet to earn the full privileges of citizenship or even Latin rights, but these troops were drafted into the legions which remained the prerogative of full citizenship.

[5] Some of the mines were, however, becoming worked out by this period, but this is a matter on which there is little information. In 205 B.C. Arretium supplied Scipio with 3,000

In the early third century Rome was faced with a new threat and for the first time faced a professional Hellenic army which was led by Pyrrhus, who came from Greece to embrace the cause of Tarentum against Rome. Pyrrhus, an able soldier, was an adventurer seeking easy conquests and soon discovered that the solidarity of Rome and her allies could not easily be broken nor her tough armies put to flight. It was a severe test for the Roman troops, faced for the first time with a well-trained Macedonian phalanx and elephants, the smell of which put the horses into a state of panic. Pyrrhus withdrew to Sicily, but this war, however indecisive its battles may have been, left Rome undisputed mistress of Italy and sealed the fate of the Greek colonies along the coast.[1] Roman engineering skill, inherited from the Etruscans, laid down the main lines of communication across Italy needed for the rapid deployment of troops in addition to the aqueducts and drains which made towns, old and new, healthier places.

In the third century Rome became involved for the first time with a great overseas power, Carthage. The first Punic War, a long-drawn out rather indecisive business, was mainly confined to Sicily and its coastline. Its most interesting aspect was Rome's emergence as a naval power. The use of ships was not entirely new, but the effort needed to face the greatest and most experienced navy of its day was prodigious. That Rome was successful was due to a new conception of the warship—as a fighting platform for her soldiers. The invention of a grapple which enabled the troops to swarm aboard gave Rome five victories at sea[2] and inflicted a severe and humiliating blow to the Carthaginians. Much of this achievement was unfortunately minimized by the inept way the Romans handled their vessels and there were severe losses due to tempests. The net result for Rome was the annexation of Sicily, Rome's first overseas possession and later, taking advantage of Carthage's weakness,

---

shields, 3,000 helmets, 50,000 Gallic spears and as many javelins and axes, buckets, bills, mattocks, etc., sufficient for forty ships (Livy, 28, 45).

[1] H. H. Scullard has pointed out that at this time Ptolemy II of Egypt began to establish diplomatic relations with Rome, clearly indicating that Rome had now become a power in the Mediterranean worthy of consideration (*A History of the Roman World, 753–146 B.C.*, 1961, p. 124).

[2] The boarding-bridge, known as the *corvus* (crow), because of its beak-like appearance, probably made the ships top-heavy and difficult to handle in heavy seas. This might explain why its use was not continued after the middle of the century (J. H. Thiel, *Studies on the History of Roman Sea-Power*, Amsterdam, 1954).

Sardinia and Corsica were also taken over, since Rome realized that islands so near to Italy could no longer be left in potentially unfriendly hands.

Towards the end of the century some of the Gauls of North Italy moved south, no doubt anticipating easy booty; but this time Rome could draw on new resources of man-power and the Gauls were trapped between two Roman armies and destroyed. To prevent further trouble from this quarter the Romans decided to take over the tribes north of the Po by treaty, conquest and the placing of colonies, so that their authority should extend as far as the Alps. The suppression of piracy in the Adriatic also led to successful campaigns along the coast of Illyria. It was at this time that the young Hannibal, extending Carthaginian power in Spain, took Saguntum, a city with an alliance with Rome, and this act led directly to the second Punic War, a fearsome struggle which almost destroyed Rome and her allies.

Hannibal became the Carthaginian leader by chance while still in his twenties. He soon showed that he was a general of outstanding ability. Until the rise of the younger Scipio Rome had no military leaders to match those of their enemy, although the elder Scipio, had he stayed in Italy, might have risen to the occasion. The details and broad strategy of the war are of significance to the military student. Unlike the earlier periods, we have reasonably full and accurate accounts of the battles, but it is more important to see how the Romans reached towards the concept of total war. While Hannibal was ravaging the lands of Italy, forces had to be spared for Spain, Sicily and Sardinia to prevent supplies and reinforcements from reaching him.[1] It was the success of this strategy and the brilliant victories in Spain which prevented Hannibal from completing his work in Italy and eventually sealed his fate.

The Army Described by Polybius

It is for this time also that we have a detailed description of the Army by Polybius.[2] This stands as the most complete and accurate account, surpassing

---

[1] But these ideas were held only by a far-seeing group led by the Scipios; the opposing faction, epitomized by Fabius Maximus, were willing to come to terms with Carthage, providing her forces were removed from Italy. There was much opposition to the plan for young Scipio to carry the war into Africa.

[2] Book vi. One of the best interpretations of the Polybian account is by General W. Roy (*The Military Antiquities of the Romans in North Britain*, 1793); for further comments see E. Fabricius, 'Some notes on Polybius's description of Roman Camps', *J.R.S.*, 22 (1932), pp. 78–87.

all others. He first describes how the four legions were annually selected from the citizens, who were obliged to serve for twenty years (ten in the cavalry) before they were forty-six years old and if their property exceeded four hundred *asses*; poorer citizens could, however, be drafted into the Navy. The legion consisted of the same three lines of battle described by Livy, the *hastati, principes* and *triarii*, but they were screened by the lightly armed *velites*, drawn from the younger and poorer citizens, and who carried swords, javelins and a circular shield three feet in diameter, and wore a helmet without a crest. The javelin was a short one, only about four feet long with a head nine inches long, well hammered, but so fashioned that it bent on impact and could not be returned by the enemy. Polybius gives us here the first description of what had by now become the characteristic throwing weapon of the legion, the *pilum*. It was basically a missile with a small hardened point, a long shaft, and was hafted into a wooden stock. Its advantages lay in its lightness, cheapness and the ease with which it could be made in comparison with other types of javelins with more elaborate heads. Polybius informs us that there were several types of *pila*, classified by their thickness,[1] and this seems to be borne out by the discovery of actual weapons by Schulten in Spain on the site of camps of P. Cornelius Scipio Aemilianus (*c.* 133 B.C.).[2] The finer sort had a wooden shaft four feet six inches long, while the iron head was barbed and the same length as the wood. Each *hastatus* had a plume of purple and black feathers standing upright eighteen inches to increase his height. They were also issued with a small

---

[1] This distinction may have marked the different kinds of legionaries; once the arms had become standardized throughout, the *pilum* would reach its final and universal pattern.

[2] *Numantia,* 1914. The sections dealing with the weapons and equipment were written by von Groller. There are two main types, one with the end of the iron shank hollowed out into a socket to house the wooden shaft and the other with the end flattened and having two holes for fitting into a split wooden shaft and being riveted to it. The longest example of the former is 95.5 cm. (iv, Taf. 21, No. 1, and 25a, No. 20), while the longest of the latter is 74 cm. (iv, Taf. 25, No. 8, and Taf. 25a, No. 11). There are, however, much shorter types of the socketed shank, a complete example being 77 cm. overall (iv, Taf. 25a, No. 3, and iii, Taf. 34, No. 6). A reconstructed comparison of these *pila* with those described by Polybius has been attempted by General Schramm (iv, Taf. 25c). None of these finds can be proved to have been associated with the campaigns of Scipio, and since much of this ground was fought over in the succeeding century, some of these weapons may be of later date. It is interesting to note that two examples published from Osma and Gormax (iii, Taf. 53a and Taf. 54a) show the bending of the shank typical of Caesar's day.

bronze breastplate (about nine inches square). The more wealthy had mail coats, presumably of chain or small plates. The other two ranks were equipped in the same manner except that they carried long spears (*hastae*) instead of *pila*. The *triarii* had only half as many men (600) as the front two ranks had and there is no mention of the complexities of the third rank given by Livy, so presumably the *rorarii* and *accensi* had all become *velites*. The *velites* did not form their own battle line, but were divided equally among all the maniples of the legion. Whether this was a tactical or administrative arrangement is not clear, but as Polybius designated the *hastati* as the second line one must presume that the *velites* normally operated in the front, drawing the enemy's fire and stinging them with their javelins before retiring through the main legionary ranks.

The *hastati* were arranged in ten maniples each of 120 men and were equipped with the *scutum* (two and a half feet by four feet, larger ones being available) made by fastening together two layers of wood by bull's-hide glue, covering it with calf's skin and binding the edges with iron. At the centre was fixed the iron boss (*umbo*), containing the handgrip on the inside. Their sword was the *gladius*, known as the Spanish sword and presumably adopted from that country at about this time. The blade was two inches wide and two feet long and was admirably suitable as a cut-and-thrust weapon for close work. The helmet was once more of bronze and the iron casque introduced by Camillus to withstand the savage blows from the Gallic swords had been replaced; perhaps the new helmets were of thicker metal. The *hastati* wore greaves (*ocreae*) and carried two *pila*.

On the subdivisions of the three ranks Polybius tells us that each was divided into ten companies called *ordines*, maniples or *vexilla*, each having two centurions, two *optiones* and two standard-bearers (*vexillarii*). This division of the maniple of 160 men produced centuries of eighty men each, as the tenting arrangements and later barrack blocks clearly show. It would seem from Polybius that the maniple was the main tactical and administrative unit. He further states, that when two centurions were both in the field, the first elected commanded the right of the maniple, the second the left. We are, however, left with the problem of the *triarii* with only half the number of men, but with ten maniples and twenty centurions. One can only presume, in spite of Polybius's statements, that the third line had the *velites* assigned to it, but perhaps only for administrative purposes. The cavalry force of 300 men was divided into ten squadrons (*turmae*), each with three *decuriones*. On the equipment Polybius is quite explicit in stating that it was all borrowed from the Greeks: in this way, he says, the lance, shield and body armour were much improved.

Polybius then describes in detail the arrangements of the military camp. The systematic way the Roman Army set out and constructed its camps set it apart from all other forces of the ancient world. Nothing so clearly illustrates the peculiar Roman genius for logical, orderly and practical arrangements. Other armies sought out places with natural defences or were billeted in towns. All the Romans needed was a reasonably level piece of ground, not too easily overlooked by near-by heights. Here the main picket and tent lines would be set up and defences constructed. The first tent to go up was the commander's, the *praetorium*, a system of tented structures about 200 feet square. Then the tribunes' tents were pitched in a single line parallel to the front of the *praetorium*, fifty feet from it and facing it. The men's tents were arranged in maniples in carefully measured lines laid out at right-angles to those of the tribunes and behind them. He also describes the roads between the blocks. In each maniple the end tents were occupied by the centurions. On one side of the *praetorium* was the market and on the other the office of the *quaestor* (quartermaster). The rear of the camp, i.e. the area behind the *praetorium*, was given to the cavalry, any *auxilia* and foreigners. We thus have the tripartite division of the camp which is so familiar to all students of Roman military establishments. In the centre along the *via principalis* were the main tents, the front of the camp (*praetentura*) was occupied by legionaries and the rear (*retentura*) by the units of lower status. The whole camp was, according to Polybius, in the form of a square, although in practice it was more like a rectangle. Between the tents and the defences was an empty space 200 feet wide which was needed for the rapid deployment of troops in an emergency, for depositing spoils of victory, including prisoners and cattle, and which also placed the tents beyond the range of missiles.

Arrangements for the guards are then given, each tribune having four drawn from the first two ranks, the *triarii* supplying guards for the cavalry. All the sections of the legions contributed sentries for the *praetorium*. Elaborate arrangements were made for the distribution of the watchword, which was written on wooden tablets (*tesserae*), presumably with waxed surfaces, and these were passed through the maniples and countersigned and returned to the tribunes. Later this must have become the responsibility of one man in each century who became known as the *tesserarius*. The task of visiting rounds was given to the legionary cavalry. Four men were selected for each watch, the beginning of which was signified by a bugle-call from the main guard station near the first maniple of the *triarii*. When their turn came the visiting party set out to inspect all the pickets and guards and if all was in order received the *tessera* from the man in charge (this may have contained the watchword, but presumably belonged to a different set from that which was used to distribute the watchword). If the sentries were asleep or absent, the four visitors noted the fact and proceeded to the next station. At daybreak the party reported to the tribune on duty and handed in the *tesserae*, and if there were any missing an immediate investigation was undertaken: the centurion in charge was ordered to parade the picket and sentries, and the tribune closely questioned each party. If serious fault was found, a court martial was then held and should a sentry have been found guilty he was condemned to the *fustuarium*. The unfortunate man was set upon by the soldiers of his own unit with cudgels and stones and usually killed. If he happened to survive, he was ignominiously discharged from the Army and was not likely to have been received kindly in civil life. This is an excellent example of Roman justice, which was scrupulously discharged, and the severity of punishment was logical, since a sentry who sleeps or absents himself without cause endangers the lives of all his colleagues. It ensured that all watches were well kept. The same punishment was meted out for theft in the camp, giving false testimony, extreme cases of immorality, and to any soldier found guilty of the same minor offence for the fourth time. It was also considered a serious offence to lose weapons on the battlefield or to retire from a given position through fear. If a whole unit was found guilty of these acts of cowardice, the men were paraded in front of the legion and every tenth man selected by lot; these were then subject to the *fustuarium*; the rest were put on a ration of barley instead of wheat and had to set up their tents outside the camp. Thus they were made conspicuous and could not have escaped the contempt of their fellow soldiers, though they may have avoided execution.[1] This mark of disgrace often extended over a long period. The survivors of the unfortunate legions which suffered defeat at Cannae were kept together and sent to Sicily. There in 205, eleven years later, they were given to Scipio for his African campaign, a deliberate slight to him by the conservative group led by Fabius and part of their efforts to frustrate Scipio's ambitions. Scipio saw this as an opportunity for these men to redeem themselves, which they did in helping him towards his brilliant victories, culminating at Zama.

Polybius then devotes a section to rewards for those who distinguished

[1] This happened to the forces of Paccius Orfitus in the reign of Nero (*Annals*, xiii, 36).

themselves. In battles and sieges every soldier was expected to fight well, but in skirmishes and other minor engagements daring actions were performed quite voluntarily. These men received cups and spears and cavalrymen horse-trappings (probably a specially decorated set). He then gives the daily pay of the soldiers at this period: five and a third *asses* for each foot-soldier, ten and two-thirds for a centurion and sixteen for a trooper,[1] but out of this there were pay stoppages for food, clothing and equipment. Rations for the legionary were a bushel of wheat per month, for a trooper ten and a half bushels of barley and three of wheat. Auxiliaries at this period received free rations of the same amounts for infantry, and seven and a half bushels of barley and two bushels of wheat for cavalry.

When the Army struck camp, and prepared to move, everything was done in an orderly sequence. At the first bugle-call the tents were struck and baggage collected, at the second the baggage was loaded on the animals and at the third the column started to march. The order of the march was determined by the potential hazards of the terrain and danger from the enemy. Normally the allies headed the column with the legions following, each with its own baggage train at its rear, and the cavalry rode on the flanks. When an attack was imminent the march was in open order, the legions in their three ranks with the baggage between them. The legions, could on the alarm, turn right or left, ranks moving in their correct order into their battle positions, leaving the baggage in the rear. When the Army intended to halt for the night one of the tribunes took a small advance party to choose a site and marked out the camp with coloured flags along the lines already given and which formed the textbook pattern. As the plan was always the same, every man knew exactly where to find a given tent once he had his bearings. This basic concept of the marching camp, described by Polybius and which had probably evolved long before his day, was maintained into Imperial times and those investigated in Britain (p. 170) vary only in detail from those of Scipio Aemilianus in Spain.

The detailed accounts of the battles of the second Punic War clearly demonstrate the great superiority of Hannibal and his army. He had brought his

---

[1] These are the figures given by Shuckburgh and are based on the *drachma* being equal to the *denarius*. The fractions, however, seem rather unlikely and round figures of five, ten and fifteen are more probable (= ½, 1, 1½ of a *denarius*). This would give the legionary 180 *denarii* a year, at the rate then in being of ten *asses* to the *denarius*. The problems presented by these figures have been discussed by G. R. Watson in 'The Pay of the Roman Army', *Historia*, 7 (1958), pp. 113–20; 8 (1959), pp. 372–8.

Plate IV
Part of the tombstone of the centurion T. Calidius Severus, at Carnuntum, showing his equipment and horse with attendant (p. 132).

Plate V
The tombstone of the centurion Q. Sertorius Festus, at Verona (p. 133).

men to a high pitch of training and endurance and he was able to outmanoeuvre the Romans with their limited view of tactics in the field of battle. The disaster of Cannae was an example of his favourite enveloping manoeuvre. The Romans relied entirely on the strength of the legions in the centre and Hannibal's crescent formation allowed the Romans to push forward, but not to break through, while the flanks closed round and encirclement was complete. Earlier at Trebia in North Italy the Roman centre broke through and escaped; similarly at Ibera on the Ebro in Spain, Hasdrubal used the same tactics against the Scipios, but his weak centre was too quickly crushed and Rome gained an important victory. The lessons to be learned were clear enough—more training, more tactical flexibility were needed once the battle was joined. The old Greek phalanx, although skilfully modified, was now thoroughly outmoded and by implication so also was the citizen army raised and trained for particular campaigns. Successful armies now required longer training and a higher degree of professionalism. Hannibal had also shown the need for a leadership in which the men had every confidence. The Roman practice of appointing the two consuls for the year to the Army command, never a very practical idea in the early days, had already many times been subject to the granting of special powers to a *dictator* as serious emergencies arose. With the granting of the *imperium* to young Scipio, all precedents and constitutional rules were swept aside. In the ladder of promotion he had only reached the level of aedileship and was thus totally unqualified by law to receive such a command. Fortunately for Rome, popular sentiment forced the hand of the reluctant Senate. Few could have foreseen the implications of this step for the future—the rise of great generals and their devoted professional armies and the bitter struggles which brought the Republic to its bloody close.

The young Scipio must have realized at the outset that sooner or later he would meet Hannibal face to face. He had therefore to learn to outgeneral him in the field and weld together a body of men, tough but flexible enough to be controlled like a precision machine. He soon showed his mettle in his seizure of New Carthage. In this brilliant feat speed and surprise were essential. His small army did a rapid march of about 200 miles in six days and arrived at the same time as the Fleet, invading the town by land and sea. This was followed immediately by an attack from an unexpected quarter while the main forces were engaged in a frontal assault. This episode set Scipio as a man apart, a commander capable of careful planning and rapid action, taking the enemy by surprise and off their guard. But he could also be cautious and wily, as was shown in the Battle of Ilipa against Hasdrubal Gisgo. Here the two armies paraded against one another, neither being anxious to take the first step. In each case the normal order was for the best troops to be in the centre and the weaker on the wings. Then one morning Scipio ordered breakfast before dawn and with a light cavalry attack roused the enemy, who tumbled into line before they could snatch any food. To his horror Hasdrubal then saw that the Roman order of battle had been changed and the legionaries were now posted on the wings. Scipio now engaged the Carthaginians with careful delaying tactics to tire them out. At a given signal the legionaries performed a complex manoeuvre, extending and wheeling on both flanks so that they enveloped both enemy flanks at the same time, thus engaging the weaker of the enemy, while Hasdrubal's better troops in the centre were not able to participate until the flanks had collapsed and were then crushed between the two Roman forces. Scipio repeated this tactic, in different circumstances, against the forces of Hasdrubal and Syphax on the Great Plains (by Souk el Kremis) in North Africa. Scipio was now ready to meet the great Hannibal himself, but when this clash occured at Zama the generals found it was impossible to outmanoeuvre each other and the infantry were locked in a stalemate. It was decided by Scipio's cavalry, the stronger of the two arms, which having chased its opposing force off the battlefield returned to fall on the rear of Hannibal's veterans.

The dreadful years of the second Punic War produced an effect on Rome similar to that on Britain after Dunkirk. At their darkest hour Rome never wavered, but with rugged tenacity and fortitude struggled on. Adversity on this scale, as in Britain, bound all citizens together in the common cause. But in the century which followed there were no serious national trials. The provinces of the East fell like a shower of ripe fruit into the lap of Rome and most of the fighting was done by the allies. Nevertheless the growing number of provinces made it necessary to maintain a standing army of at least eight legions. This permanent commitment of men serving in distant areas provided opportunities for the man who preferred being a full-time professional soldier. Such a man was Spurius Ligustinus of Sabine stock, in whose mouth Livy[1] puts these words—'My father left me an acre of land [a *iugerum* was certainly less than the required standard and Ligustinus could technically have been regarded as one of the proletariat] and a little hut in which I was born and brought up and to this day I have these. . . . I became a soldier in the consulship of Publius Sulpicius and Gaius Aurelius [200 B.C.]. In the army

[1] xlii, 34, 1–35.

which was taken over to Macedonia I served two years as a private against King Philip: in the third year for my bravery T. Quinctius Flamininus made me a centurion of the tenth maniple of the *hastati*. After the defeat of Philip and the Macedonians, when we had been brought back to Italy and discharged immediately I set out for Spain as a volunteer with Marcus Porcius the consul [195 B.C.]. . . . The general judged me worthy to be assigned as centurion of the first century of the *hastati*. For the third time I enlisted again voluntarily in the army which was sent against the Aetolians and King Antiochus [191 B.C.]. . . . By Manius Acilius I was given the rank of centurion of the first century of the main formation [*primus princeps prioris centuriae*]. . . . When King Antiochus had been driven out and the Aetolians beaten we were brought back to Italy; and twice after that I was in campaigns where the legions served for a year. Then I campaigned twice in Spain . . . [181 and 180 B.C.]. I was brought home by Flaccus along with the others whom he brought with him from the province for his triumph because of their bravery. . . . Four times within a few years I held the rank of chief centurion [*primus pilus*]; thirty-four times I was rewarded for bravery by my generals: I have received six civic crowns. I have done twenty-two years of service and am over fifty years old. . . .' At this time no one was exempted from service under the age of fifty-one. Ligustinus was made chief centurion of the First Legion and others who had protested at enlistment gave up their appeals.

Such volunteers, especially if they were experienced soldiers, would obviously be preferred to the conscript, and the principle of voluntary enlistment was long established.[1] As more of the East came under Roman control, it was inevitable that an increasing number of citizens became involved in commercial enterprises and enforced army service would have been a considerable nuisance. Rome could no longer rely on a steady and regular supply of legionaries from the simple sturdy rustics. Service in Spain in the second century was especially unpopular. The continuous series of local wars and uprisings, the ineptitude of Roman leadership and the heavy drain on man-power all meant hardship, possible death and little prospect of loot. In 152 B.C. popular pressure in Rome was such that the time-honoured method of enlistment was modified and men were chosen by lot for a limited

---

[1] Scipio was forbidden to raise troops for Africa except volunteers; the army which P. Licinius Crassus raised to fight the third Macedonian War against Perseus was largely formed of volunteers.

period of six years' continuous service.[1] Another effect of these changes was an increased use of allied forces. When Scipio took Numantia in 133 B.C. Iberian auxiliaries accounted for two-thirds of his force. In the East the critical Battle of Pydna which ended the third Macedonian War was probably won by the allies, who with elephants crushed the left wing of Perseus and enabled the legionaries to split and outflank the Macedonian phalanx.[2]

The overseas expansion of Rome had a serious effect, too, on the citizens of the upper classes. There were now opportunities of enrichment through commerce which contemporary Romans thought had a corrosive effect on the high standards of morality and simple living. Corruption in its many guises gradually seeped into these ranks and competent military leadership became increasingly difficult to find. Nor had foreign powers any reliance on the impartiality of Roman justice. The arrogance of Roman magistrates, even in Italy, became a byword and in more distant areas acts of brutality shocked people accustomed to cruelty.[3]

The difficulties Rome had in securing adequate recruitment are reflected in the attempts of the Gracchi towards agrarian reform. There had been a decline in the numbers of small landowners with the property qualification required to render them liable for military service. The removal of the young able-bodied men from the farm for long stretches of military service had helped to contribute to this situation and there had been a steady drift of these families into Rome to swell the ranks of the proletariat. The small farms had, in many areas, become amalgamated into larger ones and the large landowner found it more profitable to turn the arable land into pasturage. A new class of landed proprietors arose, enriched by the extension of Roman power in the East and able, with their new-won capital, to buy land, equipment and slaves, now plentiful and cheap. The Gracchi attempted to reverse this situation by land grants to the Roman poor; they also saw the need to extend the

---

[1] The total liability was sixteen years' service; the new regulation meant merely that men had to be returned to Italy after six years in Spain, but could be re-enlisted.

[2] Unfortunately the details of this important battle remain very obscure; Livy's account is missing and of that of Polybius there are only fragments.

[3] At Numantia, Scipio rejected the efforts of the desperate inhabitants to seek surrender terms and finally destroyed the town by fire without seeking senatorial authority; in Epirus, Aemilius Paullus systematically plundered and depopulated the countryside, deporting 150,000 inhabitants to be sold in the Roman slave-market. Polybius gives us this piece of information, but the reasons for such an act are not obvious; possibly the Roman soldiers were dissatisfied with the amount of loot they had been able to acquire in the preceding campaign.

franchise to the Italian allies. The stubborn conservatism of the aristocratic party balked these reforms and reaped in consequence the bitter harvest of the Social War. This in turn set the scene for the struggle for power between Marius and Sulla, a grim and ferocious episode in which no quarter was given and which seriously depleted the ancient senatorial families.

To Marius are attributed some of the major reforms of the Roman Army, yet he only put the final touches to changes already in being. Minor reforms of Gaius Gracchus had been to make the State responsible for the supply of equipment and clothing to the legionaries and to forbid the enlistment of youths under seventeen. This clearly indicates that neither property qualifications nor an age limit had been deterring those responsible for recruitment. Marius merely took the final step and threw the Army open to the proletariat. Marius had himself risen from the obscurity of a remote village and his sympathies remained with the lower classes; to him this method of ensuring a ready supply of recruits would have seemed an obvious one. He was a strict disciplinarian and the task of knocking such rough material into shape would not have presented a difficult problem. He had but little political wisdom or statemanship and he could hardly have foreseen the implications of his constitutional change. It may even have been a gesture to spite the nobility. Once he had obtained the consulship and Numidian command, Plutarch tells us that Marius made violent and contemptuous speeches taunting the senators and courting the affections of the masses. Sallust[1] puts into his mouth an elegant piece of rhetoric which, while embodying similar sentiments, argues that he (Marius) did not wish to force men into military service against their inclinations, but called for volunteers, promising them 'victory, spoils and glory'. Marius then promptly began to enrol volunteers, not according to the classes but from the *capite censi,* i.e. the proletariat, men without property, but who were entered on the census list. But he was careful to enlist experienced soldiers as well, by offering special inducements to veterans. According to Sallust, the Senate did not oppose these measures, because they thought that Marius would lose popularity by making the soldiers out of the plebs.[2] In effect, of course, the promises of loot appealed to some and an army career

---

[1] *Bell. Jug.*, 85.

[2] If one follows the chronology laid down by Sallust, it is probable that Marius started by recruitment in the normal manner by the *dilectus* and threw the ranks open to the *capite censi* only at the last moment when he was about to sail, thus forestalling any move of the Senate. The result was that he took to Africa a larger army than had been authorized, presumably the usual two legions of a consular army.

with its opportunities of advancement to the more far-sighted. The small property class was satisfied, too, that they could at last escape some of their arduous responsibilities.

Marius was a very able commander; lacking the brilliance and imagination of Scipio or Caesar, he nevertheless understood the basic requirements of a good army: training, discipline and leadership. His men became devoted to him because he shared their way of life and identified himself as 'one of them' as distinct from one of those others—the gilded youth of the senatorial aristocracy.[1] His ill-assorted array needed to be welded into an effective weapon, so training and discipline began immediately. Marius was careful to give his men confidence and experience by exposing them gradually to the difficulties of war. They took easy towns, engaged in light skirmishes, victories and booty were theirs as promised, but at the same time the veterans and raw recruits were brought together in mutual respect. Later in his career he could afford to be more direct. When he trained his army to meet the Germans there were long marches, each man carrying his baggage[2] and preparing his own meals. While at the mouth of the Rhône, waiting for the enemy, he used his troops to construct a large canal to bring transports and supplies more quickly to his base.

Marius did not develop any new battle tactics, but relied mainly on surprise and always showed a reluctance to engage in set-piece fighting of the traditional kind. He preferred to determine the time and place and would not be hurried. Typical of this is his great victory at Aix-en-Provence (Aquae Sextiae). He had held back his army from attacking the great body of Germans as they crossed the Rhône, but followed them carefully at a respectful distance, choosing his camp-sites with care. The first stage of the battle seems to have started almost by accident, but only involved the Ambrones.[3] The main attack was launched when half the tribe had crossed the river and when they were gorged with food and drink. This splitting of the enemy forces and causing them to fight at an inconvenient time could hardly have been an

---

[1] During the African campaign he even undertook guard duty and was also a glutton for hard work, leading his men by example. He was helped by his rough voice and bad temper; his punishments were savage but just.

[2] The expression 'one of Marius's mules' may have come from this, but Plutarch offers an alternative origin. This passage has led some to conclude that Marius devised a new form of pack which could be easily slipped on and off (Parker, p. 45).

[3] Another interpretation of Plutarch (*C.A.H.*, ix, p. 148) is that Marius got ahead of the Teutons by a short cut and attacked the Ambrones before the main body had arrived.

accident, but has the basis of some shrewd planning. The next day the main battle was engaged. Here Marius astutely gauged the temper of his enemy; knowing that they were eager for the fray, he drew up his legions on a hill slope on uneven ground and waited for the enemy to tire themselves by charging uphill. His men were ordered to stand firm, to discharge their *pila* only when the enemy were in range,[1] then to draw their swords and force the enemy back with their shields. Marius put himself in the front rank, relying on his strength and training. He relied for his main surprise on the simple tactic of hiding a small force in woods at the enemy's rear and when the battle was joined, with great noise they attacked, throwing the German army into confusion and panic—100,000 are said to have been killed or captured with immense booty. There remained the Cimbri, who had crossed the Alps and were making inroads into Italy. Marius joined Catulus and a second decisive battle was fought near Vercellae. The great general tried to emulate Hannibal at Cannae and allow the enemy to become engulfed between strong enclosing wings. But some of the Romans were drawn away by a feint retirement of the enemy cavalry and as the battle was joined a great dust cloud rising from the dry plain blotted everything from sight. It is hardly surprising that Plutarch's account is so improbable,[2] and that there was a dispute afterwards as to who was the real victor, Marius or Catulus. Whatever one may read into these accounts, Marius stands out as the architect of two great victories and saved Rome from a massive barbarian invasion.

Marius is given credit by Plutarch for changing the construction of the *pilum* by replacing one of the iron nails by a wooden pin so that the connection would break under impact and be impossible to return. As we have seen from Polybius, the *pilum* was already made to bend, but it may have been difficult to temper the shank so that it remained strong enough to be effective, yet bent under its own weight. It has also been argued that Marius was responsible for the reorganization of the legion, the abolition of the three lines and the *velites* and making all legionaries of a uniform character. This change can probably be associated with the growing appreciation of the cohort as a tactical unit as distinct from the maniple. According to Polybius (ii, 23) the cohort was used by Scipio as a tactical division against Hasdrubal, son of Gisgo. 'Scipio with the three leading squadrons of cavalry from the right

---

[1] 'Don't shoot until you see the whites of their eyes.'

[2] Marius is said to have led his troops over the plain, failing to locate the enemy, while Catulus's men took on the full weight of the attack. This seems like a highly biased account originating during the later dispute over the spoils of victory.

wing, preceded by the usual number of *velites* and three maniples (a combination of troops which the Romans call a cohort) . . .' The historian does not refer to this unit again and when describing the flight of the Macedonians at Cynoscephalae (197 B.C.), says that Flamininus sent a tribune with about twenty maniples. Livy, on the other hand, uses the term frequently and possibly indiscriminately. In the disastrous Battle of Lake Trasimene in 217 the legionaries, he says (xxii, 5, 7), could not keep in their own cohorts or maniples[1]. A more definite use occurs in the Spanish War: Lucius Marcius detailed 'a Roman cohort and cavalry to hide in a wood' (xxv, 39, 1), although it is not certain that legionaries are involved here. Sallust uses the term several times and in such a way as to imply that it was a normal tactical division.[2]

However, the three traditional battle-lines continued in use under Scipio, who faced Hannibal at Zama with this formation. By the time of the Jugurthine War there is no further mention of the *hastati, principes* and *triarii,* but precisely when they disappeared is not clear. Had Marius made this change in Africa, Sallust is not likely to have overlooked it.[3] A more likely explanation is that it was one of the unforeseen developments of the new recruitment policy. The old divisions were based on wealth and experience and once they were removed the legionaries would have been reduced to a uniformity. The Roman legion had now reached a stage in its organization in which in strength, resilience and flexibility it had no equal. Under training, discipline and brilliant leadership these units were about to perform epics which have become the envy of most of the world's successful generals of later years.

The period between Marius and Augustus saw little change in the organization and tactics of the Army, yet in two fundamental respects the legions for which Augustus eventually found himself sole *imperator* were different. These were the effects of the Social War, the pacification of Spain and the development of the colonies in the extension of the areas of recruitment. As citizenship was granted to most of the peoples of Italy even into Cisalpine Gaul, so vast new resources of manpower became available. During the Civil Wars the protagonists were not likely to inquire too closely into legalities when young able-bodied men were needed. There was also a growing appreciation

[1] *nec ut in sua legione miles aut cohorte aut manipulo esset.*

[2] *Bell. Jug.*, 51, 3; 100, 4.

[3] *C.A.H.*, ix, 146. Parker gives credit for the developed tactical use of his cohort to Sulla (p. 48).

of the effectiveness of the allies. In the closing stages of the Social War, Cn. Pompeius Strabo, father of Pompey the Great, had enfranchised thirty men of a troop of Spanish cavalry for their services in the siege of Asculum.[1] Soon that unfortunate man Sertorius was to show how the Spaniards under proper leadership and discipline could hold Roman armies at bay, and Caesar discovered in Gaul the great value of loyal Gallic cavalry. So the character of the Army changed, but the *dilectus* remained the normal method of recruitment. The officers would, however, have little difficulty, since volunteers would be plentiful and compulsory powers would only be needed to fill in the gaps. Losses by death, illness and retirement could, however, be made up by a governor in a province without reference to the consuls. Thus Caesar was able to use these powers to raise new troops in Cisalpine Gaul for his campaigns.[2]

The other important aspect was the transference of allegiance from the State to a particular commander. When Rome was a city state she could command the respect and loyalty of her citizens. The feelings of the recruits from newly enfranchised Italian peoples who had recently been engaged in war with Rome, would be different. The moral fibre of the plebs could hardly be expected to be as tough as that of the property-owning classes. As Rome began to draw from the new reservoirs of man-power, the strong ties of loyalty and responsibility would inevitably weaken. The recruit had, on enlistment, to take an oath, the *sacramentum*, binding himself to serve loyally and with obedience.[3] Sulla was probably the first commander to force his troops to take an oath to their general in circumstances which required fighting against the official representatives of the State. Soldiers had always looked towards their great and successful generals, especially Scipio, as their personal lord and protector as in the sense of the patron and his clients. Sulla and later generals in the Civil War merely took this a step further and put personal loyalty before that owed to the State. The men on demobilization could look towards their general for substantial donations or land grants.

---

[1] *C.I.L.*, i, 709 = *I.L.S.*, 8888.

[2] *Bell, G.*, vi, 1. One legion (the *Alaudae*) was raised entirely in Transalpine Gaul. Since these men were non-citizens, the legion was not given a number, Caesar enfranchised the men later, certainly by 47 B.C. when it had become known as Leg V (Suet, *Caesar*, 24). Pompey raised two legions from Roman citizens in Asia, *Bell, C.*, iii, 4, 1.

[3] This oath probably remained unchanged into Imperial times (Vegetius, ii, 5). A valuable discussion on the length of service required by the oath is found in R.E. Smith, *Service in the Post-Marian Roman Army*, 1958, Chapter iii.

It was in this period that Rome produced a most remarkable series of great commanders; following Marius and Sulla came Sertorius, driven by circumstances into a war against his own people, but showing outstanding qualities as a leader and tactician. In the same generation Pompey and Caesar reached even higher peaks of achievement. This is hardly the place to evaluate the work of these remarkable men. Caesar remains perhaps the greatest general of all time, respected and emulated by all who followed; there was little need to reshape the Army's organization or equipment, it was a tool ready for his hands. Caesar relied on speed and surprise, showing at times a daring and rashness which almost brought him to disaster, but his coolness, and the discipline of his men and their complete reliance on his leadership, normally carried the day. Had Caesar needed a model or hero, there was Scipio Africanus, and in studying accounts of these earlier battles he could not fail to be impressed by the effectiveness of speed, surprise and a small disciplined force. An outstanding example of these tactics was the crossing of the Cevennes at the end of the winter in 52 B.C. The passes still had snowdrifts six feet deep on them and the Arverni thought it impossible even for small groups of travellers to get across, but Caesar set his army to work to clear the snow with shovels and so was able to take the tribe totally by surprise (*Bell. G.*, vii, 8). It was not always easy for Caesar's troops to understand his tactics and there were occasions when they pressed forward too eagerly when they had been ordered to withdraw, as happened with serious losses at Gergovia.[1] *de Bello Gallico* contains some remarkable details of the Roman Army not only in action but also in their construction works in building bridges and siege-works.[2] But above all, the qualities of the legions in their courage, endurance and discipline shine through and these, combined with Caesar's grasp of military strategy and his audacity, make this work the finest of all military annals.

Changes under Augustus

At the close of the Civil War, Augustus found himself with several armies comprising elements of sixty legions, some of which had sworn loyalty to

[1] Caesar's own speech to his troops is significant: 'Much as I admire the heroism you have shown in refusing to be daunted by a fortified camp, a high mountain and a walled fortress, I cannot too strongly condemn your bad discipline and your presumption in thinking that you know better than your commander-in chief how to win a victory or to foresee the results of an action. I want obedience and self-restraint from my soldiers, just as much as courage in the face of danger' (*Bell. G.*, vii, 52).

[2] Those against Alesia are particularly noteworthy (see below, p. 236 *Bell. G.*, vii, 68–90).

the opposing faction. His immediate problem was to discharge veterans,[1] remove dangerous elements and put the Army on a peacetime footing. His greatest fear, and one which faced all his successors, was that of a section of the Army defecting to a popular commander. As long as the Emperor led his army on a campaign, his troops could feel a personal devotion towards him, but with a static army spread along the frontiers it is likely that few soldiers would even see the Emperor during their service. Augustus saw clearly that he and anyone who might follow him had to depend on the men placed in command of army groups. The Emperor retained for himself the *imperium* and placed it first on his list of powers; governors sent out to provinces in which there were military units were his legates and he reserved these appointments for himself, while the Senate retained the power of selecting men for the other provinces. Thus there came into being the division of the Empire into Imperial and Senatorial provinces. Care was taken in the selection of governors and it soon became the practice to select likely young men at an early stage of their careers and channel their promotion.[2] In this way these men were placed under a strong obligation to the Emperor. Another source of trouble had been the retirement grants. Successful generals, after a long campaign, normally arranged for the veterans to receive money or land grants. This obligation now had to be taken over by the State, but there was an immediate difficulty in that the actual length of service had always been based on the original obligation of the citizen to serve for sixteen seasons and thereafter only in emergencies. The rule was never changed when men began to serve continuously overseas. Augustus now established the right of a soldier to retire with his gratuity after sixteen years as a ranker and a further four as a veteran.[3] These veterans appear to have been organized in their own units in the legion under a *curator veteranorum*[4] and with their own standard.[5] It is clear that men were, in fact, forced to continue

---

[1] Augustus discharged over 100,000 veterans and many were settled in colonies, including some new ones, as it was from the sons of these men that future legions were recruited.

[2] E. Birley, *Roman Britain and The Roman Army*, 1953, pp. 3–4.

[3] At first the period for legionaries was fixed at sixteen years, but was later increased to twenty. The mutiny of A.D. 14 caused a reduction to sixteen years, but in the following year this was rescinded (*Annals*, i, 78).

[4] *Rang.*, pp. 78–79; *C.I.L.*, iii, 2817 and v, 5832 (= *I.L.S.*, 2338); Tutilius had previously been *aquilifer leg V* and the inscription is dated to A.D. 29.

[5] *Vexillarius veteranorum legionis* (*C.I.L.*, v, 4903); see also *Annals*, iii, 21 for a unit of cohort size of veterans in Africa (*ut vexillum veteranorum non amplius quingenti numero*).

their service in spite of the reform, since this is one of the burdens of complaint in the mutiny of A.D. 14.[1] It is likely that the efficient discharge of time-expired soldiers did not come into being until the Flavian emperors.[2] Regular pay must also have been a difficulty, especially in the Civil War[3] and the troops naturally looked to their general for this. Conquests often added rich booty, but with the peace of Augustus these dreams of enrichment faded. The obvious and logical step was the establishment of a permanent War Treasury (*aerarium militare*),[4] financed at first by death duties and a tax on auction sales, supplemented from the Emperor's own resources.

Seen in retrospect, perhaps the most important decision made by Augustus was the size and distribution of the Army. This had the effect of determining the extent of the Empire,[5] and the policy laid down by the first emperor was followed, with the exception of the additions of Britain and Dacia, some adjustments to the frontier between the Rhine and Danube and the abandonment from time to time of the policy of peaceful co-existence with Parthia. The number of legions Augustus decided to keep was twenty-eight, with five in Spain, five or six in Germany, five in Illyricum, three or four in Macedonia, three or four in Syria, three in Egypt, one in Africa and one or two in Vindelicia.[6] Later the garrisons in Spain and Egypt were reduced to strengthen those along the Rhine, while those in Macedonia and Illyricum moved into the new provinces of Dalmatia, Pannonia and Moesia.

It is not clear precisely how Augustus treated the *auxilia*, but some

---

[1] *Annals*, i, 17, Tacitus gives a speech by one of the mutineers, one Percennius, in which he states that men were still serving in their thirtieth and fortieth years 'and even after your official discharge your service is not finished for you stay on with the colours (*sed apud vexillum*; see above for a note on the veterans' standard) as a reserve still under canvas—the same drudgery under another name'.

[2] It is doubtful if any of the Julio-Claudian emperors ever understood the problems of the ranker except perhaps Tiberius, but he showed no serious concern. Vespasian had spent much of his life with the Army and knew how to win the loyalty of the legionary.

[3] Antony was forced to mint *denarii* with a very low silver content; these coins, especially the 'galley' type, continued to circulate well into the second century A.D.

[4] Suetonius, *Aug.*, 49.

[5] The extent to which Augustus appreciated this is seen in the document which Tiberius read to the Senate after his predecessor's death (*Annals*, i, 11). It had been written in Augustus's own hand and listed the natural resources, number of legions and *auxilia*, etc., with a clause to the effect that the Empire should not be extended, an instruction which Tiberius scrupulously obeyed.

[6] The number and distribution are still matters for discussion, but the figure twenty-eight seems to be universally accepted (Parker, pp. 78–92).

rationalization was needed. The main problem here was the reluctance of these soldiers to serve too far away from their homeland. Most of the Gallic units served along the Rhine, and the Thracians caused serious trouble under Tiberius when it was rumoured that they were to be dispatched to distant areas.[1] The problem was eventually resolved when these units ceased being recruited from the original areas, so that although the name and method of fighting remained, recruitment into the *auxilia* became general. The annual drafts from the provinces were distributed as vacancies occurred. This took place slowly and Augustus with his reluctance to extend the franchise too widely is hardly likely to have offered citizenship to all auxiliaries on discharge.[2] This privilege was probably granted by Claudius, whose desire to bring the provincials into the embrace of Rome received such derision after his death.[3]

In the organization of the praetorian guard,[4] Augustus showed an appreciation of the potential danger of such a body near Rome. He also naturally tried to avoid the impression that he maintained his position by force of arms; some future emperors seemed devoid of such fine sentiments towards the Senate and people of Rome. The guard consisted of nine cohorts, each of five hundred men, but only three were garrisoned in Rome itself, the others being distributed among the cities of Italy.[5] These men were recruited mainly from Italy and served for sixteen years, receiving considerably more pay than legionaries. The command of such a force was restricted to men of the equestrian order and it naturally became the highest and most important post a knight could hope to reach. A useful reform which can be credited to Augustus was that of the Fleet. Under the Republic the Romans had shown a surprising lack of concern over their control of the Mediterranean and at time allowed parts to become infested with pirates. Units of seamen were raised and disbanded as occasion demanded. The Civil War had taught Augustus the value of this arm and he created naval bases at Misenum in the Bay of Naples, at Ravenna in the Upper Adriatic and at Fréjus on the coast

---

[1] See p. 58 below.

[2] Dessau, *Geschichte der römischen Kaiserzeit,* 1924, p. 276.

[3] Seneca, *Apocolocyntosis.*

[4] The origins of this body are thought to be in the personal bodyguard of 500 that Scipio Aemilianus took with him to Spain in 133 B.C., because the temper and discipline of the Army in Spain at that time were so uncertain. The evidence for the grant at this period has been summarised by G. Alföldy (*Historia,* xvii, 2 (1968), 215–227).

[5] This arrangement ceased with Tiberius, who brought all cohorts to Rome.

of Provence; fleets are also found operating along the Danube and the Rhine, but much of the work of establishing the fleets as a regular branch of the armed services came later.

Under Caesar, the Army had become a highly efficient and thoroughly professional body, brilliantly led and staffed. To Augustus fell the difficult task of retaining much that Caesar had created, but on a permanent peacetime footing. The success of his policy and its gradual consolidation under the remaining Julio-Claudian and Flavian emperors can be measured by the remarkable survival of the Empire in spite of conflicts within and pressure without. It stands as a military institution without parallel in the history of the world.

Chapter 2

FRONTIER SYSTEMS

INTRODUCTION

Imperial Policy under Augustus (27 B.C.—A.D. 14)

The conception of a system of forts and supply bases, with planned communications, belongs entirely to the Empire. The idea that there was a controllable limit to the extension of Roman authority was first enunciated by Augustus. The adequate protection of the frontier areas, in the face of the movements and pressure of barbarian peoples, became an increasing preoccupation of succeeding emperors, culminating in the eventual collapse of the western provinces.

The very word *limes* (pl. *limites*) signified nothing more than a path or road in the early Empire[1] and came to be used for the ways cleared by the Army in the campaigns into enemy territory.[2] It was only later that the term came to mean the frontier between Roman and non-Roman territory and to be used as such by Frontinus[3] and Tacitus.[4] The function of defensive boundaries

---

[1] It is often the word for a path between fields, Virgil (*Aeneid*, 12, 898 and *Georgica*, 1, 126); for a full discussion see Fabricius in P-W., xiii, 572 ff.

[2] In this sense it was used by Velleius Paterculus, *penetrat interius, aperit limites* (2, 120).

[3] Referring to the new frontier constructed by Domitian in Germany . . . *limitibus per centum viginti milia passuum actis* . . . (*Strat.*, i, 3, 10), this could, however, here be interpreted as a road.

[4] When Tacitus refers to the injunction Augustus laid on Tiberius to maintain the frontiers in their present positions the word *terminus* is used as if copied from an official contemporary source, *Quae cuncta sua manu perscripserat Augustus addideratque consilium coercendi intra terminos imperii* (*Annals*, i, 11); but in later contexts the word *limes* is used, *limite acto, promotis praesidiis* (*Germania*, 29) and *de limite imperii et ripa* (*Agricola*, 41); in this terse indictment of Domitian's policy it is clear that Tacitus sees the *limes* as a strip of frontier land as distinct from the forts occupying it. In this sense it may still have been considered as the military way connecting the garrisons. Mommsen developed this idea and tried to demonstrate that it was a technical term applied to a clearly marked strip of land with a road along it (*Westd. Z.*, 13, 134, see also Oxé in *B.J.*, 114, 99).

at earlier dates by Scipio[1] and Caesar[2] in Africa is more likely to have been to check hostile cavalry or brigands.

The problems which faced the Republican governors were not so much those of protecting a frontier from external pressures as, in some cases, containing a conquered people and generally to control the movement of people and goods in both directions. Nevertheless in some areas the presence of a hostile power immediately beyond the territory under Roman authority had to be considered. One of the most serious threats was in the East, from Parthia. Rome never quite succeeded in rationalizing this problem, but the fortunes of the earlier Greek expeditions must have always inspired the Romans to further efforts in the vain hope that complete success was almost within their grasp. But even here the Parthians were never a constant and immediate threat to the rich provinces of the eastern Mediterranean. The large desert areas, crossed only by the caravan routes, afforded a measure of protection, and for considerable periods of time the only contact was by trade.

There were two methods employed in Republican times to deal with an external hostile power which threatened the peace. Invasion and conquest seemed the most obvious; the Army could then withdraw with honour and plunder, or even occupy the country. The other method was by diplomacy, an art for which the Romans displayed a considerable aptitude. Treaties could be negotiated and agreed with the hostile power by mutual arrangements or by threats and displays of force. But treaties, as many have subsequently discovered, are only 'scraps of paper' to the would-be aggressor. It was necessary to have, at some convenient point, sufficient forces to guard against the sudden appearance of a hostile force. Here the Romans developed with great skill the establishment of friendly states between their provinces and a potential source of hostility. These acted as buffers absorbing the shock of sudden invasion and allowing the Romans time to organize and deploy their own forces. Most of these friendly powers were content to play this role, since they retained their independence and knew that should they fail in their obligations the Romans could take over at any time they wished. This kind of treaty arrangement was made with the head of the ruling house, who may have owed his position to the Romans and would know that he or she could be replaced. As a further

[1] *Nat. Hist.*, 5, iii, 25. This boundary ditch was still visible in Vespasian's time (*Ann. Épig.*, 1894, No. 65).

[2] *Bell. Af.*, 51, *duo bracchia;* an unusual meaning of *bracchium* to signify an outwork or line which 'branched out' from a definite point, i.e. in this case Caesar's main camp.

Plate VI
The tombstone of the centurion M. Caelius, who was killed in the disaster under Varus (p. 52).

Plate VII
The tombstone of an *aquilifer* of XIV *Gemina*, at Mainz (p. 137).

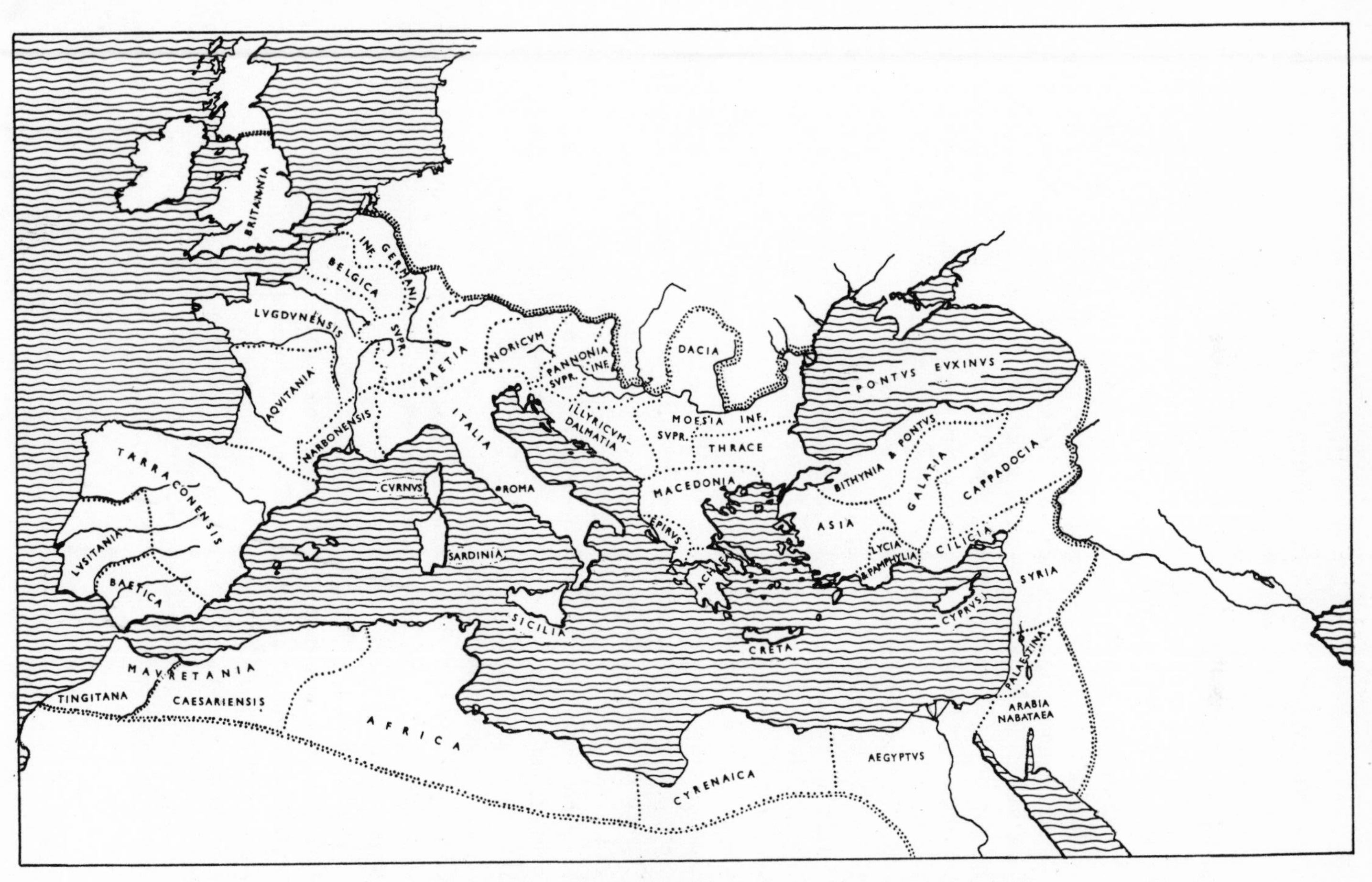

FIG. 2 *Map of the Roman Empire showing the provinces under Hadrian*

incentive to reliable behaviour, children of the ruling family were often held in Rome as hostages, ostensibly for their education.

In other areas these problems did not exist. In Spain, apart from its land frontier with Gaul, the boundary was the sea, in North Africa the desert. So apart from small groups of wandering nomads and traders, Rome had to deal only with rebellious tribes within. Campaigns were normally carried out in the summer months and as soon as the end of the season came the citizen-soldiers would be able to return to their farms and homes. But as soon as Rome began to fight in lands some distance from Italy this was no longer practicable. In addition there was usually a need to keep strong forces to hand in case of trouble. A posting to Spain, Gaul or the East inevitably meant serving several years. It became necessary to establish winter camps or *hiberna* and the selection of these sites would depend on several factors. This practice may have been necessary to keep an eye on tribes of uncertain temper and, most important, to ensure an adequate food supply. This may have been extracted as a punitive tax against enemies or by payment or treaty arrangements with friendly people. The Army always lived off the land it occupied, since supplies from Italy would present the problem of distance and bulk. Caesar gives a number of examples of his army in winter quarters in Gaul. During the winter of 57–56 B.C., the VII Legion was quartered in the north-west, where food was scarce and parties had to be sent out to secure provisions. This gave the Veneti the opportunity of seizing the officers in charge as hostages and so began the war which led to the destruction of this tribe which had close ties with Britain. The winter was also a time for preparation for the coming season and Caesar used it well in preparing, assembling and building ships for the enterprises against the Veneti and against Britain.[1] The poor harvest of 54 B.C. necessitated the wider distribution of his troops, mainly in friendly territories. This facilitated the treachery by Ambiorix which destroyed Cotta and his men and set off the attack of the Nervii on the winter camp of Cicero and further widespread revolts.

It was the subjugation of Gaul and Illyricum which began to raise in acute form the problems of frontier control. The passage of large Germanic tribes across the lower Rhine had forced Caesar to retaliation and the building of the famous bridge. Caesar's successes against the Germans led to the possibility that Roman power might be extended even further, perhaps even to the Elbe.

[1] After the difficult beach landing of 55 B.C., Caesar ordered the construction of boats of shallower draught, but with both sails and oars, the timber being obtained from Spain (*Bell. G.*, v, 1).

Caesar himself had probably begun to form ideas about the ultimate limits of Rome, but he was, up to the time of his assassination, still much occupied with the establishment of peace within the Empire. It became instead an awkward legacy for Augustus to inherit and this exceptionally able statesman was probably the first Roman to see the boundary of the Empire and its defence as a single problem. The *imperium*—the right to command absolute obedience which led to the power to raise and maintain armies—had been given only to provincial governors and dictators in conditions of emergency.[1] Augustus at first assumed the *imperium* by virtue of his provincial commands, but this was extended by the Senate to include Italy and eventually it became perpetual.[2] From 23 B.C. Augustus controlled the Army through his *legati,* who were governors of the provinces in which most of the legions were stationed. Thus there came into being the system of Imperial and Senatorial provinces. The latter were still governed by the Senate, but rarely held any troops.

In making these provisions Augustus was obliged to consider both the extent of the Empire and the number of legions required to keep the peace. In these, as in other fields, his judgement was sound. To the north-west Britain remained a problem. There is evidence that Augustus intended to complete the work of Caesar, but troubles elsewhere engaged his attention at the critical periods. He did, however establish in Britain a diplomatic foot-hold through tribes which became allies of Rome. It was an attack on one of these tribes which gave Claudius ample pretext for his invasion and native assistance was available.[3] Augustus appears to have seriously misjudged the situation in Germany. It seemed possible to him to bring the tribes east of the Rhine under Roman control and to proceed with the normal process of pacification and Romanization.[4] This seems to be demonstrated by the

---

[1] The fateful decision of Caesar to cross the Rubicon with his army rested on this constitutional restriction. This river was the boundary of Caesar's province and so of his *imperium:* for him to be in command of troops outside his province was to break the law.

[2] The precise nature of the *imperium* of Augustus has been a matter of much scholarly dispute; see Pelham's *Essays,* 1911, pp. 60–88, and A. H. M. Jones in *J.R.S.,* 41 (1951), pp. 112–19.

[3] C. E. Stevens, 'Britain between the Invasions (54 B.C.–A.D. 43)', in *Aspects of Archaeology in Britain and Beyond,* 1951, pp. 332–44.

[4] An interesting suggestion has been made (*C.A.H.,* x, 353) that there may have been a basic lack of geographic information and the case of Germany must also be viewed against the apparent ease with which Gaul had been pacified. The campaigns of Drusus and Tiberius were no more than large-scale raids.

appointment of Quinctilius Varus as governor. This man, a leading lawyer without any military qualities, may have had the special task of bringing the territory under Roman law and shaping it into a province.[1] He was led into a disaster with the loss of three[2] legions in A.D. 9, which had a profound effect on Augustus and his frontier policies. The Rhine was now to be the limit of Roman authority and was large enough to offer a physical barrier, although forts like Hofheim were established on the east side. Similarly to the north and north-west the physical barrier of the Danube imposed itself as an obvious frontier. That the conquests of Illyricum and Pannonia had been deceptively easy is shown by the great revolt of A.D. 6–9. It says much for Roman abilities that they were able to hold this difficult country in sway, and for the perspicacity of Augustus in securing the important land route across the Balkans to the east. This choice imposed problems of geography which had later to be resolved. The main task of the Army was to protect the main crossing points; legions were established at intervals, not spread out evenly, but sometimes grouped together like the three legions under Blaesus in Pannonia.

The Roman frontiers to the east and to the south were very different from those described above. The reasons for this are inherent in the geography and the cultures of the people in these areas. Apart from the coastal strips and the valleys of the great rivers there is little but desert in its various forms. Yet out of these great wastes came the riches of the East by the caravan routes bringing the exotic silks, ivory and spices which formed part of the luxury of Rome. The deserts of Parthia held the lure of conquest, and in spite of lessons of the past, Roman commanders continued to lead their armies to destruction in the wake of the legendary Alexander. To Rome, Parthia was always a special problem, at times a threat, at times a challenge, but Roman armies never really learnt how to cope successfully with desert conditions, where the legions lacked the essential speed and mobility. The other important factor was that here the Romans were dealing with cultures far older than their own. To many of the peoples of the Near East, the Romans must have seemed like upstarts and barbarians. Certainly the behaviour of some of the late Republican governors like Gabinius in Syria, Crassus in Judaea and

---

[1] Although he had married the grand-niece of Augustus and might thus be considered as a member of the Imperial family, this appointment is difficult to explain except that Augustus was misled about the state of Germany and was overanxious for a policy of peaceful absorption (Dio, lvi, 18; see also *C.A.H.*, x, p. 374).

[2] Legions XVII, XVIII and XIX (Dio, lvi, 19–21).

Antony in Armenia effectively tarnished the name of Rome. Worse was the disastrous defeat of Crassus at Carrhae in 53 B.C. with the loss of 30,000 men.[1] Caesar's experience of the East was his rapid overthrow of Pharnaces,[2] son of Mithridates, but he was planning a great expedition against the Parthians at the time of his assassination.

During the Civil War the Parthians took advantage of the situation and overran Syria. They were defeated by Ventidius Bassus, a lieutenant of Antony at Gindarus (38 B.C.), on the very anniversary of Carrhae, but Antony, under the influence of Cleopatra, lost the opportunity to follow up this advantage. It was left to Augustus to bring about a reasonable settlement, which he was able to do by skilful diplomacy. The keystone of this arrangement was the client kingdom of Judaea in the strong and capable hands of the philo-Roman, Herod the Great. Other kingdoms that were established were Cappadocia, Commagene and Chalcis, while the opportune assassination of Artaxes, King of Armenia, enabled Augustus to give the throne to Tigranes and thus ensure his loyalty. These were solid achievements, but the masterstroke which gave a greater propaganda success was the return of the standards lost by Crassus and a few surviving prisoners who had now been in Parthian hands for thirty-three years.[3] He was able to force this upon Phraates, the Parthian king, by offering his son in exchange. While the client kingdoms were expected to provide their own defence and troops for Roman expeditions, the frontier was strengthened by the presence of three or four legions and auxiliary units.[4] The boundaries of the province of Syria were, in the north-east, the upper Euphrates and, in the south, the desert.[5]

---

[1] There is a fine detailed account of this battle in Plutarch, *Crassus,* xxxi. Some 10,000 Roman prisoners were taken and a number of them may eventually have been sold to the Chinese Emperor according to an ingenious theory of Homer H. Dubs, *A Roman City in Ancient China,* 1957, published in the China Society Sinological Series.

[2] It was this campaign that he himself described in the terse words '*veni, vidi, vici*'.

[3] The significance Augustus himself gave to the recovery of these symbols is seen in its proud place in his *Res Gestae* (*Monumentum Ancyranum.*, xv, 40–43): 'I compelled the Parthians to restore the spoils and standards of three Roman armies and as suppliants to implore the friendship of the Roman people.'

[4] III *Gallica,* VI *Ferrata,* X *Fretensis* and XII *Fulminata,* all in Syria (R. Syme, 'Some Notes on the Legions under Augustus', *J.R.S.,* 23 (1933), p. 31).

[5] It may be worth noting here that the present desert and semi-desert condition of much of Syria is due partly to soil exhaustion and deforestation, but mainly to the breakdown of the elaborate system of irrigation; in Roman times these lands were under prosperous cultivation. These problems are discussed by K. W. Butzer in *C.A.H.,* 1, 1965; see also papers by

There was another important difference between the arrangements on this eastern frontier and those of the north. The garrisons, instead of having forts of their own, were normally billeted in the towns in the populated areas; official buildings for administration, stores and armaments were necessary, but they are to be found side by side with other public and private buildings in the town, like the police headquarters today. At Dura-Europos, on the upper Euphrates, the Army took over part of the town, and although some new buildings were erected, a temple was converted to military use.[1]

Egypt, once taken over by Augustus after the death of Cleopatra, always remained an exceptional case owing to its great wealth. It was too important to the Emperor to be allowed to fall into the hands of a potential rival; its governor was therefore always an equestrian and no senator was permitted to set foot in it without Imperial authority. Troops were needed

---

E. C. Semple, in the *Annals of the Assoc. of American Geographers*, 9 (1921), pp. 47–74; 12 (1922), pp. 3–8 and 19 (1929), pp. 111–48; and H. C. Butler, 'Desert Syria, the land of a lost civilization', *Geog. Rev.*, 9 (1920), pp. 77–108. Similarly there were large areas in North Africa and the Negev of Palestine which in ancient times supported large populations. An experimental farm at Avdat in the Negev has shown that the control and preservation of water from the torrential rains can be achieved and that these few inches are sufficient to maintain cultivation, especially in the loess areas of the Beersheba basin (*Arid Zone*, No. 14, 1961, published by Unesco).

[1] This town has been partly investigated by Yale University. The military buildings are published in the report on the fifth season 1931–2 (*The Excavations at Dura-Europos*, 1934, ed., by M. I. Rostovtzeff). The units here in the second century were *Cohors II Ulpia Equitata* probably brigaded with *Cohors XX Miliaria Palmyrenorum*. There were also legionary detachments of IV *Scythica* and III *Cyrenaica*. In the area explored north of Tenth Street four buildings were identified as in military use. The *principia* (described here as the *praetorium*) is of typical form, although there are interesting additions on three sides, including a row of shops. The Temple of Azzanathkona on the evidence of inscriptions, pieces of equipment and fragments of military documents, is thought to have been taken as a headquarters of the auxiliary units, while the orthodox *principia* was assigned to the legionary drafts. An example of the adaptation of earlier buildings for military purposes occurs also at Luxor in Egypt, where the Temple of Ammon, a building of the period of Amenophis III, was converted into a fourth-century fort and parts of the *principia* have been identified from the frescoes which were unfortunately destroyed by nineteenth-century Egyptologists before adequate records could be made (U. Monneret de Villard, 'The Temple of the Imperial Cult at Luxor', *Arch.*, 95 (1953), pp. 85–105). At Dura the building to the west of the *principia*, only part of which was uncovered, is identified as the commander's house (*praetorium*). To the east on a different alignment, and effectively blocking the street, a bath-house was found which was also considered to be associated with the Army, since its entrance faces the *principia*.

to maintain law and order among the inhabitants, who consisted of a native peasantry dulled into apathy by centuries of oppression and a Greek hierarchy established by the Ptolemies. The large towns had Jewish, and later Roman, traders, and disturbances between factions were common.[1] Augustus and succeeding emperors attempted to reorganize the system of land tenure and develop its economic potential, but this met with mixed success.[2] The emperors tended to regard Egypt almost as a private estate, supplying the large annual corn ration to the Roman proletariat. At least Egypt now had several centuries of peace and the presence of troops and diplomacy ensured the continuity of trade with the kingdom of Meroe, i.e. Nubia, and Ethiopia to the south which became the important meeting-place for trade from Africa and India.[3] The province has a special importance for the student of the Army, since it has produced the majority of the military documents with their often enigmatic details of pay and organization.[4]

The coastal strip of North Africa had long been civilized by Greeks and Phoenicians before the Romans created the four provinces Cyrenaica, Africa Proconsularis and the two Mauretanias, Caesariensis and Tingitana. Cyrenaica was bequeathed to Rome in 96 B.C., but did not become a province until 67 B.C., it did, however, prosper greatly under the Roman peace and an expanding agriculture. The Greeks had settled down fairly peaceably with the Libyan tribes, but there was some trouble with the Marmaridae, who had to be subdued in A.D. 2, and the Nasamones of the Syrtic Gulf were brought under control in A.D. 85–86. There is evidence of the presence of auxiliary units at this early date.[5]

---

1 These were normally riots against the Jews organized by the Greeks, but due not so much to any anti-Semitic feeling as demonstrations against the privileged status of the Jews.

2 An extreme view is given by J. G. Milne, 'The ruin of Egypt by Roman mismanagement', *J.R.S.*, 17 (1927), pp. 1–13, and a more moderate assessment by M. I. Rostovtzeff, *The Social and Economic History of the Roman Empire*, 1947, pp. 273–99.

3 Sir Mortimer Wheeler, *Rome beyond the Imperial Frontiers*, 1954, p. 116. While Rome was strong, Meroe and the trade flowing north was safe, see Rostovtzeff and his very full bibliography in Vol. 2, pp. 677–80, Notes 53–62, Chapter VII.

4 e.g. the Hunt *Pridianum*, *The Abinnaeus Archive* and *P. Gen. Lat.*, 1 (see p. 258); S. Daris, *Documenti per la storia dell' esercito romano in Egitto*, 1964.

5 E. Ritterling, 'Military forces in the senatorial provinces', *J.R.S.*, 17 (1927), pp. 28–29. The epithet given to *Cohors I Lusitanorum Cyrenaica* and *Cohors II Hispanorum Scutata Cyrenaica* is suggested as evidence of the presence of these units in the province. The situation in the third and fourth centuries was quite different and the defences this necessitated are briefly described by R. G. Goodchild, 'The Roman and Byzantine *Limes* in Cyrenaica', *J.R.S.*, 43 (1953), pp. 65–76.

The province of Africa came to Rome as the result of the defeat of Carthage and was gradually expanded to take in Tripolis after the Jugurthan War, and later, after Caesar's victory at Thapsus, Numidia was added. The fertile areas attracted settlers, including many veterans from the armies of Marius and Caesar. To the south of the coastal plain of Tripolis lies the Fezzan, which was occupied by the Garamantes. After these people had been defeated by Cornelius Balbus about 20 B.C., they settled down amicably, importing much pottery and glass from Rome, and their towns and settlements were clearly developed under Roman influence.[1]

Mauretania had come under Roman control after 33 B.C., but Augustus was unwilling to accept this large area with its difficult mountain ranges and as yet unsubdued nomadic people the Gaetuli. The solution was that of a client kingdom in 25 B.C. under Juba, then a young man. Assistance was necessary against the Gaetuli in A.D. 5–6 and although this must have been a considerable campaign very little is known about it. By this time the legionary establishment was III *Augusta*.[2]

To complete this brief survey of the military situation under Augustus it is necessary to consider Spain, which has always had affinities closer to Africa than to Europe. Although Spain had been one of the oldest and richest provinces of Rome, the nature of the country, the fierce independence of its peoples and the vacillating policy of Rome had all combined to create an almost continuous battle-ground for two hundred years, exacerbated by Roman civil strife. It was left to Augustus to end this unhappy state of affairs. His final campaigns were against the Cantabrians and Asturians in 26 and 25 B.C. But the former tribe rose again and the patient Agrippa eventually crushed these intractable people only by large-scale massacre and enslavement. Even so a considerable permanent garrison of three legions[3] was necessary, as later unrest was to prove.

---

[1] Sir Mortimer Wheeler, *op. cit.*, fn. 3, p. 55 above, pp. 97–107.

[2] It is thought that there may have been three legions in Africa at the death of Caesar and there is evidence from Thugga which may support the presence there of XII *Fulminata* (*I.L.S.*, 8966), but the situation generally is very obscure (*J.R.S.*, 23 (1933), p. 25).

[3] The campaigns had brought seven legions to Spain 27–13 B.C., but by the end of this period there were only four left, II *Augusta*, IV *Macedonica*, VI *Victrix* and X *Gemina*, and after Varus, II *Augusta* was transferred to Strasbourg. The sites of their fortresses are not known, but IV *Macedonica* may have been for part of the time on the Ebro at Caesaraugusta (Zaragoza). A coin struck here lists the three legions and shows two standards and a *vexillum* (C. H. V. Sutherland, *The Romans in Spain*, 1939, Pl. iv, No. 9; also pp. 150–1). The other two legions were presumably securing the valleys of tributaries of the Douro.

Imperial Policy under Tiberius (A.D. 14–37)

Tiberius inherited the Augustan system and received precise instructions that the Empire was not to be extended,[1] an injunction he faithfully followed. Nevertheless the Emperor, who had been one of the most experienced of the Augustan commanders, faced difficult military problems in different quarters. At the death of Augustus two serious mutinies broke out in attempts to improve pay and service conditions; four of the Rhine legions were involved and three legions in Pannonia. These events clearly showed the danger of having too many legions in the same winter camp and there was, thereafter, a tendency to spread them out along the frontiers, although it was not until Domitian that the policy of brigading two together was finally abandoned. It was perhaps as a result of these troubles that winter accommodation was improved. Living in tents through a winter on the northern frontier must have been an endurance test in itself and they began to be replaced by timber buildings, which were already in use for food storage under Augustus.[2] There was also a serious war in Africa started by Tacfarinas, a Numidian, who had been a Roman auxiliary. After deserting, he equipped and organized his native forces[3] into units on the Roman model.[4] With the help of the Mauretanians and the Cinithii he kept several Roman armies busy from A.D. 17 to 24.[5] Junius Blaesus was given IX *Hispana* from Pannonia, but the legion was returned before the revolt was over. Under Blaesus, the Army constructed a system of forts instead of retiring to winter quarters and employed mobile columns specially trained for desert conditions to keep Tacfarinas constantly engaged. This is almost reminiscent of Caesar and may well be the origin of the African *limes*. The end of this prolonged struggle certainly meant the more rapid development and urbanization of the country, as its later prosperity clearly indicates.

The revolt of Florus and Sacrovir in Gaul *c.* A.D. 21, soon suppressed, had

---

[1] *Annals*, i, xi.

[2] An example is the Augustan base at Rödgen, north of the Main, which has timber granaries, but space for tents (H. Schönberger, 'Ein Augusteisches Lager in Rödgen bei Bad Nauheim', *Saal.-J.*, 19 (1961), pp. 37–88); *Germania*, 45 (1967), pp. 84–95.

[3] He was accepted as a chief by the Musulamian tribe which lived on the edge of the desert.

[4] *Annals*, ii, 52. Further details are given in iii, 20–21; 32; 35; 73–74; iv, 23–24.

[5] Tacitus in a caustic aside (*Annals*, iv, 23) says that there were already three laurelled statues in Rome of Roman generals who had been given triumphs for victories over Tacfarinas.

as one of its causes the stamping out of Druidism.[1] The priests of this powerful nationalistic cult may have fled to Britain to strengthen the anti-Roman sentiments here until the débâcle of A.D. 60. Thrace was at this time a client kingdom, but obliged to provide troops to Rome as well as to their own king. A rumour spread that men were to be drafted into the *auxilia,* regardless of tribal affiliation, for service in distant parts of the Empire. This, according to Tacitus,[2] inflamed them into a revolt which was put down only after severe fighting. In the East the Roman position was strengthened by the annexation of Cappadocia and the selection of Zeno for the vacant throne of Armenia, but the greatest stroke was a new treaty with Parthia, engineered by the able Vitellius, governor of Syria, with Artabanus. Tiberius, while carefully following the instructions of Augustus, left the frontiers in a more stable condition, but his financial stringency did not permit the just demands from the soldiers for improved pay and shorter service to be met.

Imperial Policy under Gaius (A.D. 37–41)

The events in the reign of Gaius are difficult to disentangle from the absurdities recounted by Dio and Suetonius. He upset the balance of power by his sudden acts of both enmity and generosity. Ptolemy, who had succeeded Juba in Mauretania, was forced into suicide because of his popularity in Rome[3] and this precipitated a civil war in Africa. Only the murder of Gaius in A.D. 41 prevented a serious crisis in Judaea, where the god-Emperor was bent on erecting a large statue of himself in Jerusalem. The hatred and mistrust sown by this folly festered to burst into the terrible fury of the Jewish Revolt and destruction of Jerusalem under Titus.

Imperial Policy under Claudius (A.D. 41–54)

Claudius inherited an empty treasury, fear and dissension everywhere, but no serious trouble on the frontiers. His reign marks a break with the then

[1] *Nat. Hist.,* 30, 13. The reason for Rome's hostile attitude towards Druidism may have been the barbarous practices rather than the political sentiments (H. Last, *J.R.S.,* 39 (1949), pp. 1–5).

[2] *Annals,* iv, 46. Thrace did not become a province of Rome until A.D. 46, nevertheless Thracian auxiliary units seem to have been serving in the invasion forces in Britain in A.D. 43 (*R.I.B.,* 109; 121; 201; 291). It must be assumed that the rumour had substance and the new conscription plan was enforced after the revolt had been suppressed. There is some doubt about the status of eastern Thrace after the murder of Cotys, and it may have been the intention of Tiberius to make it into a province, but for some reason this was not put into effect.

[3] It could be argued that this was merely a ruthless preliminary to the annexation of Mauretania (Momigliano, *Claudius,* 1961, p. 55).

outdated Augustan system. The long-deferred relationship with Britain was now resolved with the invasion of A.D. 43[1]. This extension of Roman power beyond the limits of the ocean produced, as it had done in Caesar's day, a deep impression in Rome. New problems in launching such a large expedition[2] had to be faced and solved. A new fleet, the *Classis Britannica*, had to be created and transports and supply vessels commissioned.

The Roman frontier systems in Britain have a special relevance, since they exhibit considerable variety. Some of their remains have survived to a remarkable extent and, above all, they have been subjected to more intensive study than those of any other part of the Empire. The Roman military leaders were faced with two difficult problems, as it became necessary to extend control over additional areas of hostility. Firstly, there was the need to protect the English lowlands from attack from the north and west; secondly there was the need to hold down hostile populations in the highland zones of Wales and northern England. The conquest under Claudius appears to have had the limited aim of taking over the kingdom of Cunobelinus, the subjugation of the Durotriges and Belgae of the south-west and the granting of independence to the Iceni and a section of the Atrebates, the latter probably as a reward to Cogidubnus for his services. The north was protected by the creation of a client kingdom of the tribes of the Pennines under the rule of Queen Cartimandua. Of the arrangements of the frontier itself little is known.

The absence of any marked physical barriers across the Midlands forced the Romans to make use of what there was and create a fortified zone. The frontier became anchored on the Humber Estuary in the north and the estuary of the Exe in the south; full use was made of the lower Severn and Trent, while in the central area the Avon Valley with its woodland fringe and limestone escarpment may have been considered an effective barrier. The area under direct military control was the whole of the newly conquered province extending even into the client Kingdoms. Units were placed near large native settlements but otherwise spread out to form a network with

---

[1] Lessons must have been learnt about the equipping of ships and crews for heavy sea conditions in A.D. 16 when Germanicus moved his army to the mouth of the Ems and gales scattered his fleet in all directions. The account of Tacitus gives some idea of the terrors this inspired (*Annals*, ii, 23–24).

[2] The four legions were taken from the Rhine and Danube, II *Augusta* from Argentoratum (Strasbourg), IX *Hispana* probably from Siscia (Sisak) in Pannonia, XIV *Gemina* from Moguntiacum (Mainz) and XX *Valeria* from Novaesium (Neuss).

perhaps a greater concentration on the frontier where a lateral communication route was established which was to become known as the Fosse Way.[1]

XX Legion was held in reserve at the British capital—Camulodunum (Colchester)—IX Legion built its fortress on the hill-top position at Lincoln, although perhaps not immediately,[2] XIV Legion probably occupied a central site somewhere in the Midlands, possibly along Watling Street south of High Cross, while II *Augusta* may have been divided to provide garrisons for a number of forts in an attempt to hold down the peoples of the south-west, who appeared to accept Roman authority with some reluctance.[3]

The attacks of the Britons from Wales under Caratacus made it necessary to move the central portion of this frontier forward to the middle Severn, and a network of forts controlled the West Midlands. This became an unsatisfactory frontier; as it allowed the Welsh tribes to develop a relentless guerrilla warfare in a most suitable terrain, leading to considerable losses and the decline and death of the second governor, Ostorius Scapula.[4]

Although there were no land adjustments to the Rhine and Danube frontiers, the reign of Claudius saw much of the territories opened up by new roads and a deliberate policy of pacification and Romanization. Similar influences can be traced in Asia Minor and in the eastern provinces. Although King Agrippa in Judaea was ostensibly a philo-Roman, the growth of his influence by intrigue perturbed Claudius sufficiently for Judaea to be taken over at his death in 44; it was subsequent bad Roman administration which led to further trouble.

Some serious fighting was necessary in Africa following the removal of King Ptolemy. C. Suetonius Paulinus, followed by Hosidius Geta, broke the power of the nomadic tribes in the deserts and the mountains. Mauretania was thereafter divided into two provinces, Caesariensis and Tingitana, each governed by a procurator. Claudius had always been a patron of the equestrian order and it is perhaps hardly surprising to find that the new provinces here and in the East provided fresh opportunities for the

---

[1] *Arch. J.*, 115 for 1958 (1960), pp. 49–98, *Britannia*, i (1970), 179–197.

[2] Insufficient pottery has been recovered from the legionary levels as yet to determine a starting date and a fort, half legionary size, is now known from crop-marks at Longthorpe, near Peterborough and another of similar size and form at Newton on Trent (*J.R.S.*, 55 (1965) pp. 74–76).

[3] For a more detailed consideration of these and subsequent arrangements see D. R. Dudley and Graham Webster, *The Roman Conquest of Britain*, 1965; also M. G. Jarrett, 'Legio II Augusta in Britain', *Arch. Camb.*, 113 (1964), pp. 47–63.

[4] Tacitus, *Annals*, xii, 39.

advancement of the knightly class; unfortunately for Rome there were too few equestrians of the necessary calibre.

Imperial Policy under Nero (A.D. 54–68)

At the outset of this reign there were two opposing forces directed towards the frontier problems. Nero's advisers Seneca and Burrus continued the Claudian policy of consolidation and avoidance of new commitments. At first caution prevailed and it may even be that the situation in Britain caused some discussion on the abandonment of the province.[1] But the young Emperor wished for military glory and self-aggrandisement, and the selection of Q. Veranius as governor marked a change in Imperial policy[2] towards the resolution of the frontier problems of the Welsh Marches by a bold advance, and one is tempted here to see the hand of the young Nero. When Veranius died his place was taken by another experienced general, C. Suetonius Paulinus, whose rapid campaigns into North Wales were brought to a sudden halt by the revolt of Queen Boudicca in A.D. 60[3]. Little is heard of the tribes of Wales after this, so it can be presumed that the main resistance had been overcome. The rebellion merely delayed the consolidation of the new conquests.

A similar situation was resolutely faced in the East, where the Parthians had reoccupied Armenia and placed Tiridates, brother of the Parthian king, on the throne. An experienced commander was found in Cn. Domitius Corbulo, who also had a reputation as a disciplinarian.[4] At first there were difficulties with the governor of Syria, Ummidius Quadratus, who felt he had been superseded: Corbulo soon discovered that the eastern legions were unfit for serious campaigning and had become softened by the years of peace.[5]

---

[1] This is a very difficult point on which there has been much debate. The evidence comes from Suetonius (*Nero*, 18), who stated that Nero considered withdrawing from Britain, but kept his forces there, as such a decision might reflect on the glory won by his adoptive father, Claudius. Seneca was sufficiently alarmed to start calling in loans made to British chiefs, one of the factors of the revolt of A.D. 60.

[2] E. Birley, 'Britain under Nero: the significance of Q. Veranius', *Roman Britain and the Roman Army*, 1953, pp. 1–9.

[3] For a study of these events and their military aspects see D. R. Dudley and Graham Webster, *The Rebellion of Boudicca*, 1962.

[4] *Annals*, xi, 18.

[5] *Annals*, xiii, 35. Tacitus says that there were old soldiers who had never been on guard or watch, who had long since lost their armour and who had become accustomed to a completely civilian existence.

It was typical of Corbulo that he took the three legions involved,[1] reinforced by the IV *Scythica* from the Danube,[2] into the mountains to winter under tents.[3] The campaign started with a reverse suffered by a disobedient subordinate.[4] Later neither side would risk a pitched battle and there were efforts at settlement by diplomacy. In this the Romans made a real concession in recognizing the reality of the situation by offering Tiridates the throne, providing he would accept it as a gift from Rome. This he refused, so Corbulo pressed forward to the capital, Artaxata, which surrendered without a fight.[5] Next year the Roman Army undertook the difficult march of 300 miles to Tigranocerta, which also fell easily. Nero now had to find a king for Armenia and chose Tigranes, of the Cappadocian royal house, who had lived most of his life as a hostage in Rome. It was a compromise solution, since he was not accepted by all and certainly not by Parthia. As a protection the new king was given a thousand legionaries and five auxiliary units. Corbulo was now governor of Syria, and with Parthia preoccupied with wars of her own, the balance of power might have remained steady. But Tigranes invaded Adiabene and it is difficult to believe that this was done without Roman knowledge or connivance. The Parthians could no longer stand aloof and a campaign followed which is described by Tacitus.[6] Corbulo advised Nero to appoint a separate commander for Armenia, since he was fully occupied in defending Syria and the Emperor made a disastrous choice in Lucius Caesennius Paetus, who was not only totally incompetent but, unless the sources are biased, a fool and a coward. While Corbulo fortified his section of frontier and prevented any Parthian crossing of the Euphrates, Paetus was forced into humiliating surrender and Rome had lost Armenia. The Parthian king, Vologaeses, was still anxious to come to terms with Rome, and after much negotiation Tiri-

---

[1] III *Gallica*, VI *Ferrata* and X *Fretensis*.

[2] Parker, p. 135, where it is suggested that Tacitus confused IV *Macedonica* with IV *Scythica* (*Annals*, xiii, 35).

[3] Conditions were severe, but the stories of frost-bite reported by Tacitus may be extravagant rumour. The general gave certainly no quarter to deserters or shirkers.

[4] Paccius Orfitus received the indignity of having to encamp his unit outside the fortifications of the main camp as a punishment.

[5] Frontinus tells us that while the leading citizens were debating as to whether to hold out against the Romans, Corbulo executed one of the nobles he had taken prisoner and had his head shot by a ballista into the city, a barbarous but successful device which precipitated a hasty surrender (*Strat.*, ii, 9, 5).

[6] *Annals*, xv, 1–17.

dates himself went to Rome and received the crown of Armenia from Nero. Although this seemed at the time to be a great diplomatic triumph, the situation reverted to precisely that of A.D. 54; the prodigious efforts, losses and humiliation were all for nothing.[1] It created, however, a stability which remained unshaken for another fifty years, and frontier control was strengthened when Pontus, which had been a client kingdom, was annexed in 64. This brought the whole of Asia Minor under direct Roman control, together with the southern shore of the Black Sea as far east as the Caucasus. This involved an extension of the Black Sea Fleet and a new base at Trapezus,[2] which not only cut communications between Armenia and the Black Sea but meant that Roman expeditions and permanent garrisons could easily be supplied by water transport. There is a hint that at about the same time the Romans were extending their control along the north coast of the Black Sea as far as the Crimea. A Roman force under Plautius Silvanus Aelianus assisted the Greek city of Chersonesus (near Sebastopol) against the Sarmatians, who were besieging it. This activity around the Euxine Sea may be connected with Nero's idea of mounting an expedition along the northern frontier with Armenia, using the valleys of the Rioni and Kura south of the Caucasus through Tiflis (modern Tbilisi in Georgia).[3] It is difficult to imagine the real purpose of such a project or of another against Ethiopia.[4] On the Rhine and Danube the reign of Nero saw no serious problems and the Claudian policy of consolidation was pursued.

---

[1] During the final negotiations Corbulo, given a special command with *maius imperium*, mounted a display of strength with an additional legion, XV *Apollinaris*, from Pannonia.

[2] This had previously been the responsibility of the client kings and consisted mainly in the suppression of piracy.

[3] The evidence for this is in Tacitus, *Hist.*, i, 6; Suetonius, *Nero*, 19, and Dio, lxii, 8, 1. A charitable explanation would be that it was a mistaken idea of relieving the pressure of the Sarmatians on the lower Danube; these wandering nomadic tribes were on the move at this period and the Romans may have been misinformed about their origins, although according to Dio some exploration had been carried out. Troops assembled in Italy for this expedition from Germany, Britain and Illyricum were still there in 69 to add to the confusion which faced Galba. Nero had also raised a new legion, I *Italica*, the recruits for which, according to Suetonius, were Italian born and six feet tall. Nero's designation of this unit as 'the Phalanx of Alexander the Great', clearly the product of a disordered mind, may reflect the Emperor's ambition.

[4] A small exploration party of praetorians had been sent up the Nile in 61, according to Seneca, to discover its source (*Quaestiones Naturales*, vi, 8, 3–4), but it could equally well have been a military reconnaissance.

Frontier Works

At this point it may be as well to consider more precisely the nature of these early frontiers. There are two main types of problem; firstly the control of the movement of people and goods by enforcing the use only of the authorized routes which could be fully supervised; secondly that of dealing with the sudden appearance of a large hostile force. The essential task here was to prevent a frontier from being surprised and overrun. A system of some depth is needed, with watch-towers on high parts to observe the forward areas. Attached to these would be signal-towers to send information rapidly to the forts in the rear, so that, fully alerted, the main forces would have time to carry out a planned campaign to meet the attack and prevent its penetration. It is rarely possible to study such a system, since the small sites of watch-towers and signal-stations are easily lost in countries with developed cultivation. Such a system has, however, been observed in the desert conditions of the Negev by M. Gichon,[1] and he considers this may date to Flavian times, although sound dating evidence is still wanting. Full use was made by the Romans of a series of steep-sided wadis and M. Gichon draws attention to the careful siting of signal-towers on ground screened from the eyes of the enemy, so that communications could be maintained in secret. The deserts of North Africa have also produced frontier posts in a similar, excellent state of preservation. Here it became necessary to create a frontier facing the desert, and this took the form of patrol roads, a flat-bottomed ditch (*fossatum*), block-houses and forts, operated on a kind of home-guard basis.[2] These large-scale works are also associated with a complex system of water control and storage which allowed a flourishing agriculture, mainly olive-yards. The astonishing complexity of the system, as revealed by the brilliant work of Colonel Baradez in his extensive air and ground reconnaissance, is due to the long period of change and development. It should be possible, with further work on the ground, to separate out all the elements into their component historical sections. It is not easy to judge the functions of the works, especially the *fossatum*, which bears a resemblance to the *vallum* of Hadrian's Wall.

The situation on the frontier with Parthia is far more complicated and reflects the changing fortunes of the relationship between the two great powers. The main military need in this large area was to secure the main routes and river crossings of the Tigris and Euphrates. Although much work

---

[1] *Israel and her Vicinity in the Roman and Byzantine Periods*, Tel Aviv, 1967, pp. 39–47.

[2] J. Baradez, *Fossatum Africae*, 1949; see also reviews in *J.R.S.* 40 (1950), pp. 162–5.

Plate VIII
Scenes from Trajan's Column

(a) a dragon standard of the Sarmatians (p. 136).

(b) Two columns advancing towards a camp, showing the *cornucines* (p. 142).

Plate IX

(a) The standards of the Praetorians and legionaries from a scene on Trajan's Column (p. 134).

(b)
A Praetorian standard-bearer on a fragment of marble in Boston, Mass. (p. 141).
(Photograph by the Museum of Fine Arts, Boston)

has been done on these problems,[1] as in North Africa, only detailed surveys on the ground will enable the features of each period to be isolated and then the whole of each sequence studied and assessed. The position was simplified where there was a great river barrier like the Rhine or the Danube. Here additional eyes were provided by the galleys of the fleet, and the watch-towers and signal-stations, so clearly delineated on Trajan's Column, were established along the river bank.

Imperial Policy under the Flavians (A.D. 69–96)

Under the Julio-Claudian dynasty there had been a gradual extension of the Augustan Empire, but it was also a period of peaceful consolidation and the penetration of trade and Roman ideas into and beyond the barbarian frontiers. Above all, as at no other time during the Empire, the great northern frontiers were reasonably quiet. But beyond this peaceful horizon there were already forces at work which were to remain a constant and growing threat along the Rhine and Danube. Already in Nero's time the Sarmatians were pressing into the Danubian lands and other tribes were on the move; each disturbed the other in their quest for land on which to settle. But of immediate concern to the new Flavian dynasty was the unrest provoked by the Civil War. Internally this had had a grave effect on the Army, whose loyalties had been so seriously divided. Armies were assembled and pitted against each other by the four emperors Galba, Otho, Vitellius and Vespasian in rapid succession. Thus the Rhine and Danube frontiers were stripped of troops and many of the legions marched great distances according to the dictates of loyalties and circumstances.[2] Some engaged in heavy fighting and suffered

[1] In particular A. Poidebard, *La Trace de Rome dans le désert de Syrie,* 1934; R. Mouterde and A. Poidebard, *Le Limes de Chalcis,* 1945; Sir Aurel Stein, *Geog. J.,* 92 (1938), pp. 62–66 and 95 (1940), pp. 428–38; D. Oates, *ibid.,* 122 (1956), pp. 190–9.

[2] The three British legions were fortunate in this respect in remaining undisturbed, although Vitellius forced them to send a large vexillation to Italy; also the six legions which comprised the garrisons of Syria and Judaea (IV *Scythica.* VI *Ferrata,* XII *Fulminata,* V *Macedonica,* X *Fretensis* and XV *Apollinaris*) remained, since the Jewish Revolt had yet to be finally crushed. Mucianus, however, took VI *Ferrata* and a large vexillation of the other legions against Vitellius. Nor did the three legions of Egypt and Africa (III *Cyrenaica,* XXII *Deiotariana* and III *Augusta*) play any part in the Civil War. It is worthy of comment that the critical battle which helped to decide the outcome of the Civil War was fought through the night between Bedriacum and Cremona; at dawn III *Gallica,* which had been one of Corbulo's crack Syrian legions, turned in the true Eastern manner to salute the rising sun. The Vitellian Army thought they were hailing reinforcements and fled (Tacitus, *Hist.,* iii, 24–25). The complex movements of the legions are worked out by Parker, pp. 140–5.

PRAET. AGRIPP.
(Valkenburg)
Lek
Rhein
Waal
Issel
(Nijmegen)
NOVIOMAGUS
VETERA
(Xanten)
Lippe
Ruhr
GERMANIA
Maas
Wupper
NOVAESIVM
(Neuss)
INFERIOR
Rur
(Köln)
Sieg
BONNA
(Bonn)
Rhein
Vinxtbach
GER.
SUPERIOR
Mosel
THE LOWER RHINE LIMES
LEGIONARY FORTRESS
(NIEDERGERMANISCHE)
AUXILIARY FORTS
KM
MILES
BH

FIG. 3

severe losses. The net result was the disbanding of four legions[1] and the introduction of five new ones.[2]

The events of the Civil War were watched with great interest by the tribes on the Rhine and Danube. There were new and unexpected opportunities for plunder in the areas of growing prosperity. The Dacians crossed the Danube thinking that Moesia was at their mercy,[3] but fortunately for Rome Mucianus was near at hand *en route* for Italy and was able to deal with them promptly, and soon afterwards the useful task of restoring order was given to some of the defeated Vitellian legions.[4]

Far more serious was the revolt of the Batavians under Civilis.[5] Its initial success was due to the depleted state of the Rhine legions, their best men having been taken by Vitellius; also while the men were loyal to Vitellius, the officers were in sympathy with Vespasian and many had no doubts of the outcome of the Civil War. The situation was complicated by the fact that Civilis could claim to be acting under instructions from Vespasian, for it is clear that the rising was very useful to the Flavian cause in keeping the Rhine garrisons fully occupied. But the danger inherent in this was soon apparent when some of the Roman auxiliary units went over to Civilis. Among them were the eight Batavian cohorts which had distinguished themselves in Britain. Other tribes beyond the frontier joined, but the Gauls remained loyal to Rome. Civilis attacked the remnants of V *Alaudae* and XV *Primigenia*, besieged at Vetera, with engines built by Roman prisoners.[6]

---

[1] The four on the Rhine which had disgraced themselves in the revolt of Civilis, I *Germanica*, IV *Macedonica*, XV *Primigenia* and XVI *Gallica* (see also E. Birley in *J.R.S.*, 18 (1928), 56–60).

[2] VII *Gemina* was raised by Galba in Spain, I and II *Adiutrix* were created from marines of the Mediterranean fleets and Vespasian created IV *Flavia firma* and XVI *Flavia firma*, using two of the numbers of the legions he disbanded and presumably some of the troops as well.

[3] A little earlier, while Otho was still in power, the Rhoxolani, a Sarmatian tribe, had also invaded, but had been surprised by III *Gallica* when laden with booty and cut to pieces (Tacitus, *Hist.*, i, 79; the historian also gives details of Sarmatian equipment and methods of fighting.).

[4] Tacitus, *Hist.*, iii, 46. The legion responsible for this success was VI *Ferrata*.

[5] Tacitus gives us a very detailed account in *Hist.*, iv and v.

[6] *Hist.*, iv, 23. The legionaries in their turn are reported to have built a machine in the form of a tower with a projecting arm which could be lowered on to the enemy, grab individuals and spring back carrying them into the air, swing round and throw them into the fort. This must have been a swinging boom equipped with grapple hooks. Its inventor must have remembered the machines built by Archimedes in the defence of Syracuse (Polybius, 8).

A relief force under Dillius Vocula was attacked and suffered heavy losses[1] but eventually fought its way to Vetera and raised the siege. By now the news of the Battle of Cremona had reached the combatants and Civilis was forced to throw off all pretence that he was merely acting for Vespasian. The legions accepted the new Emperor with reluctance. Their supplies were low and they were very short of men. At this critical juncture Civilis was joined by some of the Gallic leaders who must have considered that Roman control of the lower Rhine could now be ended and the vision of an *Imperium Galliarum* lured them forward. Vocula attempted to restrain his men, but was murdered, and then followed the disgraceful spectacle of the legionaries taking an oath of allegiance to a foreign power, the Gallic Empire which only existed in the imagination of the rebels. The garrison at Vetera, reduced now to starvation, surrendered—only to suffer massacre. Civilis and his allies now had complete control of Lower Germany, but not for long. Mucianus, at this time ruling in Rome, in the name of Vespasian, organized a force of nine legions[2] under the command of Q. Petilius Cerialis and Ap. Annius Gallus. The rebels showed a complete lack of cohesion and strategy and the outcome was never in doubt. Cerialis on reaching Mainz sent his Gallic levies back to their homes with the message that the legions were sufficient to restore order. The rebellious Treveri took a stand in the strong hill position at Rigodulum, but Cerialis had no difficulty in taking it by storm, and the Romans entered Trier, where they met the wretched remnants of I *Germanica* and XVI *Gallica* to whom Cerialis showed wise clemency.[3] Civilis now drew together all his forces and made a surprise attack on the Romans before they had been able to fortify their camp: a fierce battle was won by the steadiness of the legions and in particular XXI *Rapax* distinguished itself. But the Germans withdrew to Vetera, now skilfully defended by diverting the flood waters of the Rhine. Cerialis, informed by a deserter of a way round, forced a victory. Civilis now retired to his island between the Waal and the old course of the Rhine and although he had some successes was eventually forced to capitulate. XIV

---

[1] IV *Macedonica* and XXII *Primigenia*.

[2] II *Adiutrix*, VIII *Augusta*, XI *Claudia*, XIII *Gemina* and XXI *Rapax* from Italy (all but the last being part of the victorious army), I *Adiutrix*, VI *Victrix* and X *Gemina* from Spain and XIV *Gemina* from Britain.

[3] Tacitus puts into his mouth the words, 'Now the soldiers who revolted are once more soldiers of their country. From this day you are enlisted in the service and bound by your oath. The Emperor has forgotten all that has happened and your commander will remember nothing' (*Hist.*, iv, 72).

Gemina crossed the North Sea to reduce the Tungri and Nervii of Belgica. It had been a long and bitter struggle with many lessons for the Romans on frontier policy. It was unwise to allow auxiliaries to serve near their homelands and under their own leaders; there was henceforth a tendency, which was soon reversed, to disperse these levies more widely into the *auxilia*. The legions on the Rhine were redistributed lest old sympathies should remain,[1] while X *Gemina* was established in a new fortress at Noviomagus (Nijmegen) to watch the Batavians.

It could be argued that the weakness of the Rhine frontier system in Upper Germany was the absence of legions, for as the *limes* moved progressively forward, taking in the Agri Decumates between the Rhine and Danube, the legions' fortresses remained on the west bank of the Rhine. Two of them in particular at Strasbourg and Vindonissa were almost a hundred miles in the rear by the end of the first century. On the lower Rhine, however, Noviomagus and Vetera remained on the frontier, since the river itself continued to provide the barrier.

To Domitian must be given the credit of strengthening the defences of the middle Rhine by subduing the Chatti and building a new frontier in the Taunus and Wetterau between the Lahn and the Main, some thirty miles east of the Rhine. The campaign of 83 was dismissed by Suetonius as having been unjustified,[2] ridiculed by Tacitus,[3] but merited a serious comment from Frontinus.[4] It is likely that there were no pitched battles, but a grim war of attrition in the forest-covered hills. A large army was needed and vexillations were called from all four of the British legions.[5] The result was a broad salient pushed into Germany, fortified with a line of timber watch-towers and signal-posts,[6] and in communication with the forts now established in the broad plain of the Main. Domitian and his advisers showed a firm grasp of the

---

[1] VI *Victrix* went to Novaesium, XXI *Rapax* to Bonn, I *Adiutrix* and XIV *Gemina* to Mainz, XXII *Primigenia* to Vetera, XI *Claudia* went to Vindonissa and VIII *Augusta* to Argentoratum (Strasbourg). For studies of the legionary dispositions and movements in A.D. 69 see Parker, pp. 140–5, and generally under the Flavians, R. Syme, *J.R.S.*, 18 (1928), pp. 41–55.

[2] *Dom.*, 6.

[3] *Agricola*, 39, is an echo of the events of the reign of Gaius.

[4] *Strat.*, i, 1, 8; Domitian disguised his intention of leading his troops by announcing that his departure from Rome was for the purpose of taking a census in Gaul, 'under cover of this he plunged into sudden warfare and crushed the ferocity of these savage tribes . . .'

[5] The senior tribune of IX won decorations (*I.L.S.*, 1025).

[6] These small posts were about ninety yards square and 400 to 700 yards apart.

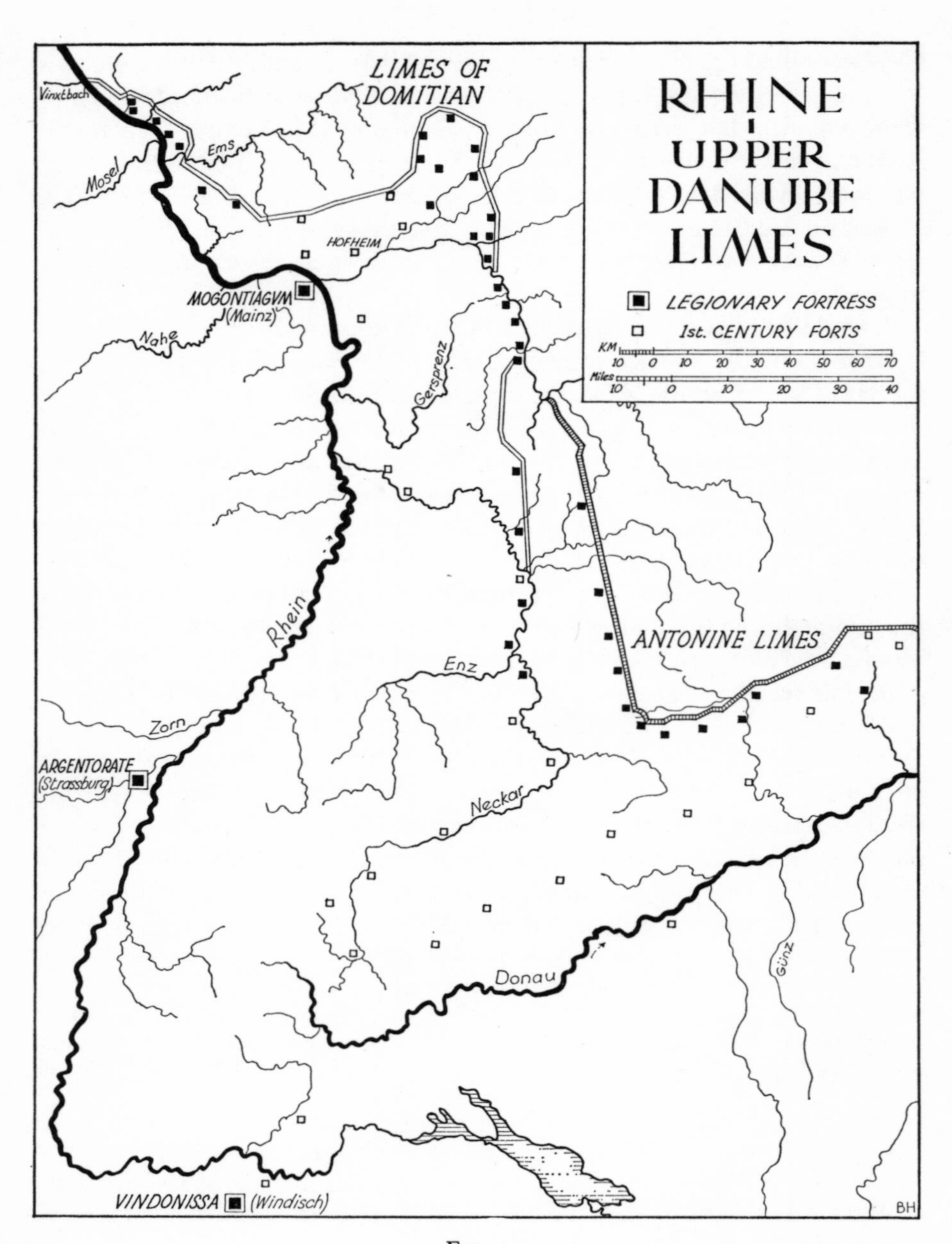

RHINE
UPPER
DANUBE
LIMES
LEGIONARY FORTRESS
1st. CENTURY FORTS
KM
10 0 10 20 30 40 50 60 70
Miles
10 0 10 20 30 40
LIMES OF
DOMITIAN
Vinxtbach
Ems
Mosel
HOFHEIM
MOGONTIAGVM
(Mainz)
Nahe
Gersprenz
Rhein
ANTONINE LIMES
Enz
Zorn
ARGENTORATE
(Strassburg)
Neckar
Donau
Günz
VINDONISSA
(Windisch)
BH

FIG. 4

strategical importance of the lands thus acquired, for Rome now controlled some of the main routes taken by tribes from the north and east advancing to the Rhine, in addition to completely straddling the principal north-south link. Furthermore it paved the way for the next logical step, a similar advance to the south up the Neckar to cut across the awkward re-entrant angle joining the Danube at Faimingen and thus shortening the frontier by 150 miles.

The situation on the Danube is less certain, but forts may have been established on the north side before the time of Domitian. By the end of his reign the line had been advanced beyond the Danube some twenty to thirty miles, joining the river near Eining a little way above Regensburg. The two new *limites* probably joined at Köngen.

The frontier to the east of Eining is basically a simple situation, since the great river itself was the *limes* and its garrisons sited at intervals along the southern bank. There are, however, serious exceptions to the very generalized picture. The three provinces Noricum, Pannonia and Dalmatia, with their difficult mountainous terrain, continued to be held with permanent garrisons. Legio IV *Flavia* was at Burnum and XIII *Gemina* at Poetovio under Vespasian, while only one legion was on the Danube itself, XV *Apollinaris* at Carnuntum. During this period there were four legions on the lower Danube in Moesia, all along the river bank.[1] The balance of power along this frontier suddenly changed in A.D. 85, when the Dacian tribes became united and hostile under Decebalus and there were twenty stormy years culminating in the campaigns of Trajan and the annexation of Dacia. A reorganization of the Danubian frontier may be assigned to this period as a result of this grave threat. At the mouth of the river where it swings to the north before flowing into its delta across the flat plain of the Dobrudja, the *limes* was shortened by the construction of a bank between Tomi on the shore of the Black Sea and Raşova on the Danube.[2] The first campaign mounted by Domitian in Dacia under Cornelius Fuscus, the praetorian *praefectus*, ended in disaster in A.D. 86. A

---

[1] VII *Claudia* at Viminacium (Kostolac near Belgrade), V *Macedonica* at Oescus and I *Italica* at Novae (Swisjtow), while the fourth legion, V *Alaudae*, may have been the one destroyed or disbanded in 86 after the loss of its eagle (Dio, lxviii, 9, 3), but this may refer to a praetorian standard (*C.A.H.*, xi, p. 171, fn. 1).

[2] This may have been the first continuous frontier structure erected by the Roman Army, and although only about forty miles in length it was probably the genesis of all later works, including Hadrian's Wall in Britain. There are actually three different lines to be observed and two sets of forts (C. Schuchhardt, *Die sogenannten Trajanswalle in der Dobrudscha*, 1918).

further campaign was mounted leading to a great Roman victory at Tapae under Tettius Julianus in A.D. 88.[1] If this advantage could have been immediately developed, Decebalus could hardly have survived to give Trajan so much trouble, but he was spared for the moment by two events. The first was the revolt of Antonius Saturninus on the Rhine, which collapsed as swiftly as it had begun,[2] and secondly there was trouble with the tribes which occupied the deep southern re-entrant between the Danube and the Theiss, the Sarmatian Iazyges, the Marcomanni and the Quadi. As long as these tribes were in friendly alliance with Rome the Pannonian frontier was safe, and prior to A.D. 87 only one legion had been placed on the Danube in the whole of its long length to Viminiacum, XV *Apollinaris* at Carnuntum. The situation was not only drastically altered, but if these tribes had combined with the Dacians it could have become extremely serious. Domitian resolved the problem, rapidly coming to terms with the Dacian king and even lending him skilled artisans of different trades, presumably in an attempt to divert his interests into peaceful channels. Domitian was at least now able to concentrate on the other troublesome tribes. Accounts of these campaigns are very unsatisfactory, but it is most probable that XXI *Rapax* was destroyed. A peace was eventually gained without any dramatic victories, but by a show of force and skilful diplomacy. The effect of all this was to strengthen the Danubian frontier of Noricum and Pannonia by four legionary fortresses instead of one.[3]

In Britain, the Civil War had brought about a change. The lessons of the great revolt had not been lost on the Roman government, and after 60 the governors were chosen for their administrative skills rather than any military abilities. Peaceful reconciliation and the creation of an urbanized province became the main aims. With the removal of Caratacus and the crippling of the

---

[1] By now three more legions may have been involved, IV *Flavia* from Dalmatia, I *Adiutrix* from Germany and II *Adiutrix* from Britain.

[2] It had an important consequence for frontier policy, since neither Domitian or subsequent emperors ever again allowed two legions to occupy the same camp (Suetonius, *Dom.* 7), except in Egypt, where two legions still shared a fortress at Alexandria as late as A.D. 119 (P-W., xii, 1510).

[3] The situation by 96 appears to have been as follows: XV *Apollinaris* at Carnuntum; XIV *Gemina* which was brought from Mainz to replace XXI *Rapax* *c.* A.D. 93 at Aquincum (Budapest); I *Adiutrix* moved to Brigetio *c.* A.D. 87 prior to the Dacian campaign of 88, and the XIII *Gemina* moved from Poetovio to Vindobona (Vienna) somewhat later under Trajan, all in Pannonia.

Druids,[1] the main hopes of the anti-Roman sentiment were now centred on Venutius, who had already turned against his spouse, the Roman nominee Cartimandua. In the time of Didius Gallus she kept her throne only because of the presence of Roman forces. The Civil War with its demands on the British legions necessitated a withdrawal of troops and with this the position of Cartimandua was untenable. Venutius was left in full possession of the north and the hostile frontier was merely guarded by Roman forces now without a strong and decisive leader. This was a situation which Vespasian could hardly tolerate and as soon as the Batavian revolt had been settled Petilius Cerialis became governor of Britain and was dispatched with an extra legion, II *Adiutrix*, to advance into the north. Cerialis naturally selected his old legion, IX *Hispana*, as his main force, advancing into the Vale of York, but XX *Valeria Victrix*, now commanded by Cn. Julius Agricola, moved in conjunction up the western side of the Pennines. Venutius decided to make his stand at Stanwick, by the division of the routes to the north near Scotch Corner. Sir Mortimer Wheeler has shown by his excavation of these remarkable earthworks[2] how they became the rallying-point of the gathering tribes, but once more the Britons failed, in a pitched battle against the superior Roman forces. The extent of the new conquests is more difficult to determine, but the northern frontier of Brigantia appears to have extended as far as the Solway -Tyne line[3] and finds from Carlisle point to Roman occupation at about this time.[4] The client status of Brigantia thus ended and the territory was incorporated into the province.

Cerialis was the first of three great military governors sent to Britain by the Flavian emperors. All the old doubts and hesitations which may have remained as the legacy of the Claudian conception of a limited conquest were finally swept away in bold decisions to advance and consolidate. Sextus Julius Frontinus, one of the most distinguished men of his day, appears from the brief note by Tacitus[5] to have completed the conquest of the Silures and

---

[1] Pliny (*Nat. Hist.*, 30, 13) stated that Druidic practice continued in Britain to his time of writing. This work was completed in A.D. 77, but it is impossible to be sure that this means that the cult survived in any strength after the destruction of the sacred groves in A.D. 60.

[2] *The Stanwick Fortifications*, 1954 (Research Report of the Society of Antiquaries of London, No. 17).

[3] E. Birley, *T. Dumfries and Galloway Nat. Hist. and Antiq. Soc.*, 29 (1952), pp. 46–65, reprinted in *Roman Britain and the Roman Army*, 1953, pp. 31–47.

[4] *Arch.*, 64 (1913), p. 311, and *T. Cumb. and West.*, 17 (1917), pp. 235 ff.

[5] *Agricola*, 17.

consolidated the military gains in Wales.[1] He was followed by Cn. Julius Agricola, whose career was almost entirely devoted to Britain and who was fortunate in having Tacitus as his son-in-law and a subsequent biographer. In spite of this detailed account there are many difficulties and problems over the campaigns in the north,[2] and one of the most perplexing is how Agricola managed to garrison such an enormous new area without weakening the hold on Wales and the southern Pennines, yet have troops to fight his great battle at Mons Graupius and supply Domitian with reinforcements for his German campaigns.[3]

Agricola's strategic reserves must have been cut down to a minimum and this may well account for the use of *auxilia* in the battle, the depleted legions being his reserve. It may also help to explain the gradual withdrawal from Scotland for which Domitian received a typical Tacitean condemnation.[4] The new conquests could not be held through lack of man-power. Agricola's scheme had been based on four legions, II *Augusta* now at Caerleon, IX *Hispana* at York, II *Adiutrix* at Chester[5] and the new fortress of XX

---

[1] The date of A.D. 75 can hardly be placed against every Roman site in Wales, as has been the tendency of antiquaries of the past.

[2] Among the considerable literature on the subject one might single out the contributions by I. A. Richmond, 'Gnaeus Julius Agricola', *J.R.S.*, 34 (1944), pp. 34–45, and 'New Evidence upon the Achievements of Agricola', *Carnuntina*, 1956, pp. 161–167, and in his introduction to *de vita Agricolae* with R. M. Ogilvie, 1967.

[3] Legionary vexillations are known to have been taken from all four legions (*I.L.S.*, 1025 and 9200).

[4] *perdomita Britannia et statim omissa, Hist.*, i, 2. The retreat to the Clyde–Forth line was probably rapid, but the area to the south of this, relinquished more slowly.

[5] The lead water-pipe from Chester stamped with Agricola's name (*E.E.*, ix, 1039; Wright and Richmond, 1955, No. 199) shows that construction of the fortress must have started by A.D. 79. At this time Agricola was under instructions from Titus to find a suitable defensive line and not proceed beyond it. It became evident in 80 that this was to be the Clyde-Forth line (*Agricola*, 23); the obscurity of this passage of Tacitus is due to his need to avoid giving Domitian credit for initiating a further advance which gave Agricola the opportunity for his fine victory at Mons Graupius (Frere, p. 109). The logic of these events is that Agricola intended to establish *Legio* XX in a new fortress in the north-west (Carlisle?), since the founding of Chester implies a withdrawal from Wroxeter. The change of policy in 83 and the further advance and conquests forced a change of plan and a new fortress was built at Inchtuthil, presumably after 84, even perhaps after Agricola's withdrawal. If the building of Inchtuthil had begun as early as 84, it is difficult to understand why it should still be in an unfinished state in 87, the earliest possible date on coin evidence for its abandonment (*J.R.S.*, 45 (1955), p. 123). Somewhere in the north-west there must be a legionary fortress to house XX between *c.* 79 and 84 or later. The evidence from Wroxeter suggests that the fortress

Valeria Victrix at Inchtuthil.[1] But it was a compromise solution: the eastern lowlands were to be occupied and the mouths of the glens to be blocked. No doubt this would have been successful, but Domitian had need of troops elsewhere and a withdrawal became imperative. Faced with the mass of Scotland left to the Caledonian tribes, the solution of finding a suitable frontier line was dictated by geography and the forces available. There were only two possible lines where estuaries cut deep into the land mass, the Clyde-Forth line and the Solway-Tyne line. Between these was about 7,000 square miles of land, much of it hilly, even mountainous, and heavily forested. To have kept this area under close control, assuming that one of the tribes, the Votadini, remained friendly to Rome, may have required at least thirty-five auxiliary units. If the units had not been available, the Roman high command would have had no choice but to fall back on the southern of the two lines. It was the high demands of manpower created by the Dacian Wars that made this decision inevitable. This left a large reservoir of hostile tribesmen, still smarting from the defeat imposed by Agricola. To them the withdrawal may have been seen as a sign of weakness. Domitian withdrew II *Adiutrix* from Chester, where it was replaced by XX *Valeria Victrix*. Just how gradual was the withdrawal from the North it is difficult to say. The destruction at Newstead dates from *c.* A.D. 100 and some of the forward area may still have been held. It is possible that a sudden native uprising at this time took the Roman command by surprise.[2]

In the East there were no major wars or disturbances under the Flavians once the Jewish Revolt had been crushed by Titus. But the strengthening and consolidation of the frontiers must have been in process here as elsewhere. At least the realities of the situation were understood in Rome not only by the two Emperors who had seen tough service there, but also by Domitian who was unlikely to embark on lost causes or expensive gestures against the Parthians. The lessons so painfully learnt under Nero were still clearly understood. When trouble seemed to threaten, M. Ulpius Traianus, father of the later

---

there was kept in commission at least until *c.* 80. It seems unlikely, however, that the unit would have been erecting a large stone bath-house there at this late date, as Frere has suggested (p. 117, fn. 2).

[1] When the devoted labours of Sir Ian Richmond recovered the total plan of the timber fortress (details of which have appeared progressively in the annual summaries in *J.R.S.*), it was found that a number of buildings including the legate's house had not been erected before the order to demolish was given.

[2] *Roman Britain*, 1963, pp. 45–46.

Emperor, used diplomacy rather than force. Parthian suspicions may have been aroused by the steady process of consolidation and road-building.[1] Finally Domitian is credited with raising the pay of legionaries from 225 to 300 *denarii* per year.[2]

Imperial Policy under Trajan (A.D. 98–117)

With Trajan, the Empire was once more under the command of a great military figure, regarded, in fact, by many as the greatest of imperial times. But his position has been enhanced by the adverse reports of Domitian, for behind the barrage of propaganda rested the very solid achievements of the Flavian dynasty. The immense care taken in readjusting and consolidating the frontiers enabled Trajan to concentrate entirely on two problems, Dacia and Parthia. The Parthian campaigns impressed his contemporaries, but in the sobering light of historical reassessment they appear grandiose and without serious justification.

Domitian's solution of the Dacian question had been only a temporary one to enable him to gain time to deal with the other tribes. The Dacians, united under their great leader, remained a serious potential threat to the security of the Danubian provinces. But Trajan proceeded with caution along the other frontiers. The keynote was consolidation and at many points there is evidence of the conversion of the old earth and timber forts into more durable stonework. By A.D. 101 Trajan was ready: two new legions, XXX *Ulpia* and II *Traiana*, had been raised to replace the two which had been lost, and by now at least thirteen legions were available on the lower Danube. In the first campaign[3] the route used by the successful Julianus was followed towards Tapae. The Dacians fell back, avoiding a pitched battle, and the Romans consolidated their gains. During the winter Decebalus launched a counter-attack somewhere in Lower Moesia which was repulsed. Having surveyed the difficult terrain in the approach from the west to the enemy capital Sarmize-

---

[1] A milestone found near Palmyra is dated to A.D. 75 (*Ann. Épig.*, 1933, No. 205). Vespasian had annexed Commagene in A.D. 72 (Josephus, vii, 7, 1).

[2] Suetonius, *Dom.*, 7, 3; Dio, lxvii, 3, 5.

[3] The only account of the Dacian campaigns is that of Dio, very fragmentary and confused. Trajan wrote his own commentary, but only one sentence, *Inde Berzobim, deinde Aizi processimus*, has survived in a contemporary grammar-book, indicating the route of the first season. Attempts to turn the scenes on Trajan's Column into a continuous narrative by Cichorius and Lehmann-Hartleben have never been very satisfactory, since the Column was never intended to do more than illustrate events well known to most contemporaries.

getusa, Trajan for the second season decided to advance from the east towards the Red Tower Pass.[1] Once inside the Carpathians he would need to reduce the Dacian fortresses on the Mühlbach before advancing towards Sarmizegetusa. So eventually the capital was reached, but Decebalus capitulated to save his town from destruction. Garrisons were left behind while Trajan returned to Rome for his triumph and assumed the title *Dacicus*. Soon, however, Decebalus felt strong enough to break the peace and the struggle was renewed in 105. A grim campaign followed, with much hard fighting, until the Dacian capital fell; the King fled to the north pursued by Roman cavalry, to be brought to bay and driven to commit suicide. This if depicted in a beautifully composed scene which is one of the most striking os all those on Trajan's Column.[2]

The incorporation of the new province necessitated a new arrangement of the frontier and its garrisons. It appears to have been of consular status under Trajan with XIII *Gemina* at Apulum, and possibly also *I Ad* and IV *Flavia* in the province[3] while there were now three on the lower Danube east of the river Aluta (Oltul).[4] The two previous legionary fortresses at Oescus and Ratiaria became *coloniae*, while in Upper Moesia the VII *Claudia* remained at Viminacum, IV *Flavia* and II *Adiutrix* at Singidunum. The two Pannonian provinces probably had five legions at the close of the wars, but the situation is far from clear.

There are further problems over the position of the Dacian frontiers. It has been assumed, since there is a series of forts along its course, that the Aluta became the eastern frontier of the new province. There is, however, evidence of garrisons and civil settlements to the east in Wallachia and a road was built up the river Sereth to Poïana. Thus it would seem that it was the Roman intention to hold this area, although it may never have become as pacified as the western part of Dacia.[5] Nor is the western boundary defined,

---

[1] G. A. T. Davies, 'Topography and the Trajan Column', *J.R.S.*, 10 (1920), pp. 1–28.

[2] Cichorius, Taf. cxlv.

[3] R. Syme, *Laureae Aquincenses*, i (1938), pp. 267–86.

[4] I *Italica* at Novae; XI *Claudia* at Durostorum (Silistra) and V. *Macedonica* at Troesmis (about thirty miles south of Galaţi).

[5] That the Sereth was held as early as 100–5, probably prior to the outbreak of the second war, is shown by the Hunt *Pridianum* of *Cohors I Hispanorum* (Robert O. Fink, *Roman Military Records on Papyrus*, 1971, No. 13). This interesting military document lists troops on garrison duty at Piroboridava which can be identified as Poïana.

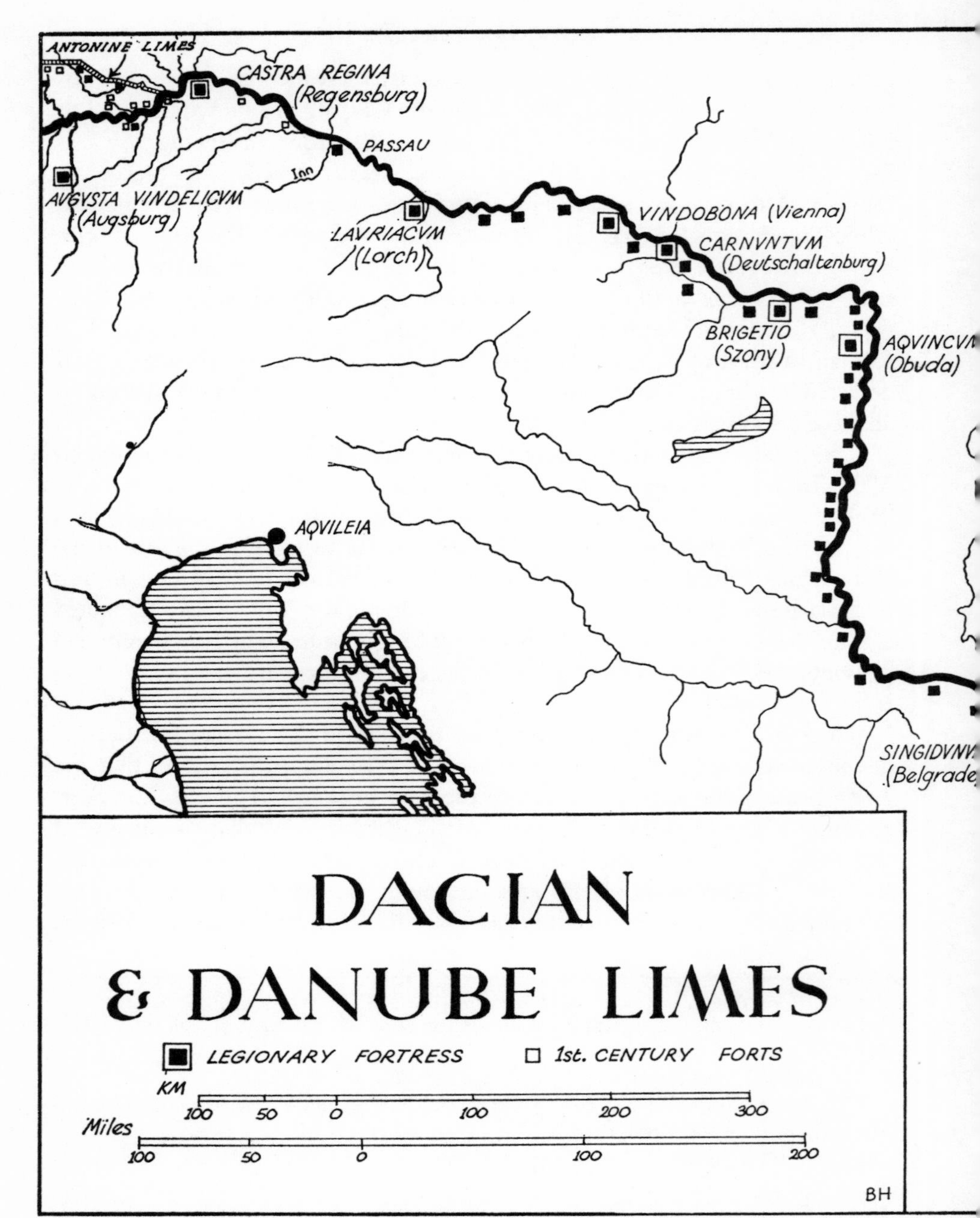
ANTONINE LIMES
CASTRA REGINA
(Regensburg)
PASSAU
Inn
AVGVSTA VINDELICVM
(Augsburg)
LAVRIACVM
(Lorch)
VINDOBONA (Vienna)
CARNVNTVM
(Deutschaltenburg)
BRIGETIO
(Szony)
AQVINCVM
(Obuda)
AQVILEIA
SINGIDVNVM
(Belgrade)
DACIAN
& DANUBE LIMES
LEGIONARY FORTRESS
1st. CENTURY FORTS
KM
100
50
0
100
200
300
Miles
100
50
0
100
200
BH

F

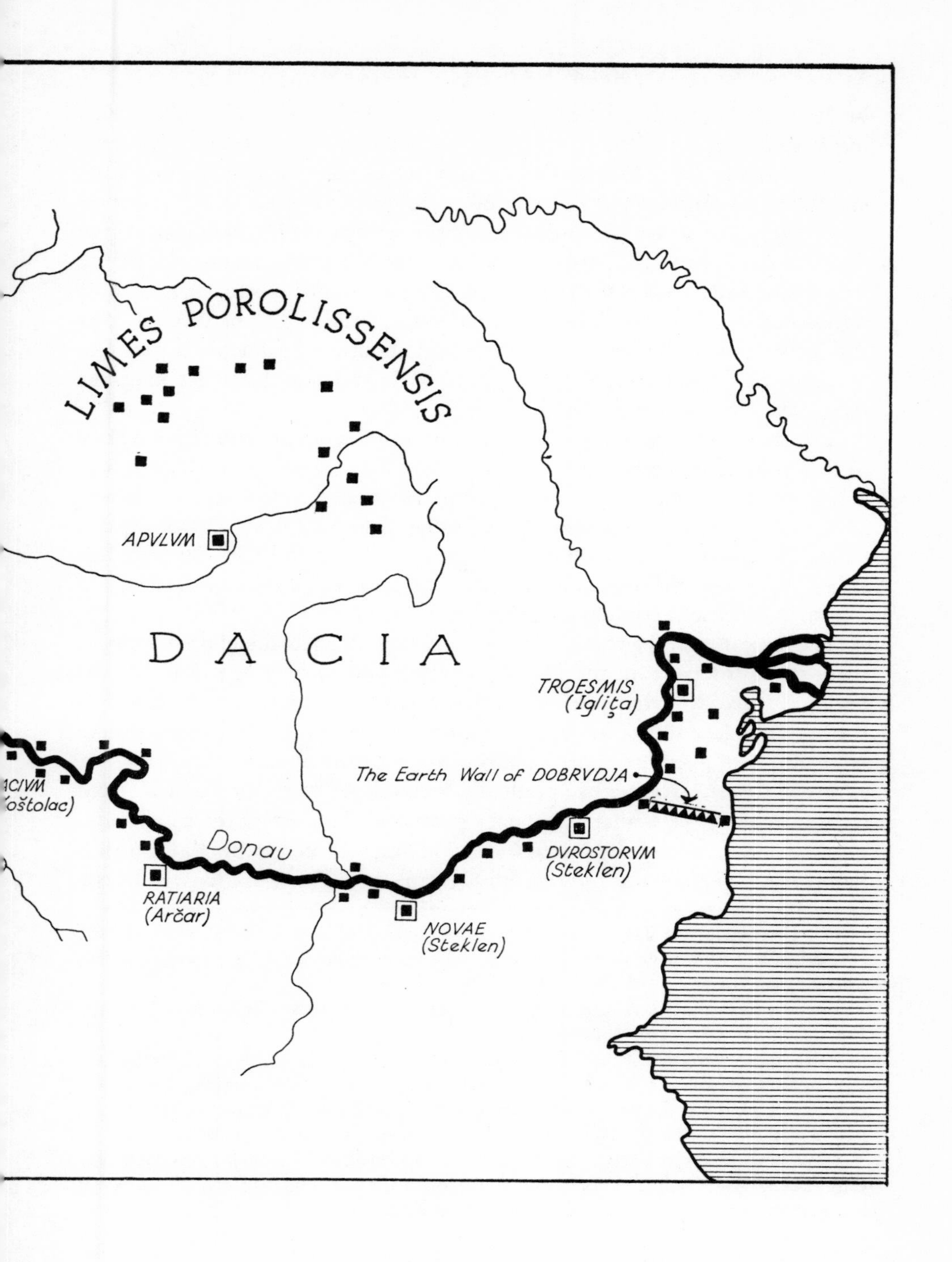
LIMES POROLISSENSIS
APVLVM
DACIA
TROESMIS
(Iglița)
The Earth Wall of DOBRVDJA
Donau
DVROSTORVM
(Steklen)
RATIARIA
(Arčar)
NOVAE
(Steklen)

but it would seem unlikely that the area between the Theiss and Danube, forming a deep inlet into Roman territory, would have been left entirely in the hands of the Iazyges. The northern limits can be seen by a network of forts and their garrisons defined with the aid of *diplomata*.[1] In 1965 an investigation was made of the area in the south-west of the province of Dacia Porolissensis.[2] The frontier follows the narrow summit of the Meses Mountains, where the steep-sided valleys obviate the need for any continuous barrier. A system of signal- and watch-towers has been found with a series of forts between Bologa and Teháu seven kilometres in the rear and about ten kilometres apart. The *municipium* of Porolissum itself appears to be defended by linear features, but these have yet to be firmly dated.

The effect of the extension of the Empire was a rapid growth of prosperity in the lands of the lower Danube, its delta and the shores of the Black Sea, and this is even reflected in a new upsurge of urban development in Thrace. More important to the student of the Roman Army are other durable results of the Dacian wars, the Column of Trajan in Rome and the monument at Adamklissi[3] with their remarkable series of reliefs which portray the army of Trajan in its varied activities in such splendid detail.

The unfortunate Parthian expedition which clouded the end of Trajan's reign drew troops from the Danube frontier and some reorganization was required there. XV *Apollinaris* was taken from Carnuntum and remained in the East: its place was taken by XIV *Gemina* moved from Vindobona;[4] there were also considerable vexillations from some of the other legions.

In the East the frontier had gradually been moved up to the Euphrates and strengthened by permanent legionary garrisons. The last of the small independent states, that of the Nabataeans, was annexed in 106 by A. Cornelius Palma and VI *Ferrata* was now established at Bostra. This was followed by the

---

[1] C. Daicoviciu, 'La Transylvanie dans l'Antiquité', in *La Transylvanie*, 1938, and C. Daicoviciu and D. Protase, 'Un nouveau diplôme militaire de Dacia Porolissensis', *J.R.S.*, 51 (1961), pp. 63–70.

[2] Report forthcoming. A brief account was presented by Stephen Ferenczi at the Limes Congress in 1967.

[3] This great memorial dedicated to Mars Ultor (Mars the Avenger) stands behind the *limes* of the Dobruja near the altar thought originally to have been erected after the disaster of Fuscus in 86. It is in the form of a *trophaeum*, and its panels, while of crude provincial work, have a vitality of their own (Florea Bobu Florescu, *Monumentul de la Adamklissi*, 1959).

[4] The new dispositions appear to be as follows: X *Gemina* at Vindobona, XIV *Gemina* at Carnuntum, XXX *Ulpia* at Brigetio, II *Adiutrix* at Aquincum, and the IV *Flavia* at Singidunum; the legions in Lower Moesia remained unchanged.

Plate X
Standards on a relief from S. Marcello, Rome (p. 139).

Plate XI
The tombstone of an *optio* Caecilius Avitus of XX *Valeria*, at Chester (p. 141).
(Photograph by the Grosvenor Museum, Chester)

building of a great north-south highway linking the Red Sea, where there was now a squadron of the fleet, with Syria, and controlling the great caravan trade routes and facilitating the sea link with India. The speed of Roman penetration must have been observed with growing alarm by the Parthians, but there were no hostile acts. The war was the result of a seizure of the throne of Armenia by the Parthian King Osroes for his own nominee, thus causing a break with the Neronian conception of this throne being under the control of Rome. Trajan might have intervened with diplomacy and a show of force, but lured by the vision of Alexander, chose to mount a vast expedition with a view of settling the Parthian problem once and for all.[1] Although there were eight legions already in the East, some of them perhaps of dubious quality, it soon became necessary to call in XV *Apollinaris* from Carnuntum and vexillations from several others. The first moves were into Armenia, but merely to meet the King, who was informed that he no longer enjoyed that position. A Roman governor was appointed to the new province, which included Cappadocia. Trajan was still thirsting for conquests and as yet there had been little fighting. He advanced south into Mesopotamia and occupied Nisibis again without loss. He then wintered in Antioch, where he was injured by an earthquake. In the following spring the Army was ready for another advance, this time towards Ctesiphon, the Parthian capital, where the Tigris and Euphrates are only some forty miles apart. The Parthian king, troubled by internal dissension, was unable to offer serious resistance and the capital fell after a short siege. It is possible that Trajan now travelled down to the Persian Gulf and created a client-kingdom there, bringing the whole of the great trade route to the Far East under Roman control.[2] This was the peak of Trajan's achievements; the great conquests so easily won proved more difficult to hold. Serious revolts broke out and, spreading to several provinces, were put down with difficulty. It was very obvious that the grand design of the Eastern Empire was completely impracticable. The weakening of the Army on other frontiers began to show with troubles in Britain and on the Danube and a serious revolt in Judaea which spread to Egypt. At this point the Emperor's health gave way and he died in 117 after handing over to Hadrian an Empire now expanded to its fullest extent, and the Emperor, faced with the task of maintaining this vast frontier, abandoned Mesopotamia. As he moved from province to province with restless energy the predominating theme was consolidation.

---

[1] The sources for this war are very poor; cf. F. A. Lepper, *Trajan's Parthian War,* 1948.

[2] R. P. Longden, 'Notes on the Parthian campaigns of Trajan', *J.R.S.,* 21 (1931), pp. 1–35.

It was during the reign of Trajan that there was instituted a general programme for the steady replacement of timber buildings and turf ramparts with stone, dated by inscriptions from all three legionary fortresses in Britain.[1] The Trajanic frontier in Northern Britain yet remains somewhat uncertain, but the troubles of the early years of the second century must have necessitated a new arrangement. A pre-Hadrianic frontier has been postulated along the Stanegate[2] with the features associated with frontier works, forts, fortlets and watch- or signal-towers[3] between Carlisle and Corbridge, but there is so far no evidence of the eastern section.

Imperial Policy under Hadrian (A.D. 117–38)
Hadrian's Wall (Fig. 6)

It was in Britain that Hadrian left his most impressive mark, the great Wall that bears his name. The Trajanic solution proved to be inadequate as pressures from the North increased during the early years of the second century. Nothing is known of these events and speculation has filled the void. A romantic aura of mystery and imagination has surrounded the disappearance of IX and a great Roman defeat has been visualized, leading to its replacement by VI *Victrix*. However, the recent discovery of traces of this unit on the lower Rhine opens up alternative possibilities.[4] The situation in Britain had become

---

[1] York, A.D. 108 (*R.I.B.*, 665); Chester, A.D. 102–17 (*R.I.B.*, 464), and Caerleon, A.D. 100 (*R.I.B.*, 330).

[2] E. Birley, *Studien zu den Militärgrenzen Roms*, 1967, pp. 8–10.

[3] Summarized in *Research on Hadrian's Wall*, 1961, pp. 132–50. The main forts fairly evenly spaced are Carlisle, Old Church, Nether Denton, probably Carvoran, Chesterholm, possibly Newbrough and Corbridge. In between there are fortlets at High Crosby, Castle Hill, possibly at Birdoswald, Throp (195 ft by 210 ft) and Haltwhistle Burn (210 ft by 170 ft), and signal- or watch-towers like Pike Hill, a stone structure 20 ft square.

[4] The evidence from Nijmegen is given by J. E. Bogaers, 'Die Besatzungstruppen des Legionslagers von Nijmegen im 2. Jahrhundert nach Christus', *Studien zu den Militärgrenzen Roms*, 1967, pp. 54–76. Professor Frere follows Professor Bogaers's argument that Platorius Nepos brought VI with him from Germany in 122 and that IX was at Nijmegen for a few years before going to the East. There are difficulties about this. The stamped tiles of IX from Nijmegen imply more than a passing phase and the presence of the legion there has not been securely dated. A suggestion made by Dr John Mann in correspondence is that IX may have been sent to the lower Rhine during the second Dacian War when the legionary strength had been reduced to the VI Legion at Vetera. IX however was back in strength at York by 108 to undertake the reconstruction of the fortress in stone (*R.I.B.*, 665). VI was sent to start work on Hadrian's Wall, the building of the Tyne bridge being one of their first tasks (*R.I.B.*, 1319 and 1320). It is not impossible that four legions were engaged on the Wall and IX was work-

serious enough by A.D. 122 to warrant a visit by the Emperor himself. His answer was impressive even for the Roman Army.[1] A man-made barrier 73½ miles long, 8 ft thick[2] and at least 15 ft high to the rampart walk, with a parapet and merlons adding another 6 ft. In front was a 20 ft berm and a ditch with an average width of about 27 ft and a depth of about 9 ft; the upcast was heaped on the outer lip into a counterscarp shaped into a glacis. Behind the wall was the Vallum,[3] a remarkable feature which has been the subject of much controversy. This consists of a flat-bottomed ditch 8 ft wide at the bottom, 20 ft at the top and 10 ft deep. The upcast was carefully made into two mounds on both sides and equidistant from the ditch. Each of the mounds was 20 ft wide and 6 ft high with turf revetments placed so that from crest to crest the distance was 100 ft, but the whole work was 120 ft or a Roman *actus* wide. The only crossings were at the forts where there were 20 ft causeways and stone-framed gateways. The Vallum was clearly intended to control access to the Wall zone and the idea put forward by Collingwood of a fiscal control[4] has now been discounted.[5]

An integral part of the wall, or curtain, the word by which it is known, were the milecastles and turrets. The former, provided every Roman mile, were fortlets varying from 50 to 60 ft wide and 65 to 75 ft long internally. They contained one or two barrack blocks for about thirty to fifty men.

---

ing on the Turf Wall, hence the absence of records which would have been in timber (Frere, p. 138). This suggestion was originally made by Professor Birley, who has also indicated that the careers of two senatorial military tribunes (*I.L.S.*, 1070 and 1077) show that IX must have still been in existence after *c.* 120 (*Roman Britain and the Roman Army*, 1953, p. 26).

[1] There is a vast literature on the Wall; fortunately there is also a very full and scholarly summary of all that has been written up to 1961 in E. Birley's *Research on Hadrian's Wall.* The most important contribution since this has been C. E. Stevens, *The Building of Hadrian's Wall*, 1966, published in an extra series, Vol. 20, by the Cumberland and Westmorland Ant. and Arch. Soc. This is a revision of a Horsley Lecture given in 1947 and first published in *A.A.*, 4th ser., 26 (1948), pp. 1–46.

[2] i.e. the narrow wall; the original scheme was for a turf wall 20 ft thick at its base built west of the Irthing, then a broad wall 9 ft 6 in. thick from the Irthing to Newcastle. This was changed during the course of the construction to a narrow wall throughout, so that in places elements of the broad wall survive in the eastern sector where the work started. In certain places the foundations of the broad wall were used for the narrow wall.

[3] The word *vallum* should refer to the Wall itself, but it has been used by antiquaries for this ditch for so long that the misuse of the word has become hallowed by tradition.

[4] *Roman Britain and the English Settlements*, 1937, p. 133.

[5] I. A. Richmond, *J.R.S.*, 40 (1950), pp. 52–53.

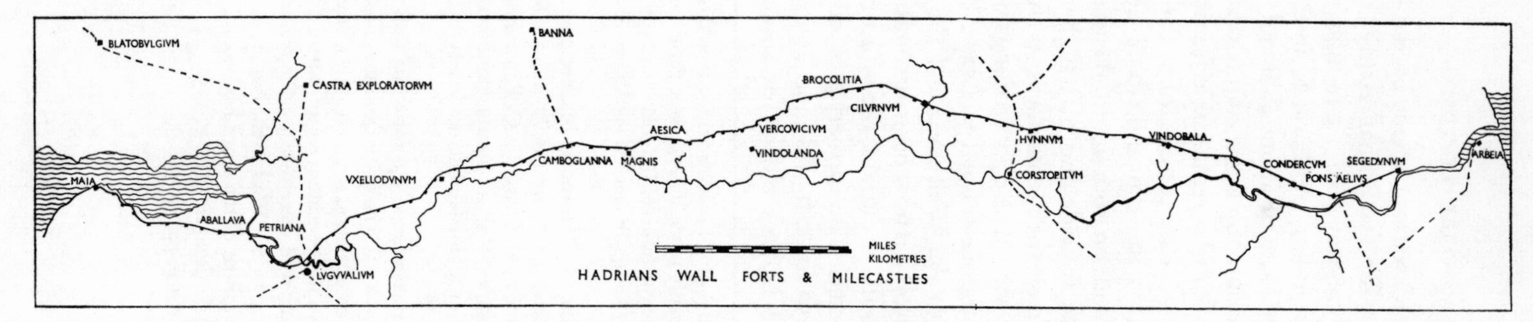

FIG. 6. *Hadrian's Wall, excluding the coastal defences on the west coast*

There were also two gateways, one into the milecastle and one through the curtain itself. There was also a stairway which led up to the patrol track on the Wall. The turrets, spaced out to form three intervals between the milecastles, were recessed into the curtain and measured 14 ft square internally. They could have been used for signalling as well as to provide shelter for the patrolling force.

The final feature which was part of the system was the Military Way, a metalled and cambered road behind the curtain linking the forts, milecastles and turrets. This road was not part of the original scheme, but a later addition, and the earliest milestone is dated to A.D. 213. It overlies the north *vallum* mound, showing that this feature had by then ceased to function.[1] The west end of the curtain was at Bowness, but shore defences continued as far as Skinburness in the form of fortlets and towers with similar spacings to the milecastles and turrets. It seems probable that the system extended as far as St Bee's Head, adding another forty miles to the Wall from Bowness, and garrisoned by four forts, at Beckfoot, Maryport, Burrow Walls and Moresby. At the eastern end the situation is more difficult to interpret, since riverside development has obliterated traces of earlier structures. The curtain turns at Wallsend towards the Tyne, and there were probably fortlets and forts along the south bank as far as South Shields, where the Wall system ended.

In the original scheme for the Wall it was intended to place the forts in the rear, having garrisons only attached to milecastles and turrets on the Wall itself. But during the construction it was decided to incorporate the forts into the Wall. This change of plan has been demonstrated by the finding of the Wall and turret foundations under some of the forts where the latter project in front of the curtain. In some cases the ditch (where it had already been excavated) had to be filled in to enable the fort to be built over it. Although the forts vary in size, according to the unit for which they were designed, they are all of the same basic plan in relationship to the curtain. In six cases about a third of the fort projects beyond the curtain, so that three of the gates give access to their northern side to facilitate rapid forward movement. The nature of the ground dictated the siting of Castlesteads behind the curtain. Stanwix, Carrawburgh, Greatchesters, Bowness and possibly Carvoran may represent another change in the plan. Two of these, Carrawburgh and Greatchesters, were part of the narrow wall and lie wholly behind the curtain and all have

[1] The road diverges towards Turret 29b, which has been shown to have been occupied up to the end of the second or the early third century (*A.A.*, 3rd ser., 9 (1913), pp. 56–67; pottery reassessed by J. P. Gillam, *A.A.*, 4th ser., 35 (1957), p. 215, dated group No. 47).

only one gate through it. Housesteads occupies a position on the very edge of a cliff and this determined that its long axis should be parallel to the curtain. It was built after a turret had been built, but before the narrow wall had reached this stretch. The north-east angle tower built in the correct corner position had to be dismantled and another erected to fit the junction of the fort corner with the curtain.

Thus the ideas on frontier control which had been steadily evolving over the previous hundred years reached their final, logical solution, a continuous man-made barrier containing all the required elements for watching, signalling, controlling and grappling with the enemy, all together in a single co-ordinated system. The Wall was crossed by three roads, the westernmost one through Carlisle proceeding north to Netherby and thence towards the fort at Birrens, while another road passed through Birdoswald to Bewcastle. Dere Street had a special gate (Portgate) through the Wall near Halton Chesters, towards Risingham and High Rochester. Three of these outpost forts, Birrens, Netherby and Bewcastle, have produced Hadrianic inscriptions, so it would appear that they were concerned with a part of the Wall system.

The command headquarters of the Wall would have been at Stanwix, where the senior officer led the milliary *Ala Petriana* and it is clear from this and the presence of the outpost forts that the most serious threat was likely to come from the Novantae and Selgovae.

The function of this great mural barrier was to impose a very severe control on traffic passing into and out of the province. Its precise military purpose is less clear. The gateways in the milecastles and forts served as access points for units to advance to deal with any hostile forces. Richmond saw the further possibility of using the curtain as a screen, so that detachments could move rapidly to selected points unseen by the enemy, and then suddenly appear on their flanks. An important feature needed to facilitate this tactic would have been a metalled road linking all the milecastles, but the Military Way has been shown to have been of much later build. It is doubtful if the Roman Army command saw the Wall quite as a means to such a sophisticated manoeuvre. The advance of large forces would have been by the main routes already held by the outposts. The Wall garrison would have been alerted and moved forward to battle in the open conditions favoured by the Army at this period. Small parties of raiders could be watched and rounded up or dispersed by the cavalry. The Wall system was a remarkable conception and its main advantage was undoubtedly the constant line of communication from one end to the other. Action could be planned at the main headquarters. There was a naïve idea in the minds of earlier antiquaries of a speaking tube running along the Wall. A highly efficient signalling system based on the turrets and milecastles, and placing all parts of the barrier in rapid telegraphic communication, would be almost as effective as the Victorian speaking tube.

Work on the Wall must have begun soon after Hadrian's visit to Britain in 122, under the direction of Aulus Platorius Nepos.[1] Building was still being carried out after 128 when Hadrian assumed the title *Pater Patriae* which appears on the inscription from the east gate of Greatchesters.[2] The completion of the forts may have taken longer. Building stones from Carvoran show that this fort was not converted to stone until 136–8.[3]

On the Rhine and Danube

On the Rhine and Danube, as in Britain, Trajan seemed to have embarked on a policy of consolidation without many large-scale changes in the system. The solution imposed by Hadrian was the continuous barrier, but beyond the Rhine it was not such a massive undertaking as his Wall in Britain. It consisted of a wooden palisade set in a steep-sided trench 3–4 ft deep; the tree trunks, normally 1 ft in diameter, were set at about 1 ft intervals. The trench was packed solid sometimes with stones, and it has been estimated that the timber fence stood some 9 ft high above the ground; but in some stretches of stony terrain an unmortared wall was substituted. The carefully planned tactical arrangements of Domitian in Upper Germany were abandoned and the whole line moved forward, and some of the auxiliary units which had hitherto remained in the rear were brought up to the frontier. At this period, too, the development of towns and agriculture was encouraged in the rearward areas, which was possible only with the establishment of the frontier barriers.[4] The effect of this policy was to provide an increasing man-power potential for the Army and the beginning of the conception of frontier peoples as *limitanei*.[5]

There was little change on the Danubian frontier. XXX *Ulpia* was moved

---

[1] C. E. Stevens has, however, argued on the basis of an inscription from Milecastle 17 (*R.I.B.*, 1419), that work started before 122, but the absence of the name of Nepos need not mean that it predates his governorship.

[2] *R.I.B.*, 1736.

[3] *R.I.B.*, 1816, 1818 and 1820.

[4] *S.H.A.*, *Hadrian*, 11, on his Wall in Britain—*qui barbaros Romanosque divideret,* which may imply an acceptance of Romanization within the frontiers.

[5] E. Birley, 'Hadrianic frontier policy', *Carnuntina*, 1956, p. 30.

to Vetera in Lower Germany and its place at Brigetio taken by I *Adiutrix*, which had taken part in the Dacian Wars and may still have been in Dacia; a campaign under Marcius Turbo was conducted against the Iazyges. Although the detachments which Trajan had taken with him to the East were by now returned, further vexillations from the Danube legions were needed in the Jewish Revolt of 132, the most serious campaign which Hadrian had to lead. The losses on this were severe and probably included XXII *Deiotariana*.[1]

Elsewhere the hand of Hadrian fell heaviest in Judaea, where trouble had started under Trajan. Hadrian decided to build a new colony on the site of Jerusalem, Aelia Capitolina; the temple, destroyed by Titus, was to be replaced with a new structure dedicated to Jupiter, linked with the Emperor himself. Although this was the only logical course the Romans could pursue, nothing could have been more calculated to arouse the fury of the Jews than the further despoliation of their most sacred site. A brutal war of the utmost savagery followed, demanding eventually the presence of the Emperor himself. The destruction and slaughter was immense; over half a million died. Finally the smouldering embers were stamped out by the end of 135 and only gradually was it possible to begin peaceful reconstruction.

Hadrian emerges not as a brilliant general but as a master tactician in defensive warfare. His decisions, like those of Augustus, affected the physical limits of the Empire. Unlike the first Emperor, he journeyed everywhere seeing for himself and settling the actual *limes* and types of structural barrier, and looking into everything in the greatest detail.[2] Although much concerned with discipline and training, he could ameliorate a measure which seemed too harsh as in the case of allowing soldiers' children to claim inheritance rights.[3]

[1] The last notice of this legion is in 119 at Alexandria. Talmudic sources hint at a battle in the southern foothills of Judaea.

[2] Dio summarizes his activity (lxix, 9): 'He personally viewed and investigated absolutely everything, not merely the usual appurtenances of camps, such as weapons, engines, trenches, ramparts and palisades, but also the private affairs of everyone, both of the men serving in the ranks and of the officers themselves—their lives, their quarters and their habits—and he reformed and corrected in many cases, practices and arrangements for living that had become too luxurious. He drilled the men for every kind of battle, honouring some and reproving others, and he taught them all what should be done.' (Loeb trans.) An interesting example of luxurious living arrangements has been found by Dr D. Baatz in the fort at Echzell. The timber buildings of this period had been demolished to make way for those of stone and Dr Baatz recovered from the quarters of a *decurion* some remarkable decorated wall plaster which one does not normally associate with the military life (report forthcoming).

[3] Berlin Papyrus No. 140; Lewis and Reinhold, pp. 519–20.

The feelings of contemporaries towards Hadrian were echoed, although fulsomely, by Aristides in his famous oration.[1] But one is rather nearer with the Greek Flavius Arrianus, a friend of the Emperor, who became governor of Cappadocia. His military textbook, *Tactica*, shows a keen interest in the older fighting formations, but also indicates ways of adapting the army of his own day to the new threats posed by the highly mobile horsemen from the Steppes.[2] From Africa comes the inscription recording Hadrian's inspection and report[3], and it is significant, too, that from the very boundary of the Empire, Flavius Arrianus was writing of the state of the naval squadrons patrolling the eastern shoreline of the Black Sea.

But Hadrian's most important contribution was the stabilization of the Army itself. He encouraged provincials to make a career of the Army and civil administration and many of the officers are found in their own territory. This seems especially so in Africa, where there is abundant epigraphic evidence.[4] He was also responsible for the introduction of *numeri* as permanent official units of the Army. Hitherto native levies had often been used in campaigns and dismissed at their close. Now they became part of the frontier garrisons and are found from this time on the Rhine and Danube. Like the *auxilia*, in their original state they were led by their own tribal chiefs as officers and were equipped, and fought, in their own native style. They are to be found in forts adjacent to the *auxilia* and it must be presumed that they were used as frontier police and customs guards.[5] This was a logical development of the permanent barriers. Access to the Empire was now strictly controlled, the regular auxiliary forces being kept as reserves in case of real trouble, while the legions, often deep in the rear, acted as the strategic reserve.

---

[1] *To Rome;* an English translation has been published by Saul Levin, The Freepress, Glencoe, Illinois, 1950.

[2] F. Kiechle, 'Die "Taktik" des Flavius Arrianus", *Bericht der Römisch-Germanischen Kommission,* 45 for 1964 (1965), pp. 87–129; a fragment has also survived of Arrianus's expedition against the Alans, *de acie contra Alanos;* see also *Essays* by Henry F. Pelham, 1911, pp. 212–33. Hadrian may also be credited with the introduction of units of cataphracts into the Army (J. W. Eadie, 'The Development of Roman Mailed Cavalry', *J.R.S.,* 57 (1967), p. 167).

[3] *I.L.S.*, Nos. 2487 and 9133–5, see p. 149.

[4] R. H. Lacey, *The Equestrian Officers of Trajan and Hadrian: Their Careers with some Notes on Hadrian's Reforms,* 1917, and R. Cagnat, *L'Armée romaine d'Afrique et l'occupation militaire de l'Afrique sous les empereurs,* 1912; see also M. G. Jarrett, 'The African contribution to the Imperial Equestrian Service', *Historia,* 12 (1963), pp. 209–26.

[5] E. Birley, 'Hadrianic frontier policy', *Carnuntina,* 1956, p. 27.

The pattern of frontier garrisons thus becomes firmly established under Hadrian and the important aspect is its permanency. Recruitment became more and more dependent on the frontier peoples, until the Army forming the garrisons became almost a hereditary caste.

Imperial Policy under Antoninus Pius (A.D. 138–61)

Although Pius did not come to the throne until he was almost fifty-two, he reigned for twenty-three years. During this long period he endeavoured to continue the policies of Hadrian, but, unlike that wandering Emperor, he remained at Rome with a firm grasp on the many intricate reins of government. Consolidation and conciliation continued to be the watchwords along the frontiers. There were troubles here and there; in Africa part of VI *Ferrata* was dispatched from Palestine with detachments from the Rhine and Danube. The Dacian frontier required to be altered and strengthened, but it is in Britain that there was the only major advance. The Wall of Hadrian was abandoned in favour of an advance and the construction of yet another barrier, this time a shorter one between Clyde and Forth, to become known as the Antonine Wall.[1]

What caused Antoninus Pius, normally a cautious Emperor content to let the decisions of his predecessor remain, to make this bold advance and give up such a solid monument? As Richmond has written, 'It is the prerogative of Great Powers to change their minds and the habit of military problems to change also.'[2] But the key must lie in the second part of this statement. The tribes beyond the Wall had by now recovered their strength and they saw the Wall as a permanent barrier sealing them off from former allies, as yet far from pacified,[3] and the opportunities of raiding the rich lands to the south.

The extension of the Wall defences along the Cumberland coast and the presence of the milliary *Ala Petriana* at Stanwix may be indicative of the

---

[1] The major study of this frontier was by Sir George Macdonald, *The Roman Wall in Scotland,* 2nd ed., 1934; much work has been done since then and has been summarized by Dr. Anne S. Robertson in *The Antonine Wall,* a handbook published by the Glasgow Archaeological Society, 1960. The latest reappraisal of the evidence is by Dr Kenneth A. Steer, 'John Horsley and the Antonine Wall', *A.A.,* 4th ser., 42 (1964), pp. 1–39. Detailed reports of excavations, etc., have appeared in the *P. Soc. of Ant. Scot.*

[2] *History,* 44 (1959), p. 9.

[3] cf. the complaints of the Tencteri against the *colonia* (Cologne) of having to surrender their arms to visit their kinsmen and pay an entry tax, *inermes ac prope nudi sub custode et pretio coiremus* (Tacitus, *Hist.,* iv, 64).

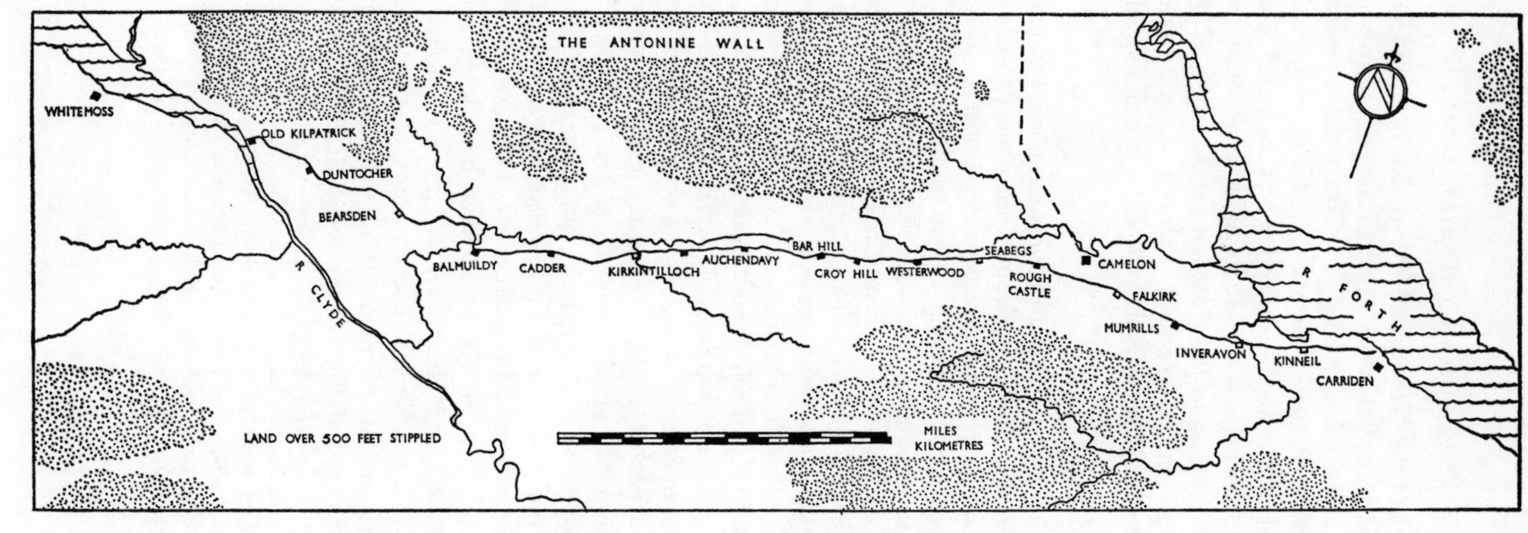

FIG. 7 *The Antonine Wall*

Roman assessment of the Novantae.[1] The withdrawal under Domitian had left the pro-Roman Votadini at the mercy of their aggressive neighbours, the Selgovae. The sudden advance may have been made necessary by tribal warfare to which the Romans, because of their treaties, could not remain indifferent. Probably also Roman discipline had slackened after the death of Hadrian and the hostile tribes may have thought they could act with impunity. The only possible solution to this situation was to bring all the lands under occupation and tight military control. Coins of 142 and 143 commemorate victories and there is historical authority[2] for the building of a new Wall.

The Antonine Wall (Fig. 7)

The Antonine Wall was a turf wall (*murus caespiticius*) standing on a stone base 14 ft wide. With its sloping back and front it was probably no more than 10 ft high with a 6 ft patrol track protected by a timber breastwork. On the north side there was a ditch, leaving a berm of about 20 ft. The dimensions of the ditch vary from point to point, but the average is about 40 ft wide and 12 ft deep. There was an 80 ft length on Croy Hill where the tough nature of the rock defeated the legionaries. The profile was V-shaped and in places there was the typical rectangular slot in the bottom. Behind the Wall ran a military road 16–18 ft wide. The whole of the system was about thirty-seven miles long and like Hadrian's Wall built by detachments of all three legions in Britain under the governorship of Lollius Urbicus. As many as seventeen of their distance slabs have been found and at least fourteen marching camps which the legions probably used during the construction work.[3] Unlike the frontier to the south, the Antonine Wall had no milecastles or turrets, but the forts were more closely spaced. The thirteen forts known to date are at about two mile intervals, and if this pattern was followed throughout there must be several forts waiting discovery. They are not all for complete auxiliary units; only five or six could have been large enough. Rough Castle is only an acre and Duntocher half an acre in size. None of the forts projects beyond the barrier. The economy of effort and man-power, which seems clear on the Antonine Wall when compared with that of Hadrian extends to the buildings of the forts. Here only the principal ones had stone foundations, the barracks being wholly in timber, and this

[1] J. P. Gillam, in *Roman and Native in North Britain*, 1958, pp. 75–76.

[2] Capitolinus, *S.H.A.*, *Pius*, 5, 4; Pius became Imp. II in 142 (*I.L.S.*, 515).

[3] An attempt has been made by R. W. Feachem to relate these camps to actual lengths attributable to legions, *P. Soc. Ant. Scot.*, 89 for 1955–6 (1958), pp. 329–39.

accounts for their plans not being fully recovered by the excavators in the first three decades of this century. Collingwood considered the work as slight and makeshift,[1] but this view was based on the assumption that Hadrian's Wall was still held as the main barrier. But this was not so. There may have been some units left in the forts, but the doors were removed from the mile-castles and the Vallum systematically breached. The Antonine Wall itself terminated at Old Kilpatrick at its western end, but, as in the case of the Cumberland coast, forts and fortlets continued the line. At Bishopton there is one guarding a ford across the Clyde and a fortlet at Lurg Moor above Greenock, and there must be more to come. Similarly on the east side the barrier ends at Bridgeness, but there is Carriden, a fort to the east, and there may be others apart from Cramond and Inveresk. The main control routes between the two Walls were held, but there is at present an absence of Roman sites in Carrick and Galloway and the status of the Novantae remains uncertain. The Agricolan strategy of a fortified line across Strathmore was adopted and a chain of outposts once more controlled the territory between the Highlands and the coast. Ardoch was now occupied, as probably also were Dalginross, Bertha, Strageath and Cardean. The whole system seen in its entirety, as far as the present evidence allows, was far from a temporary or makeshift arrangement; the northern frontier was moved forward and Hadrian's Wall abandoned and this was to be the new permanent arrangement for Britain.

But it failed, as the previous barrier had failed. In A.D. 155 a serious uprising led to the deliberate withdrawal from the Antonine Wall and the systematic dismantling and destruction of the forts.[2] But there was also destruction by the enemy at Newstead. The breaches in Hadrian's Wall had enabled the anti-Roman tribes to work in concert and it seems likely that the Selgovae and the Brigantes planned their revolt jointly and to telling effect. After the revolt had been put down with the help of legionary reinforcements sent direct to the Tyne from Germany,[3] the situation of the frontier is uncertain. However, there is evidence that Hadrian's Wall was once more

[1] *Roman Britain and the English Settlements*, 1937, p. 148.

[2] This policy accounts for the remarkable preservation of so many fine distance slabs, the assumption being that they were carefully buried, as may also have been the fate of the collection of altars from Auchendavy.

[3] *R.I.B.*, 1322. A dedication was set up at Corbridge by an officer of VI Legion to Mars Ultor, Mars the Avenger, under Julius Verus, who was responsible for some of the reorganization at this time (*R.I.B.*, 1132; the letters VL [TOR] had previously been read as VE[XILLATIO], *A.A.*, 4th ser., 21 (1943), pp. 177–8). There is, however, still some doubt about this reading.

brought into commission with the reoccupation of milecastles and turrets. Some of the old abandoned fort sites in the Pennines[1] and Peak District[2] were rebuilt, while forts beyond Hadrian's Wall continued in commission at this time, as evidence from Birrens and Newstead indicates. It is doubtful if the Antonine Wall forts were rebuilt and reoccupied as soon as the Brigantian Revolt was crushed, although Dr Steer has produced an ingenious argument to support this theory.[3]

At one time it was thought that the garrison of Wales was seriously reduced in the Antonine period to provide troops for the new frontier,[4] but this has now been shown to be an exaggerated view, since there is ample pottery evidence for the continued occupation of many of the Welsh forts at this time,[5] although some regrouping may have been carried out. By this time some of the tribes may have been considered sufficiently pacified to have been allowed to develop as normal *civitates*. An example may be the Silures, the one-time bitter enemy of Rome, for their tribal centre at Caerwent and the scatter of villas and peasant communities clearly indicate peace and a growing prosperity. The absence of forts, in the territories of the Demetae and the Deceangli, may have a similar significance.[6] Thus the situation in Wales is

---

[1] Brough-on-Noe has produced a dedication *c.* 155–9 (*R.I.B.*, 283), and Ilkley has a dedication to Marcus and Verus (*R.I.B.*, 636) which must fall between 161 and 169 and shows that the fort was occupied at this time, and this has now been amply confirmed by excavation (B. R. Hartley, 'The Roman Fort at Ilkley, Excavations of 1962', *Leeds Phil. and Lit. Soc.*, 12 (1966), pp. 23–72).

[2] There is evidence of this as far south as Littlechester (*Derbyshire Arch. J.*, 81 (1961), pp. 85–110).

[3] In his paper 'John Horsley and the Antonine Wall', *A.A.*, 4th ser., 42 (1964), pp. 1–39, and later accepted by Professor Sheppard Frere, *Britannia*, 1967, pp. 155–61. The argument is based on auxiliary inscriptions of the reign of Pius which are associated with the second occupation and also the pottery dating from Mumrills (*P. Soc. Ant, Scot.*, 94 for 1960–1 (1961), pp. 100–29). A detailed study of the samian from northern sites by Mr. B. R. Hartley (*Britannia*, 3 (1972), pp. 1–55) has shown how difficult it is to postulate on archaeological grounds any occupation of the Antonine Wall after 160. This means a very brief second period (c. 159–160) when Hadrian's Wall may have been held by legionary detachments, on the basis of the inscriptions from Chesters (*R.I.B.*, 1460–1, hitherto placed somewhat earlier, see M. G. Jarrett and J. C. Mann, *B.J.*, 170 (1970), p. 188).

[4] R. E. M. Wheeler, *Prehistoric and Roman Wales*, 1925, pp. 230–2; V. E. Nash-Williams, *The Roman Frontier in Wales*, 2nd. ed., 1969, pp. 22–27.

[5] Grace Simpson, *Britons and the Roman Army*, 1964, pp. 85–96.

[6] Michael Jarrett, *Bull. Board of Celtic Studies*, 20 (1963), pp. 206–20.

highly complicated and until much more work is done on a large number of sites it will not be possible to summarize the position at any given period. The great highland massif of Central and North Wales was probably under continuous military surveillance with forts controlling all the passes and river crossings linked by splendid roads giving rapid access to any trouble spot. One is tempted to ask why such lavish expenditure of money and man-power was necessary in a land where the population could never have been great and trouble-makers could easily have been rounded up and deported. The Army may have seen the situation from a different viewpoint. The soldiers, perhaps thinner on the ground than we suppose, for the most part carried out police work and the area presented a fine training-ground, while the wild Celtic tribes no doubt produced their quota of recruits from their tough mountain stock.

The two legionary fortresses were well placed with estuarine harbours on the Dee and Usk allowing them to be provisioned from the sea. Chester occupied a key position between the two highland zones and XX Legion was able to watch both (Fig. 8). These sites were so well chosen that there was no need for alterations in either case. However, for the greater part of the occupation the great fortresses remained mere depots, their best troops serving as front-line units or as construction corps in the north,[1] or supplying vexillations for campaigns elsewhere.

In the middle of the second century *numeri* of Britons appear in the Odenwald in Germany and there are building slabs indicating their work in constructing stone watch-towers. The latest dated stone is A.D. 148, and this has led to the theory that these Britons were moved from the area between the two Walls after the campaign of Lollius Urbicus,[2] but recent work in Germany throws considerable doubt on the suggestion.[3] On the Upper German and Raetian *limes* under Antoninus Pius, there was considerable

---

[1] As Dr Jarrett has indicated (*Arch. Camb.*, 113 (1964), p. 54), while vexillations of VI and XX are attested in construction work on the Antonine Wall, it would appear that the whole of II *Augusta* was involved.

[2] Collingwood (*Roman Britain and the English Settlements*, 1937, p. 146) was somewhat reluctantly followed by Richmond (*Roman Britain*, 1963, p. 52). The theory always had difficulties; the units, for example, are named after districts in Germany and not northern Britain. The Germans have compared some decorative details on their towers with those on the Antonine Wall, but the latter was built by legionaries.

[3] Excavations by Dr D. Baatz on the small *numerus* fort at Hesselbach have shown earlier timber phases with very similar plans of buildings, which suggests that these *Numeri Brittorum* may have been sent to Germany at an earlier date either by Trajan or Hadrian.

FIG. 8 *The Legionary Fortresses in Britain*

building activity and in places the frontier advanced to a new position. A remarkable feature of the new line is the straight lengths from Walldürn to Weltzheim, a distance of eighty kilometres, with the deviation of only a metre, a fine testimony to Roman surveying skill. The forts along the Antonine *limes* fall into several groups according to size; those of the *numeri* are about one and a quarter acres, then there are some from half to three-quarters of an acre, and a series of very small ones from 240 to 700 sq. yds in area. This would seem to demonstrate an economy of manpower similar to that on the Antonine Wall in Scotland. The reasons behind the new forward move in Germany seem to be economic rather than military. The shortening of the line and the diminution in fort sizes may indicate a stretching of resources, but it also had the effect of including good farming land to help to support the population in the rear expanding under the *pax Romana*. This peace which had endured so long was soon shattered by the Chatti in 162. There is evidence of forts being overrun, destroyed and later repaired.

Imperial Policy under Marcus Aurelius (A.D. 161–80)

Marcus had a long apprenticeship, learning and absorbing the art of government from Pius. It is one of the ironies of history that Marcus Aurelius, a man of contemplation, was obliged to spend much of his life at the head of his armies. In the East the peace was suddenly shattered by the Parthian Osroes, who put a Parthian nominee on the throne of Armenia and then advanced into Syria, scattering the legions before him, for the troops, as in the days of Nero, had become demoralized by a long peace. Marcus had to detach legions from the other frontiers. The campaign was led, at least in theory, by Lucius Verus and was very successful, the Parthian Army melting before the solid and well-planned advance of the Romans. By 166 the situation was almost the same as that when Trajan had reached the pinnacle of his success, but once more the Romans were denied the consolidation of their triumph. A fearful plague spread from Seleuceia and spread rapidly through the army of the East, and as the troops returned to their home stations they took it with them and the distress and devastation swept through the whole Empire. Its effect on the Army was disastrous, for the frontier garrisons were sadly depleted and the watchful tribes beyond soon became aware of this serious weakness. For some time the northern frontiers had been under indirect pressure from the tribal movements which were probably set in motion by the trek of the Goths from the lower Vistula towards south Russia in search of new lands. The weakness of the Danubian frontier was suddenly exposed by a group of tribes led by the Marcomanni and Quadi, which broke through and crossed the Julian Alps into northern Italy while Marcus was with his army on the Save. Their main target was the important trading port at the head of the Adriatic, Aquileia, which held out in a siege, but it seemed for a time as if the rest of Italy and Rome itself was at their mercy. But the tribes seem to have been content with their plunder and did not press south. Troops were called in from the more peaceful areas,[1] and Marcus led his army against the tribes as they were making their escape across the Danube, laden with booty, probably late in 171, and won a hard-fought victory. But troubles broke out everywhere at once. The threat to Italy, withdrawal of troops, and loss of men by the plague revealed to the lawless and adventurous the serious weakness along the frontiers, which they promptly attempted to exploit. From the Nile delta to Britain, from Armenia to Dacia, there were revolts and incursions.[2] The picture is confused and the information very fragmentary.[3] Marcus took to the offensive with resolute energy and sense of purpose. First the provinces behind the frontier were cleared of the enemy; then the stability of the frontiers themselves restored. The coin issues record victory after victory as the years roll by. From 171 to 173 Marcus was at Carnuntum pressing deep into barbarian lands and his purpose became clear —he was determined to bring to submission all peoples bordering the Danube, and under direct control rather than rely on treaties. All the tribes were reduced one by one.[4] The work was now almost complete and groups of conquered barbarians were settled in frontier districts depopulated by the plague and the wars. In 175 it was the turn of the Sarmatians, but the task was not finished, since it was interrupted by the revolt of Avidius Cassius in Syria. While this was soon crushed, it demanded the presence of the Emperor in the East. The Danube remained restless and further campaigns of 177 and 178 were needed; the end of the long struggle was now in sight, but so was that of Marcus himself. On his death-bed in 180 he called upon his son Commodus to complete his task. But to the young Emperor this was a disagreeable duty when the pleasures of Rome beckoned, so treaties were

---

[1] There is evidence of parts of X *Fretensis* from Jerusalem and III *Augusta* from Africa; even brigands were turned into soldiers (*S.H.A.*, *Marc.*, 21, 7.)

[2] One tribe, the Costoboci, crossed the Danube and managed to make its way as far as Eleusis in Greece and plunder the Temple of the Mysteries.

[3] Probably the best chronology yet devised is that of Anthony Birley, *Marcus Aurelius*, 1966.

[4] A rock-cut inscription of II *Ad.* at Trenčin in Czechoslovakia is evidence of the army sixty miles beyond the Danube (*C.I.L.*, iii, 13439).

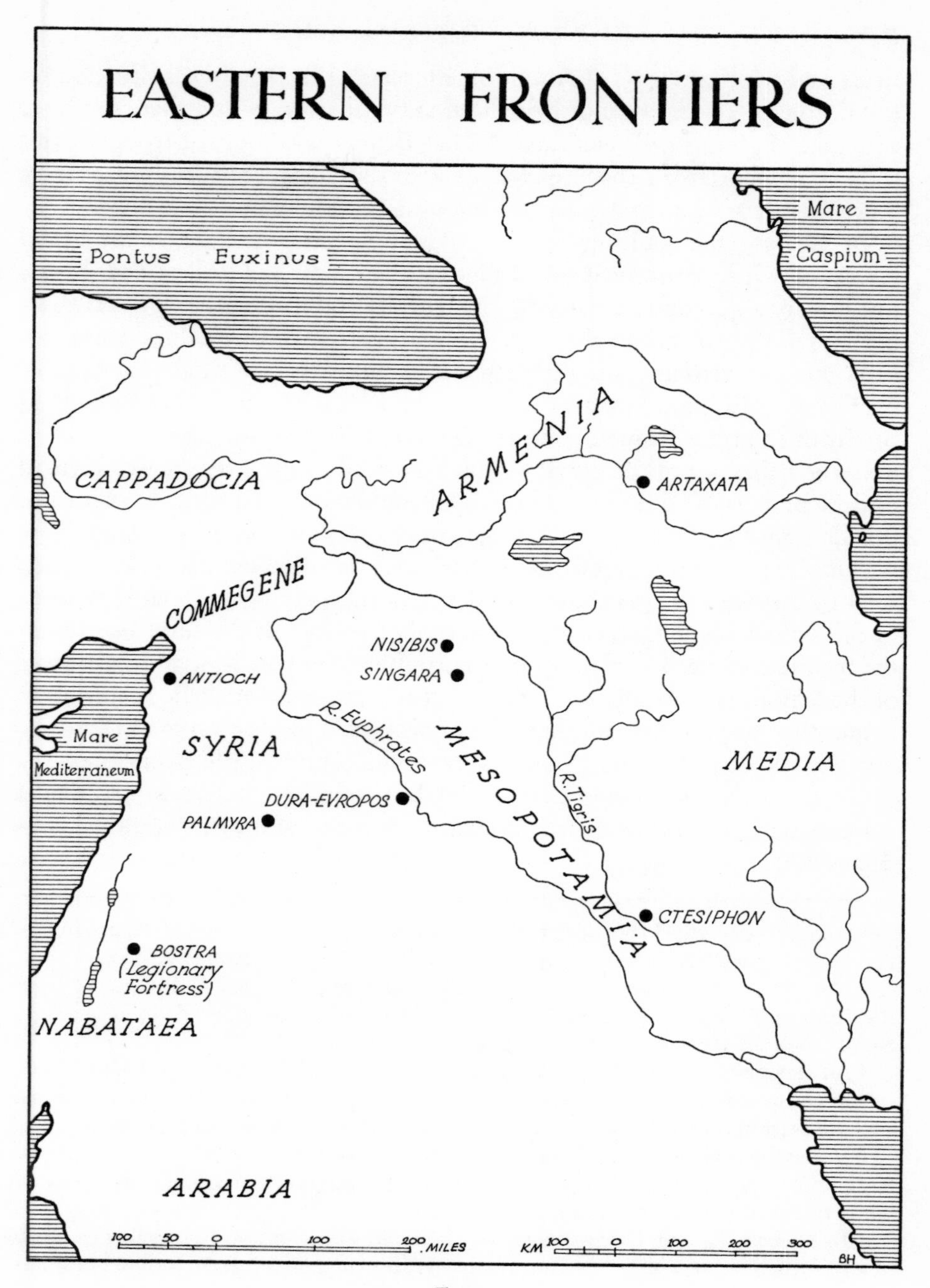
EASTERN FRONTIERS
Mare
Caspium
Pontus Euxinus
ARMENIA
ARTAXATA
CAPPADOCIA
COMMEGENE
NISIBIS
SINGARA
ANTIOCH
R. Euphrates
Mare
Mediterraneum
SYRIA
MESOPOTAMIA
MEDIA
R. Tigris
DURA-EVROPOS
PALMYRA
CTESIPHON
BOSTRA
(Legionary
Fortress)
NABATAEA
ARABIA
100 50 0 100 200 MILES
KM 100 0 100 200 300
BH

FIG. 9

made and the frontier was left to look after itself.[1] In Britain the situation on the northern frontier cannot have been very satisfactory and there are hints of further fighting and rebuilding.[2] That Britain was short of troops at this time is evident from the dispatch of 5,500 auxiliary cavalry, by Marcus Aurelius in 175, at a time when other frontiers were in desperate need.[3]

In viewing the achievements of Marcus Aurelius one senses the grand design which he conceived when the threat to Italy had to be faced. It was not merely a question of restoring the frontier and its defences, but preventing other similar outbreaks. To this end garrisons themselves were not sufficient and Marcus made little attempt at any large-scale redeployment of his forces.[4] His main aim was to bring the barbarians immediately beyond the frontier under Roman control. The army he had assembled in Italy to clear the tribes out of the districts round the head of the Adriatic was formed of detachments of all those legions which could spare them and the success of this demonstrated the need for a permanent mobile force. Although there was nothing new about this, since legionary *vexillationes* had always been used in this way, it was shown up dramatically here against the ineptitude of the depleted static garrisons. The idea of creaming off the best troops from the frontiers to form such new army corps began to take root and the power of the legions was already on the wane. It is perhaps unfortunate that Marcus bequeathed to posterity not the story of his wars and ideas on grand strategy but a book of philosophical meditations which bring little comfort to the historian. In another respect he did at least emulate Trajan in erecting a column with a spiral of reliefs modelled on those of his distinguished predecessor.[5]

---

[1] Some scholars who do not accept this view may overstress the strength of the barbarians, Pavel Oliva, *Pannonia and the Onset of Crisis in the Roman Empire*, 1962, p. 299–306.

[2] Associated with the governor Sextus Calpurnius Agricola *c.* 163–6, *R.I.B.*, 589 from Ribchester; J. P. Gillam, 'Calpurnius Agricola and the northern frontier', *T. Archit. and Arch. Soc., Durham and Northumberland*, 10 (1953), pp. 359–75.

[3] An alternative view is that these Sarmatians were virtually captives and their dispatch to a distant province helped to solve an embarrassment; nevertheless it also shows the real need for reinforcements in Britain (Anthony Birley, *Marcus Aurelius*, 1966, p. 260).

[4] V *Macedonica* was transferred from Troesmis to Potaissa in Dacia in 167 on its return from Parthia; the new legion II *Italica*, was posted to Lauriacum (Lorch, near Linz), in Noricum, but at first was established in a new fortress at Ločica near Poetovio.

[5] E. Petersen, A. von Domaszewski, G. Calderini, *Die Marcus-Säule*, 1896, the reliefs lack the details on the earlier Column and the events they illustrate are even more obscure; the Column stands in the Piazza Colonna. The best modern studies of the difficult chrono-

Imperial Policy under Commodus (A.D. 180–92)

The son of the great Marcus had none of his father's qualities. He preferred to spend his time and energies on the pursuit of pleasure and left the government to favourites who took full advantage of such a situation. Fortunately for Rome there were still competent governors and legionary legates who kept a firm control of the frontiers and there was, except in Britain, little trouble. In Britain the Caledonian tribes rose and crossed Hadrian's Wall. A successful campaign was launched against them with a victory recorded on a coin issue and important enough for the Emperor to take the title *Britannicus*. It was at this time, however, that a revolt in the Roman Army broke out in Britain. This was not against the Emperor, but his favourite Perennis, the praetorian prefect who had absolute power and may have been niggardly with soldiers' pay.[1] The trouble in Britain was settled in 186 by Pertinax, who was later Emperor for three months after the murder of Commodus.

There are many inscriptions recording repairing and rebuilding,[2] but no basic changes are evident and the frontier tribes remained quiet, slowly recovering from the onslaughts of the previous reign. The assassination of Commodus in 192 set in train events comparable to the year of the four Emperors. The peace and security of the Empire was split from end to end by a tragic Civil War which brought the Severan dynasty to the fore.

Imperial Policy under Severus (A.D. 193–211)

The death of Commodus on the last day of 192 precipitated a long and fearful Civil War out of which Severus rose as final victor. The struggle was one

logical aspects of the campaigns and the Column are W. Zwikker, *Studien zur Markus-Säule*, 1941, and a renewed assessment by Dr J. Morris in his valuable paper 'The Dating of the Column of Marcus Aurelius', *J. Warburg and Courtauld Institutes*, 15 (1952), pp. 33–47. A recent photographic record has been published by P. Romanelli, *La colonna Antonina*, 1942, and with a commentary, Caprino—Colini—Gatti—Pallotino—Romanelli, *La colonna di Marco Aurelio*, 1955.

[1] According to Dio, lxxii, 9, 1,500 men were involved in a march to Rome to denounce Perennis, but it is more likely they had real grievances to air.

[2] Inscriptions from the Danube between Aquincum and Intercisa refer to the erection of watch-towers (*burgi*) and guard points (*praesidia*) to protect the province from secret crossings by brigands (*ad clandestinos latrunculorum transitus*) (Fülep, Epigraphie, *Intercisa, i*, 1954, pp. 248 ff., also *Arch. Ert.* 1955, pp. 67 and 78). Some scholars have considered that the inscriptions are merely Roman attempts to disguise more serious barbarian raids, but the term may only have reflected an official attitude when enemies of the State are often thus denigrated.

between three groups of legions, those of the East under C. Pescennius Niger, those of Britain under D. Clodius Albinus, and those of the Danube under L. Septimius Severus. The last of these started with definite advantages. He was the nearest to Rome and quickly established himself there. He dismissed the praetorians and filled their places with picked men from his own legions. Severus had the largest army group, since the Danubian Wars had led to a concentration of twelve legions along that frontier, whereas there were only four on the Rhine, and as soon as the new Emperor was established in Rome the legions of the Rhine came over to his side. The nine legions of the East which Niger could command were never quite of the same calibre as those of the northern frontiers. With much internal dissension and wavering of loyalty Niger's position was hopeless and his forces fell before the advancing Army of the North in 194; only Byzantium held out, suffering a protracted siege. The severity of the penalties of Severus on the Syrians drove many soldiers into Parthia to seek protection.[1] Although the Parthians had taken little advantage of the Civil War, some Roman outposts had been attacked and taken. This justified Severus in undertaking an expedition to recover lost ground and consolidate the frontiers; he was wise enough however, not to threaten Parthia itself and to hand over Osrhoëne to the local ruling house. Any further activities had to be curtailed to deal with the threat from the west. Severus had recognized the third claimant, Albinus, as his heir, but this was only a gesture allowing him time to deal with Niger. Albinus was a popular candidate with the Senate and many of the influential families encouraged him to claim the purple, and he crossed to Gaul with his three legions and began recruiting troops.

The governor had to take a calculated risk in removing too many troops from Britain and one can only guess at the measures he took for the defence of his province. It seems possible, however, that the towns were given their defences, perhaps only in the form of ditches and a bank, but adequate protection against marauding bands from the north. The Britons themselves probably made these precautions a condition of their support. Sufficient strength must have been left to stave off any attack and possibly one might guess at some kind of treaty arrangement with the northern tribes involving rewards had Albinus been successful. The final battle took place at Lyons in 197, each side, according to Dio, having 150,000 troops, but the larger body

---

[1] Herodian, *Severus,* iv, 7–9, notes that special skills were thereby placed at the disposal of the Parthians, so making them stronger enemies of Rome.

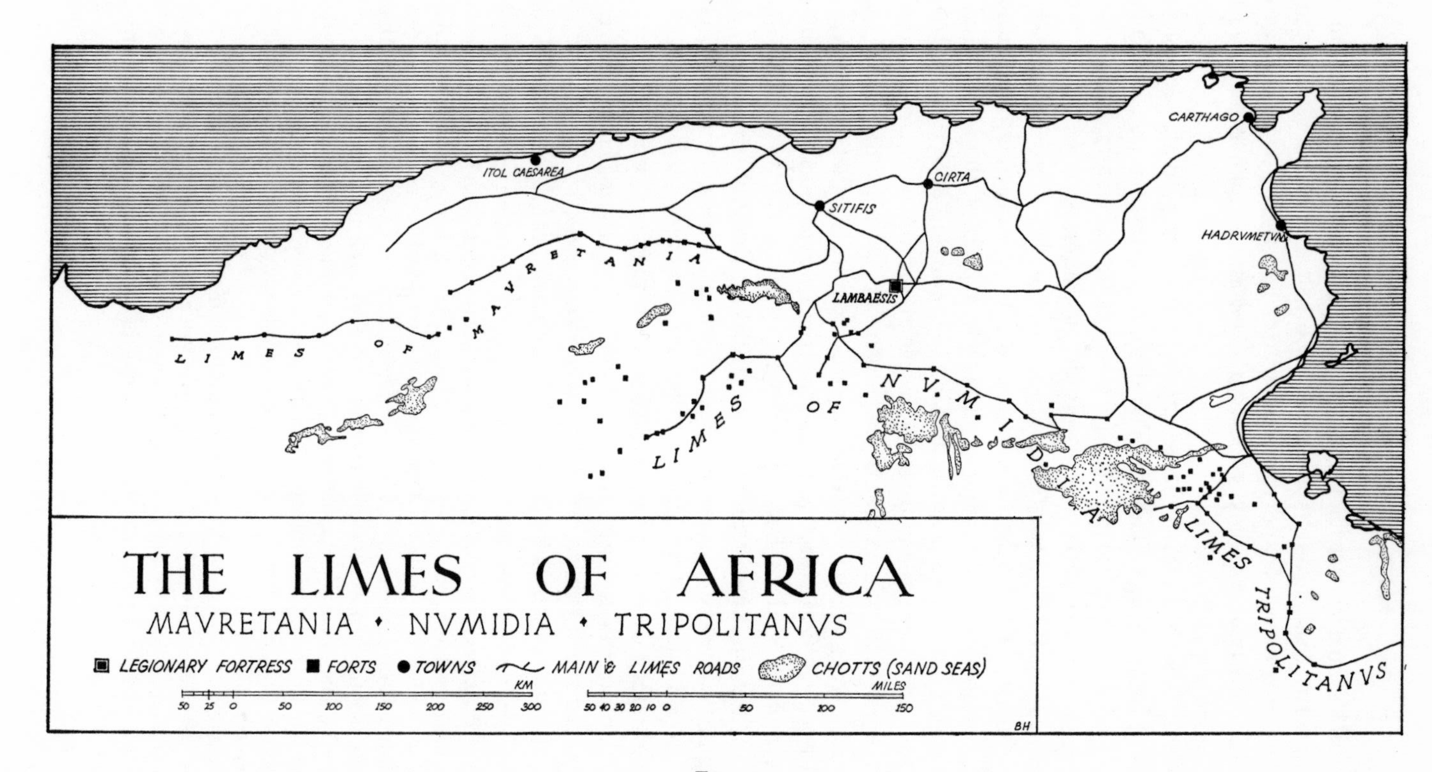
THE LIMES OF AFRICA
MAVRETANIA • NVMIDIA • TRIPOLITANVS
LEGIONARY FORTRESS
FORTS
TOWNS
MAIN & LIMES ROADS
CHOTTS (SAND SEAS)
KM
50 25 0 50 100 150 200 250 300
MILES
50 40 30 20 10 0 50 100 150
BH
CARTHAGO
HADRVMETVM
CIRTA
SITIFIS
ITOL CAESAREA
LAMBAESIS
LIMES OF MAVRETANIA
LIMES OF NVMIDIA
LIMES TRIPOLITANVS

FIG. 10

of seasoned legionaries were Severan and there was little doubt over the outcome.[1] Severus dealt heavily with those who had supported his opponents. Many were put to death and they and others had their estates and wealth confiscated. This was a policy which for Gaul and Britain must have had serious consequences.[2]

The Civil War had clearly demonstrated the effect of the policy of local recruitment begun under Hadrian. The three legionary armies each had not only their own leaders but possessed a unity which contrasts strongly with the confused situation of 69 with its divided loyalties. Severus had not only the largest group but the toughest, and his use of this solid core of experienced legionaries, first in the East, then in the West, and then once more in the East, foreshadowed the great mobile field armies of the later centuries. Another result of the wars was the division of command in Syria and Britain. Severus had no wish to see potentially hostile governors with large forces. So each of these two provinces was divided into two, splitting the legions, each now under two governors instead of one. Loyalty to the throne was reinforced by large donatives[3] and allowing the men to make legal contracts with their wives and to live with them in the *canabae*.[4] The Emperor's troubles on the frontiers were by no means over; the Parthians began to be active again while the two armies were locked in their struggle in Gaul. Severus carried out a demonstration in strength and reorganized the defences of the Euphrates;[5] by relaxing the severe measures against Syria he was able to bring back economic stability. It was in Britain that Severus faced his greatest challenge, for the northern tribes had taken advantage of the situation and overrun the weakened defences, causing sufficient damage to the Wall of

---

[1] Dio's account (lxxvi, 6, 1) indicated that it was a close thing, with Severus almost losing his life, but this may have been from a biased source. There is an interesting practical detail given about some of the troops of Albinus. They constructed concealed trenches and pits (*lilia*) in front of their lines and by a feint retreat lured their opponents into them with disastrous results. Another significant detail is the use of cavalry as an independent arm.

[2] It is perhaps significant that samian from Central Gaul ceases to be imported into Britain at this time and this may have been due to severe economic dislocation.

[3] *S.H.A.*, *Severus*, xii, states that he gave his soldiers greater sums of money than any previous emperor.

[4] Herodian, iii, 8, 5. Although there is no evidence that this was put into practice, the distinction between the *canabae* and the civil settlement which could receive the status of a *colonia*, as at York, is brought out by Sir Ian Richmond (*Eburacum*, 1962, pp. xxxiv–xxxix).

[5] This involved the creation of two new legions, I and III *Parthicae* and a new province, Mesopotamia.

Hadrian and its forts to make large-scale reconstruction necessary.[1] Roman forces appear to have been adequate to protect the civil part of the province, since there is no evidence of destruction there. Clearly to set matters aright, campaigns had to be mounted, ground regained and forts rebuilt, and the governors appointed, Virius Lupus, Valerius Pudens[2] and Alfenus Senecio, were at work from A.D. 197 to A.D. 205. But their work did not resolve the frontier problem, which had now remained a constant worry for fifty years. Severus decided to settle matters with an imperial invasion of Scotland and was for three years (A.D. 208–11) active in the field with Caracalla. When busy preparing for what may have been his final campaign the Emperor died, but Caracalla may have completed the work.[3]

There seems little doubt that the Romans had at last found a satisfactory solution to the northern frontier, since the tribes beyond gave no trouble for almost a century. The Severan campaigns must have been savage indeed for such a long-term effect. It is more likely that a new policy had developed. Hadrian's Wall, now rebuilt was once more in commission, but there were strong outposts holding the main routes to the north. Some of the units in these outlying forts were trained and equipped for patrol work[4] which ranged over the whole of the Cheviots and probably beyond. But this is only part of the arrangement, for the Romans had at last realized that the solution to a stable frontier lay in the control of the great fertile triangle between the Highlands and the sea and the Forth. Recent work has shown that Carpow on the south bank of the Tay became a permanent fort with massive stone buildings garrisoned

---

[1] The view that the Wall and forts were thoroughly demolished is far too sweeping and based on unwarranted assumptions. The actual evidence of destruction is very limited and it is not always possible to determine whether the reconstruction was made necessary by enemy action or dilapidation. There certainly was much work done on the structures of the Wall at this time and the reorganization of the garrisons suggests that the barrier was thoroughly overhauled by Severus.

[2] *J.R.S.*, 51 (1961), p. 192; this governor is now known from an inscription found at Bainbridge in 1960.

[3] According to Dio, Caracalla made peace immediately and departed for Rome; however, he did not arrive until late in 211 and coins of 212 celebrated victory; *Roman Britain*, 1963, p. 59, and Frere, p. 175.

[4] These were the scouts (*exploratores*), I. A. Richmond, 'The Romans in Redesdale', *Hist. of Northumberland*, 15 (1940). pp. 95–97. The significance of two other aspects is also brought out in this important study: the use of skilfully mounted artillery at forts like High Rochester and the effective political control over the Lowland tribes.

with detachments of men from VI and II *Augusta*.[1] This site and Cramond on the Forth, where there is also evidence of Severan work, could both be provisioned from the sea, but it seems unlikely nevertheless that these two posts would stand in such isolation so remote from Hadrian's Wall. The Severan solution to the problem of the northern frontier proved to be the only successful one, but it is only partially understood. Effective control over the area beyond Hadrian's Wall must have been maintained by strong garrisons deep in the territory, a vigilant patrol system and new treaty arrangements with the tribes which opened up a better and more peaceful relationship with Rome.

---

[1] *P. Soc. Ant. Scot.*, 96 for 1962–3 (1964), pp. 184–207; *Scottish Hist. Review*, 42 (1963), pp. 126–34; *J.R.S.*, 52 (1962), p. 163; 53 (1963), p. 127; 55 (1965), pp. 200–2.

Chapter 3

THE COMPOSITION OF THE ARMY

THE LEGIONS

Introduction

The legions were the principal force in the Roman Empire in early Imperial times. They had always been recruited from amongst the Roman citizens by annual draft, and the way this gradually broke down into a more voluntary system in the late Republic has been described above. Caesar had recruited extensively during his Gallic campaigns from Cisalpine Gaul, and the *coloniae* of this province continued to supply men for the legions into the second half of the first century. Recruitment from Rome and Central Italy into the ranks ceased almost certainly from the time of Augustus, except into the praetorian and urban cohorts. There was, however, a place for citizen volunteers in the *Cohortes Civium Romanorum Voluntariorum ingenuorum,* during the period when the *auxilia* were recruited entirely from non-citizens.[1] That they held a special place is clear from the bequests of Augustus when they all received three hundred *sestertii* on an equal footing with legionaries,[2] and their commanders held the title *tribunus.*[3] This privileged position may, however, have ceased at a later date, since there is evidence of unenfranchised recruits by the end of the first century.[4]

It is clear from a study of the *origines* of legionaries given on their tombstones that the recruitment areas gradually spread outwards towards the frontiers in the first two centuries of the Empire. Italy fell from the position

---

[1] Velleius, ii, 111; Suetonius, *Aug.,* 25. The earlier view, initiated by Mommsen (*Die Conscriptionordnung der römischen Kaiserzeit,* 1884) based on Dio (lv, 31; lvi, 23) that these units were raised by Augustus from freedmen at the time of the Pannonian revolt has now been superseded (see K. Kraft, *Zur Rekrutierung der Alen und Kohorten an Rhein und Donau,* Berne, 1951). By the time of Trajan the title *civium Romanorum* could be awarded to an auxiliary unit for valour in the field.

[2] *Annals,* i, 8.

[3] Cheesman, *Auxilia,* p. 67.

[4] *Ibid.* The diploma cited here, however, is probably a soldier recruited in the Civil War.

		Numbers of actual examples from inscriptions											
	% of Italians compared to Provincials	*Italy*	*Spain*	*Gaul*	*Macedonia*	*Asian Provinces*	*Syria and Egypt*	*African Provinces*	*Noricum, Dalmatia and Raetia*	*Germany*	*Pannonia and Moesia*	*Dacia*	*Thrace*
Augustus to Caligula	65	207	8	35	14	50	14	7	—	—	—	—	—
Claudius and Nero	48.7	117	18	59	9	11	3	4	12	—	—	—	—
Vespasian to Trajan	21.4	73	16	51	6	43	56	22	23	20	15	—	2
Hadrian to end second century	·9	17	17	34	10	21	79	786	56	35	237	50	99

of supplying 65 per cent of the recruits in the period from Augustus to Caligula down to less than 1 per cent in the second century.[1] The figures on the facing page are taken from the Tables given by Forni[2].

The Numbering and Naming of Legions

In Republican times the legions were given serial numbers (I, II, III, etc.) each year as they were recruited, or by a commander raising troops in his province,[3] and this method persisted. It is therefore hardly surprising to find that in the Civil War there were several having the same number. Distinctions were made by the use of nicknames (*cognomina*) which often reflected the circumstances of their creation like XIV *Gemina* (or Twin), where one legion was made by merging two together.[4] Or the name may indicate the person under whom it was created (the title *Augusta* may also mean that it was reconstituted after the Civil War). Other names are true nicknames like *Ferrata, Fulminata,* or *Rapax,* or may show the province in which they were raised or had served (*Hispana, Gallica,* etc.). Others have the name of the deity particularly favoured by the Emperor who founded the legion (*Apollinaris* and *Minervia*). Later, legions received titles imposed by the ruling house, such as *Antoniniana* (A.D. 212–22) or *Gordiana* (A.D. 238–44), which were held only under the relevant emperors. Augustus retained the numbering system, as legions would have been reluctant to abandon numbers and names hallowed by tradition, just as recently some of our county regiments took unkindly to their regrouping.

Legionary Names and Origins[5]

	Origin of Name	*Date of Creation*
Legio I *Germanica*	Province where it gained distinction under Tiberius[6]	pre-Augustan (?)

[1] An exception to this general rule appears to be in the case of new legions, as Dr J. Mann has indicated (*Hermes*, 91 (1963), pp. 483–9).

[2] *Il reclutamento delle legioni da Augusto a Diocleziano*, 1953, Appendix B, pp. 157–212.

[3] Caesar, for example, had his own series which went at least as far as XII; he was, however careful to avoid the numbers I to IV which were reserved for consular armies.

[4] Caesar thus explained the title *Gemella* (*Bell. C.*, iii, 4), and it has been generally assumed that *Gemina* has the same meaning.

[5] Mostly compiled from Parker.

[6] One of the four legions disbanded after its disgraceful conduct during the revolt of Civilis. The two new legions, IV and XVI, contained two of the same numbers, but with different titles.

	Origin of Name	*Date of Creation*
I *Adiutrix pia fidelis*	Assistant, i.e. raised to supplement the legionary strength[1]	Nero
I *Italica*	Raised in Italy[2]	Nero
I *Macriana*	Raised by Clodius Macer[3]	Nero
I *Flavia Minervia*	After Minerva[4]	Domitian
I *Parthica*	Raised for campaigns in the East	Severus
II *Adiutrix pia fidelis*	Assistant, i.e. raised to supplement the legionary strength	Vespasian
II *Augusta*	Raised by Augustus	Augustan
II *Italica pia*	Raised in Italy	M. Aurelius in 165
II *Parthica*	Raised for campaigns in the East	Severus
II *Traiana fortis*	Strong, raised by Trajan	Trajan
III *Augusta pia fidelis*	Formed by Augustus	Augustan
III *Cyrenaica*	Province where it gained distinction	Pre-Augustan (?)
III *Gallica*	From the veterans of Caesar's Gallic legions	Pre-Augustan
III *Italica concors*	United, raised in Italy	M. Aurelius in 165
III *Parthica*	Raised for campaigns in the East	Severus
IV *Flavia firma*	Steadfast, raised by Vespasian	Vespasian in A.D. 70

[1] I and II *Adiutrix* were both raised from marines of the Fleet, I by Nero and II a little later, but their precise origins are obscured by the confusion of the Civil War.

[2] According to Suetonius (*Nero,* 19), all the recruits had to be Italian born and six feet tall. He called the force 'The Phalanx of Alexander the Great', an example of his strange romantic delusions.

[3] Clodius Macer, governor of Africa, set himself up as Emperor on the death of Nero, but was executed by the order of Galba (*Hist.*, i, 7); the legion was disbanded by the new Emperor, but came once more into commission under Vitellius (*Hist.*, ii, 97), only to disappear finally with his death.

[4] Minerva was particularly favoured by Domitian (Suetonius, *Dom.*, 15); further titles *pia fidelis Domitiana* were given for its loyalty in the revolt of Saturninus, but *Domitiana* was erased when Domitian incurred the *damnatio memoriae*.

	Origin of Name	*Date of Creation*
IV *Macedonica*	Province where it gained distinction[1]	Augustan
IV *Scythica*	Region where it gained distinction[2]	Pre-Augustan
V *Alaudae*	The Lark[3]	Pre-Augustan
V *Macedonica*	Province where it gained distinction	Pre-Augustan[4]
VI *Ferrata fidelis constans*	'Iron-sides',[5] a nickname indicating endurance	Pre-Augustan
VI *Victrix*	Victorious, given after an outstanding victory	Pre-Augustan[6]
VII *Macedonica Claudia pia fidelis*	For its loyalty to Claudius in A.D. 42[7]	Pre-Augustan
VII *Gemina*	One legion made out of two[8]	Galba
VIII *Augusta*	Reconstituted by Augustus	Pre-Augustan
IX *Hispana*	Province where it gained distinction	Pre-Augustan
X *Fretensis*	From the naval war between Octavian and Sextus Pompeius	Pre-Augustan

[1] See fn. 6, p. 109 above.

[2] *Ibid.*

[3] This legion was raised by Caesar (Suetonius, *Caesar,* 24) and its name is the Celtic word for the crested lark, but it also means 'the great', from the allusions to the lark as the great songster. This legion was probably lost in A.D. 86.

[4] The two emblems, the bull and capricorn, associate it with both Caesar and Augustus, and probably refer to its creation and later reorganization. It seems possible that this legion was at one time known as V *Gallica*(*J.R.S.*, 23 (1933), pp. 18–19).

[5] To use a Cromwellian parallel; *ferrata* means iron shod.

[6] It is possible that both the VI Legions originated from Caesar's VI. It was also known as *Hispana* from its service in Spain (*Rang.*, p. 187).

[7] In the mutiny of Scribonianus in A.D. 42, the two legions (VII and XI) in Dalmatia were ordered to march behind the self-appointed 'Emperor', but according to Suetonius some divine intervention prevented the troops from dressing their eagles with garlands and perfumes and then the standards resisted all attempts to lift them from the ground (*Claudius*, 13). These omens, the loyalty felt towards Claudius and the skilful handling of the situation by the officers, dissuaded the men from revolting.

[8] It originally had the title *Galbiana* and suffered serious losses in the battle of Cremona (*Hist.*, iii, 22).

	Origin of Name	Date of Creation
X *Gemina*	One legion made out of two	Pre-Augustan
XI *Claudia pia fidelis*	For its loyalty to Claudius in A.D. 42[1]	Pre-Augustan
XII *Fulminata*	'Lightning-hurler', probably gained under Caesar	Pre-Augustan
XIII *Gemina pia fidelis*	One legion made out of two	Augustan
XIV *Gemina Martia Victrix*[2]	One legion made out of two	Augustan
XV *Apollinaris*	After the god Apollo[3]	Augustan
XV *Primigenia*	After Fortuna Primigenia[4]	Caligula or Claudius
XVI *Flavia firma*	Raised by Vespasian	Vespasian in A.D. 70
XVI *Gallica*	Province where it gained distinction[5] possibly under Drusus	Augustan
XX *Valeria Victrix*[6]	Gained distinction under Valerius Messalinus[7]	Augustan
XXI *Rapax*	'Greedy'—in the sense of sweeping everything before it[8]	Augustan

[1] After the abortive revolt of Furius Camillus Scribonianus in 42. This legion probably derived from Caesar's XI, and the title *Actiacus* it held under Augustus indicates that it fought at Actium (*I.L.S.*, 2243 and 2336).

[2] The title *Martia Victrix* was granted after 60 for the legion's part in the defeat of Boudicca.

[3] Apollo was considered by Augustus to be his protecting deity.

[4] An ingenious but inconclusive argument has been put forward (P.W., xii, p. 1246, under 'Legio') that Caligula raised the two *Primigenia* legions. Whether it was Caligula or Claudius, clearly the need for additional troops for the invasion of Britain was the reason for their creation.

[5] See fn. 6, p. 109 above.

[6] The title *Victrix* was granted after 60 for the legion's part in the defeat of Boudicca.

[7] An alternative possibility put to me by Dr A. Birley, is that this title was bestowed by Claudius for some special distinction it may have earned in the invasion of Britain. Just as he gave the title Britannicus to his son, so this legion was named after his wife Valeria Messalina. It had a boar as its emblem, the significance of which is not understood.

[8] This legion was lost under Domitian and the erasure of its name on two tombstones at Vindonissa (*C.I.L.*, xiii, 5201 and 11514) may indicate that it seriously disgraced itself.

Plate XII
A battle scene from Trajan's Column showing a stone-thrower, a native ally and auxiliary infantry.

Plate XIII
The tombstone of a *tubicen*, from Köln, now in the Museum at St. Germain-en-Laye, France (p. 141).

	Origin of Name	Date of Creation
XXII *Deiotariana*	Raised by Deiotarus[1]	Augustan
XXII *Primigenia pia fidelis*	After Fortuna Primigenia	Caligula or Claudius
XXX *Ulpia victrix*[2]	Raised by Trajan and called victorious presumably after distinguished conduct in Dacia	Trajan

The Number of Legions in Service—a Summary

The actual number of legions at any given time has always been a matter of serious dispute. The only certain fact is given by Tacitus in his review of the military situation in A.D. 23—eight legions on the Rhine, three in Spain, two in Africa, two in Egypt, four in Syria, two in Pannonia, two in Moesia, and two in Dalmatia,[3] making a total of twenty-five. Since none had been added or lost under Tiberius, this is the strength left by Augustus. But exactly how Augustus reorganized the fragments of sixty or so legions after the Civil War into a rationalized army is not very clear.[4] It was Parker's view, however, that by 16 B.C. Augustus had an army of twenty-eight legions to be reduced by the Varian disaster to the figure of Tacitus. Two new legions were raised either by Caligula or Claudius, one by Galba and three more under Nero or Vespasian. Then four were disbanded by Vespasian after the Revolt of Civilis and replaced by two re-formed units. Thus the total by now was twenty-nine, increased to thirty by Domitian in 83. V *Alaudae* was lost probably by 92, so that when Trajan raised XXX for the Dacian Wars, as its number implies, there were now thirty legions. However, Trajan also created II *Traiana* and this probably replaced XXI *Rapax*, which had disappeared earlier.

---

[1] Deiotarus, tetrach of Galatia, was a client king whose domains had been extended by Pompey. He raised two units and had them trained in the Roman manner, as if they were legions, to assist Caesar in his lightning campaign against Pharnaces (*Bell. A.*, 34, 4). When the kingdom was bequeathed to Rome 25 B.C. this regular Roman legion was formed out of the troops of these two units.

[2] The number XXX would seem to indicate that at the time of its formation there were already twenty-nine legions.

[3] *Annals*, iv, 5.

[4] Parker recites the arguments of Mommsen and Hardy, pp. 78–90.

An inscription in Rome, datable to the time of Marcus Aurelius, lists twenty-eight legions in a west to east order with IX *Hispana* and XXII *Deiotariana* missing.[1] The former may have been lost on the Danube or in Cappadocia in 161 (see p. 97), and the latter in the Jewish Revolt under Hadrian. The only additions in the second century were two new legions raised in Italy by Marcus Aurelius in 165. The total remained remarkably constant for over a hundred years.

Organization

The smallest unit in the legion was the century which may originally have been a hundred men, but by the time of Polybius contained eighty. It was divided into ten sections of eight men each (*contubernia*) sharing a tent and a mule in the field and a pair of rooms in permanent barracks; it seems likely that this was also a mess-unit. In Republican times there is greater reference to the maniple which consisted of two centuries. The pairing of centuries in the camp and fort continued and this old unit probably had some administrative and tactical significance. The troops probably paraded and marched in this order. The century was, however, the basic unit of the Imperial legion. Six centuries made up a cohort (480 men) and ten cohorts the legion, except the first cohort which had double centuries and was in effect a milliary unit. That there were only five centuries in the first cohort seems to be implied by a list of *optiones* on an inscription from Lambaesis of the time of Severus.[2] This agrees with the plans of legionary fortresses where the five double blocks, each with a centurion's house, appear along the *via principalis* at Novaesium (Fig. 35) and Inchtuthil (Fig. 34) and are implied by the spacing at Caerleon (Fig. 33)

Vegetius is our only authority for the arrangement of cohorts in their battle order and he also gives us some indication of their relative importance.[3] In the front line the first cohort was placed on the right, the third in the centre and the fifth on the left flank, while between them were the second and fourth. Behind the first, in the second line, was the sixth, which he says

---

[1] *I.L.S.*, 2288.

[2] *C.I.L.*, viii, 18072.

[3] The *antiqua legio* of Vegetius (ii, 4–14), Parker suggested (*Class. Q.*, 26 (1932), pp. 137–9), describes the organization of the legion between A.D. 260 and 290 after Gallienus had increased the size of the cavalry (Zosimus, i, 52, 3), but the arrangement of the battle order of the cohorts most probably remained unchanged from earlier times.

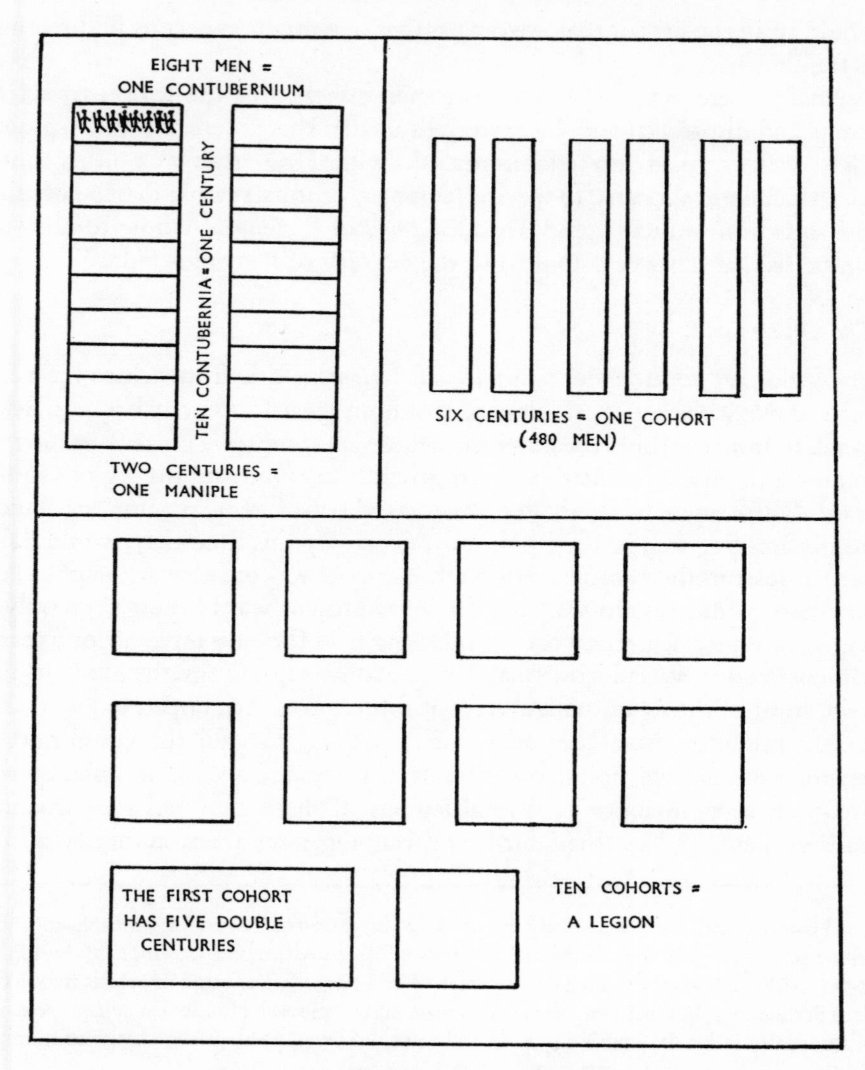

FIG. 11 *The organization of the legion*

should consist of the finest of the young men. The eighth was in the centre with selected troops and the tenth on the left flank also with good troops, the other two, the seventh and ninth, coming between. The weakest cohorts thus appear to be the ninth and seventh and the fourth and second, and it would be in the first of these two pairs that one might expect to find recruits in training.[1]

Finally there was a body of horsemen attached to the legion to act as scouts and dispatch riders. Josephus tells us that there were 120 to a legion.[2] They were ranked 'on establishment', with H.Q. staff and other non-combatants and allocated to specific legionary centuries rather than belonging to a squadron of their own.[3] Thus the *eques* of II *Adiutrix* whose tombstone can be seen at Chester belonged to the century of Petronius Fidus.[4]

The Officers

The legionary commander was a *legatus legionis*, a title from which has been derived our word legate, as someone to whom special authority has been delegated. In Imperial times the Emperor, whose *imperium* gave him official power to raise an army, appointed his own generals to command the legions in his place. These men were usually senators who had been praetors or senior magistrates as a step in their political careers. By this time they would have been at least in their thirties, although Agricola was only twenty-eight. The weakness of this system was that this appointment was in most cases only a stage in a career, legionary command being held for only three or four years. Military ability varied greatly and while some eagerly sought glory in the field, most of the *legati*, with a view of future prospects, hoped to avoid any serious mistakes. At a later stage, after passing through the consulship, a senator would have become a provincial governor and, if in an Imperial province, a commandor of several legions. If these men had no particular military bent, at least their birth and training gave them authority and a

---

[1] Professor Birley has drawn my attention to the career of T. Flavius Virilis a centurion whose tombstone has been found at Lambaesis in North Africa (*C.I.L.*, viii, 2877 = I.L.S., 2653; *J.R.S.*, 2 (1912), pp. 21–24); after forty-five years' service, some of which may have been in the ranks, but including the centurionate in six legions, he finished as *hastatus posterior* of IX; surely in his sixties (his age at death is given as seventy) he was in charge of training young recruits or given a staff post.

[2] iii, 6, 2. An inscription from Lambaesis (*C.I.L.*, viii, 2562, discussed by von Domasewski in *Neue Heidelberger Jahrbuch*, 9, p. 150) of the time of Severus Alexander implies that the commander of the cavalry was an *optio equitum*. This means that the strength of this unit had not materially altered by then. It probably remained the same size until Gallienus.

[3] D. Breeze, "The Organisation of the Legion: The First Cohort and the *equites legionis*" (*J.R.S.*, 59 (1969), pp. 50–55).

[4] *R.I.B.*, 481, Catalogue No. 29, Wright and Richmond, 1955, p. 21, and Pl. x.

sense of responsibility, and any weakness was only revealed by an unforeseen crisis.[1]

The next in rank was the senior tribune. Another senior officer was the *praefectus castrorum* or camp prefect, normally a man of fifty or sixty years of age who had risen through the centurionate to become a *primus pilus* (chief centurion) and who had spent his whole life in the Army.[2] He assumed command when the *legatus* and senior tribune were away, as in the case of the ill-fated Poenius Postumus of II *Augusta* who disobeyed the command of Suetonius Paulinus in the Boudiccan Revolt and afterwards fell upon his sword. It was for the ex-centurion the very peak of his army career and he could hope for no further advancement in this field, but held the position as long as he was able. For example, M. Aurelius Alexander, a Syrian of Commagene of XX *Valeria*, died at the age of seventy-two, as his tombstone at Chester indicates, still apparently in harness.[3] Although a senior quartermaster, mainly responsible for engineering and building works, munitions and equipment, the *praefectus* could have independent command in the field.[4]

There were six military tribunes, each of whom held this commission as a step in his career. The senior tribune was a senator designate (*tribunus laticlavius*, i.e. entitled to a broad purple stripe on the *toga*, the senatorial distinction), who served a short term as a tribune before he was twenty-five, prior to entering the Senate as a *quaestor* (a junior magistrate in Rome or a junior financial officer in one of the senatorial provinces). He could look forward to receiving the full command of a legion later in his career and this was his apprenticeship. The

---

[1] Governors of provinces were in the early Empire chosen with care for their specific abilities and the particular situation. Thus Quinctilius Varus was a distinguished lawyer and may have been selected as governor to integrate the laws of the inhabitants into the Roman code, clearly a premature decision in view of his loss of three legions!

[2] Tacitus gives an effective pen-portrait of Aufidienus Rufus, who was the target of the legionaries' abuse during the mutiny on the Rhine in A.D. 14. The soldiers removed his carriage, loaded him with equipment and forced him to march at the head of the column. He had been a private, then a centurion and finally *praefectus castrorum*, and was in favour of the introduction of tougher methods. Having known hard toil himself, he was all the more hard-bitten because of what he had endured (*Annals*, i, 20).

[3] *R.I.B.*, 490; Catalogue No. 36. The alternative reading OS(ROENVS) seems less likely. This is the latest attested example of the rank and his name indicates that he could hardly have died before the time of the Caracalla.

[4] As in Britain (*Annals*, xii, 38).

other tribunes were, most of them, young men of equestrian rank and designated as *tribuni angusticlavii* (i.e. having a narrow stripe). They sprang from municipal magistracies or prefectures of auxiliary cohorts, and a few of them might later expect high positions in the civil service as procurators or in the Army as prefects of *alae* (cavalry regiments). In the legion, these officers held staff appointments and had administrative and judicial responsibilities rather than those of direct command.[1]

The most responsible officers in the legion were undoubtedly the centurions, of which there were sixty of graded seniority.[2] The senior was the *primus pilus*, who, although actually commander of the first century of the first cohort, had much wider responsibilities. The post was held for only one year, a practice continued from Republican times when all centurionates were held for each campaigning season. A senior centurion could thus have held six or seven *primipilates* at this early period (see p. 35). The rank was highly paid and a substantial grant was available on discharge, sufficient to acquire equestrian status. If the man wished, he could continue to serve in the Army, providing there was a vacancy, or there were now attractive posts open to him in the civil service. In the early Empire there are examples of both *primus pilus bis* and *primus pilus ter*, but under Claudius the practice changed with the appearance of the *primus pilus iterum*. This appointment was reserved for officers who had been tribunes in the *vigiles*, urban and praetorian cohorts at Rome, after which they were posted to a legion. Their functions are not clear but they ranked immediately below the senior tribune. They may have had the duties of, and even held the title of *praefectus castrorum*, but enjoyed a higher scale of pay. If they advanced further up the promotion ladder, they could be appointed to procuratorships senior to those open to the ordinary *primipilares*.[3] Next in seniority to the *primus pilus* was the *princeps*, who was responsible for the H.Q. staff and training. The other centurions of the first cohort, known as the *primi ordines*, were in order of seniority, *hastatus, princeps posterior* and *hastatus*

---

[1] Tacitus mentions their duties as arranging the discharges of veterans (*Annals*, i, 37) and after the mutiny of A.D. 14, reporting on the reliability of centurions (*Annals*, i, 44).

[2] In the revolt in Germany, the legionaries turned against their centurions and gave them each sixty strokes of the lash, one for each centurion in the legion (*ut numerum centurionum adaequarent. Annals*, i, 32).

[3] i.e. they qualified for a ducenarian grade of the equestrian order (with an annual salary of 200,000 *sesterces*) instead of a centenarian one (*The Later Roman Empire*, pp. 5 and 525). These difficult points are discussed by Dr B. Dobson in his edition of *Rang.*, p. xxxiii, and I am grateful to him for further elucidation.

posterior. In the other cohorts, the centurions were ranked as *pilus prior, pilus posterior, princeps prior, princeps posterior, hastatus prior* and *hastatus posterior*, but were of equal status, varying only in service seniority.[1]

The senior positions were filled by outstanding or especially favoured men from the lower centurionate, but direct promotion was common from the ranks of the praetorian guard and also direct commissions were given to municipal worthies in the provinces. Special circumstances might disrupt the normal promotion pattern. Gnaeus Calpurnius Piso, when governor of Syria, pursued an arrogant course of ill-judged hostility towards the popular prince Germanicus, apparently under the illusion that he was carrying out the will of Tiberius. To curry favour with the Army, Piso demoted unpopular tribunes and senior centurions and replaced them with men of his own choice who allowed discipline to become lax.[2] Support for the winning side in a civil war might also bring its reward in promotion and *vice versa*. Tacitus records changes in the centurionate of the British legions by Vitellius[3] when II *Augusta* was earnestly espousing the Flavian cause, but doubtless the tables were turned when Vespasian came to power; he was not a man likely to forget old loyalties. One of the remarkable features of the centurionate is the way in which officers were posted from legion to legion and province to province. An example is Petronius Fortunatus;[4] he probably came from Africa, whither he eventually returned, and was recruited to the ranks of I *Italica* in Lower Moesia. After four year's service during which he was in turn *librarius, tesserarius, optio* and *signifer*,[5] he became a centurion and was then transferred to twelve other legions: VI *Ferrata* in Syria Palaestina, I *Minervia* in Lower Germany, X *Gemina* in Upper Pannonia, II *Augusta* in Britain, III *Augusta* in Numidia, III *Gallica* in Syria, XXX *Ulpia* in Lower Germany, VI *Victrix* in Britain, III *Cyrenaica* in Arabia, XV *Apollinaris* in Cappadocia, II *Parthica* probably at Albano in Italy, and finally I *Adiutrix* in Pannonia; a total service of forty-six years, receiving decorations in one of the Parthian campaigns, yet never reaching the senior ranks. Centurions were not normally discharged but died in service.

[1] von Domaszewski in his basic study, *Rang.*, followed Vegetius (ii, 21) in suggesting an over-elaborate system whereby each centurion held a different status; a more realistic approach has been made by T. Wegeleben, *Die Rangordnung der romischen Centurionen*, 1913.

[2] *Annals*, ii, 55.

[3] *Hist.*, iii, 44.

[4] *C.I.L.*, viii, 217 = *I.L.S.*, 2658.

[5] Professor Eric Birley, to whom I am indebted for this and other references in this section, considers that this rapid promotion was due to Fortunatus being literate in a province which was probably deficient in *literati homines*.

Thus the postings involving travel over long distances can hardly be regarded as promotions; there must be other factors involved. The most obvious would be the mounting of campaigns, when it was the usual practice to create the forces needed by vexillations from legions in those parts of the Empire where conditions were peaceful. In some cases these detached units may have been returned, but losses in the field could best be made good by absorption of the vexillations into the legions established in the campaign area. Equally well, losses could be made good by dispatch of further units. After its defeat by Boudicca in the early stages of the revolt of A.D. 60, IX *Hispana* had its complement made good with 2,000 men from the legions of the Rhine.[1] Fortunatus was serving during the second half of the second century, a time of trouble certainly in Britain, which could account for his transfer to II *Augusta* and VI *Victrix* in turn. Vexillations from the Rhine were drafted to all three of the British legions under the governorship of Cn. Julius Verus;[2] but the date, *c.* A.D. 155–8, is too early to fit into the pattern given by Fortunatus.[3]

Each centurion had an *optio,* so called because originally he was nominated by the centurion. The title *optio ad spem ordinis* was given to an *optio* who had been accepted for promotion to the centurionate, but who was waiting for a vacancy. There is at Chester the memorial of an officer of this rank who was lost in a shipwreck.[4] Another officer in the century was the *tesserarius,* who was mainly responsible for small sentry pickets and fatigue parties, and so had to receive and pass on the watchword of the day.[5] Finally there was the *custos armorum* who was in charge of the weapons and equipment. There was no office staff at this level. The cohort had no particular officers or staff, as it was a tactical rather than administrative unit, and when occasion required one of these units to act independently a commanding officer and H.Q. staff would have been specially appointed.

---

[1] *Annals,* xiv, 38.

[2] *R.I.B.,* 1322, *E.E.,* ix, 1163, now in the Museum of Classical Antiquities, at the University of Newcastle upon Tyne.

[3] His last post was probably held *c.* A.D. 200–10 if his Parthian decorations were won under Marcus and Verus.

[4] *R.I.B.,* 544, Catalogue No. 92, *qui naufragio perit.*

[5] See p. 30 above.

The organization of the legionary headquarters was complicated. There was a staff of clerks and orderlies which formed the *tabularium legionis* under the *cornicularius*. Among its officers were *actuaii*, *librarii* and *exacti* (literally one who exacts payment). There were clerks with special duties, such as those who kept granary records (*librarii horreorum*), the soldiers' compulsory savings bank (*librarii depositorum*) and those who were concerned with the properties of the men killed on active service (*librarii caducorum*). There must have been, as in a modern army, a considerable amount of paper work and the archaeological residue of this survives only in very exceptional circumstances, such as in Egypt[1] and Dura-Europos, where fragments of papyrus give us tantalizing glimpses of army life.[2]

The senior officers had their own orderlies (*beneficiarii*), who were aides rather than personal servants and took their seniority from their respective officers. Legionaries of this rank might also be found on the staff of the provincial governor, as in the example of C. Mannius Secundus of XX Legion, whose tombstone, found at Wroxeter,[3] has given rise to the erroneous idea that this legion was stationed there at the same time as XIV, from which there are tombstones of serving soldiers.

As the legions were responsible for all the major engineering and building projects of the Army, they included a number of specialists such as surveyors (*agrimensores*) and *metatores*, who were sent ahead to select a site for a camp and set out the main lines. There would also be the various armament workers making and repairing different kinds of weapons. The technical officers in charge of this kind of work do not appear to have had any special rank, but were from the time of Hadrian given immunity from heavy fatigues and other routine duties.[4] The reason for this exemption was simply that these men, because of their specialists duties, were not available for the normal

---

[1] *The Abinnaeus Archive*, 1962.

[2] *The Excavations at Dura-Europos*, Final Report V, Part i, The Parchments and Papyri, 1959.

[3] *R.I.B.*, 293; Collingwood suggested that the two legions advanced into the area together, *Roman Britain and the English Settlements*, 1937, p. 90. It would seem likely, however, that XX succeeded XIV on its withdrawal from Britain in A.D. 64.

[4] An almost complete list of *immunes* of the late second century is recorded by Tarruntenus Paternus (*Digest*, 50, 6, 7); there is a useful discussion on the precise meaning of this immunity by G. R. Watson, *The Roman Soldier*, 1969, 75ff.

work of the *miles gregarius*. Although Hadrian gave official recognition to the *immunes*, there is little doubt that the practice had been recognized much earlier. This immunity may have been a greater privilege than it would appear to us, since it was customary for centurions to receive payment for release from these responsibilities.[1] An unscrupulous centurion could organize a lucrative business out of this simple practice and official immunity would have been a considerable advantage and certainly carried with it an important legal status.

Among these specialists were the priests (*haruspices*), but their duties were not like those of our service chaplains, to help soldiers with personal problems, but to officiate at the sacrifices and other ceremonies. There is a vivid scene on Trajan's Column (Pl. IIa) which shows the long procession with the animals, in appropriate order, winding its way round the fort towards the altar, where the veiled priest stands with the horn players, ready to sound the fanfare which scared away the evil spirits. There were also the doctors (*medici ordinarii*) and medical orderlies (see below, p. 251).

The Legionaries' Equipment

The equipment issued to the legionaries was remarkably uniform throughout the Empire and it is possible that there were large centres in Gaul and North Italy for the mass manufacture of helmets, armour and weapons as well as the kettles and mess-tins[2]. One can detect changes in style at different periods

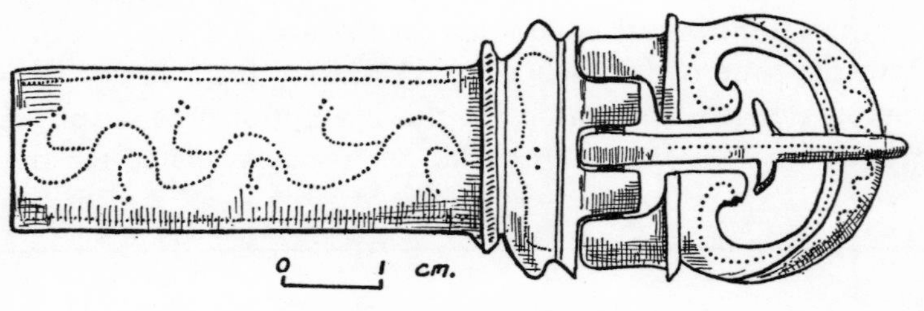

FIG. 12 *A buckle from London (actual size)*

[1] *Annals*, i, 17; Tacitus puts into the mouth of Percennius, the leader of the mutiny of A.D. 14, the words: 'Life and limb are reckoned at two and a half *sestertii* a day and out of this wretched pittance, he must pay for his clothes, tent and weapons as well as bribe his centurion to secure exemptions from fatigues and punishments.'

[2] There is, however, no direct evidence of this (R. MacMullen, 'Inscriptions on Roman Armour and the Supply of Arms to the State', *American J. Arch.*, 64 (1920), pp. 23–40). In the East armaments were supplied, on demand, by the cities (*Hist.*, ii, 82; Dio, lxix, 12).

and there seems to be a tendency in the first two centuries gradually to simplify and reduce the over-elaboration on more decorative pieces. In the middle of the first century, for example, the buckles (Fig. 12), belt-plates (Fig. 13) and apron terminals were not only silvered and occasionally gilded, but also decorated with black niello inlay. By the end of the first century this practice had ceased and in the second the tendency is for rather lighter metal with open 'fretwork' designs placed over coloured cloths or leather. It may even be possible to trace a similar change of style in the humble mess-tin.[1]

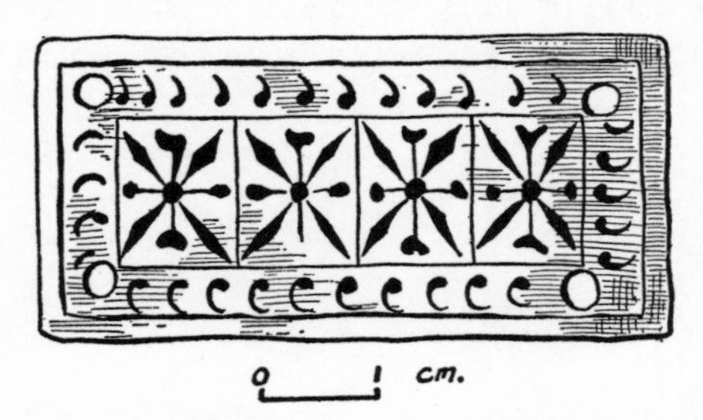

FIG. 13 *A belt-plate from Leicester (actual size)*

The soldiers wore a linen undergarment next to the skin and over it a short-sleeved woollen tunic which came down to the knees. The latter garment can be clearly seen on the Column, worn by soldiers who have stripped off their body armour while in working parties in friendly areas (Pl. IIb).[2] Although the Romans had originally considered the wearing of trousers (*bracae*) a foreign and effeminate habit, legionaries in cold climates were allowed to wear ones made of leather which were skin-tight and reached just below the knee.[3] On their feet they wore the elaborate military boot

[1] An attempt to do so (*A.A.*, ser. 4, 13 (1936), p. 139) now requires revision in the light of a number of well-dated mid-first-century examples (e.g. from Gloucester, in the British Museum (*Arch. J.*, 115 for 1958 (1960), Pl. ix B), from Broxtowe, Nottingham (*Ant. J.*, 19 (1939), Pl. lxxxvii), Caves Inn, Warks (*T. Birmingham Arch. Soc.*, 81 for 1963–4 (1966), pp. 143–4), and Doorwerth (*Oudheidkundige Mededeelingen*, Leiden 12 (1931 Afb. 12 and 14)). Differences in shape may be associated with workshops rather than with a chronological significance.

[2] The arrangement of this garment is difficult to understand. It appears to be gathered at the waist, where it must have been fastened with brooches, allowing it to fall in semicircular folds back and front. This formalized treatment is also seen on tombstones.

[3] On Trajan's Column, *bracae* are seen worn by legionary and praetorian standard-bearers and all auxiliaries, but not by the praetorians and legionaries themselves. This also seems to be the case on the Column of Marcus.

(*caliga,* from which the Emperor Gaius received his nickname, as a child, of Caligula or Little Boots). They were heavy sandals with several thicknesses of sole studded with hollow-headed hob-nails. The leather thongs were continued half-way up the shin and tied there, and in the cold weather could be stuffed with wool or fur.

The type of body armour seems to have varied from time to time. In the first century legionaries had hardened leather jerkins with metal or additional leather shoulder-pieces, but on Trajan's Column all the legionaries are equipped with body armour made of metal strips and plates.[1] This is a complex system in several parts. There are the pairs of front and back-plates (Fig. 14) covering

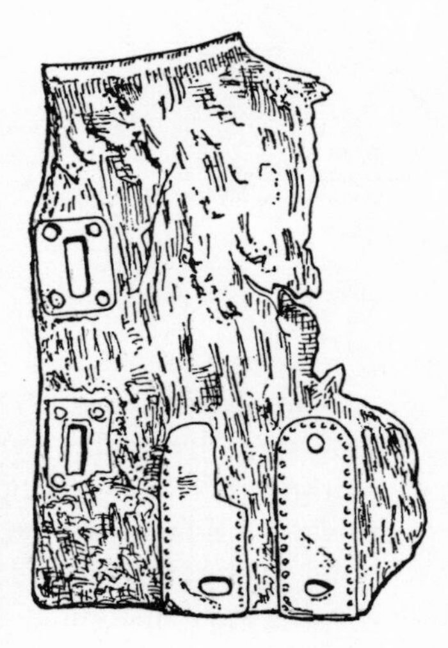

FIG. 14 *A plate from a* lorica segmentata *from Newstead (half-size)*

the upper part of the chest and back, while the trunk is protected by six or seven overlapping metal strips hinged at the back and fastened at the front with hooks laced together (*lorica segmentata.*) They were arranged so as to give complete freedom of movement and so must have been held together with leather strips to allow the necessary articulation, or riveted to a leather jerkin, so that

[1] A problem arises when the legionaries on the Column are compared with those on the Adamklissi monument. One group on the latter are without helmets and wear cloaks which conceal their body armour (Metopes xxviii, xxxviii and xliii). But the other group, identified by their rectangular shields and the *gladius* carried high on the right side, wear mail, and their right arms are covered with reticulated armour, presumably to protect them from the long knife the Dacians attached to a shaft Metopes, xvii–xxiii).

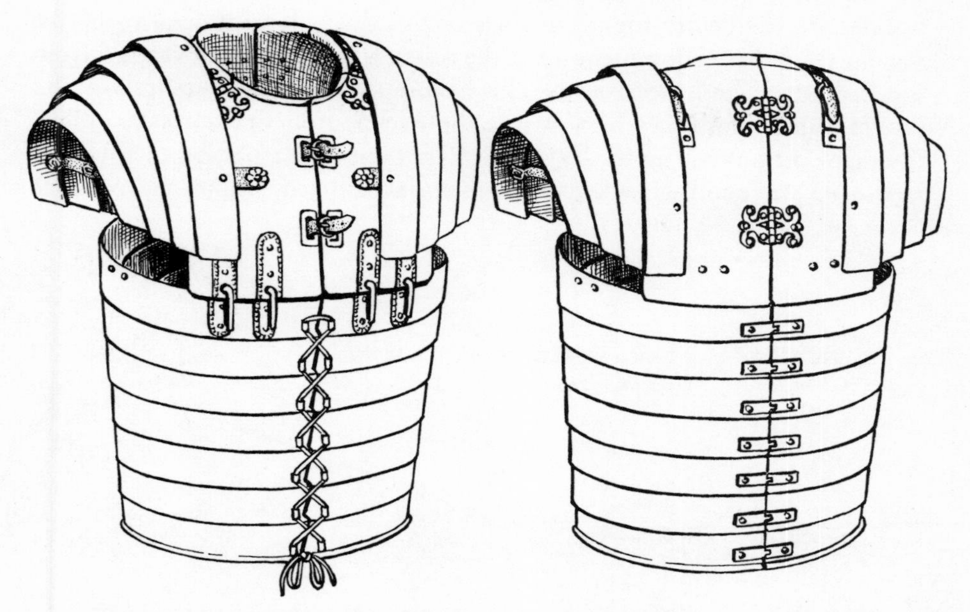

FIG. 15 *The* lorica segmentata, *front view*

FIG. 16 *The* lorica segmentata, *back view*

each strip had independent movement.[1] Over this was a pair of shoulder-pieces each with five or six strips carefully shaped and presumably buckled on to the plates or strips.[2] The *lorica segmentata* (Figs. 15 and 16) was in use in the Army of invasion of Britain, since fragments of hinges (Fig. 17) hooks and

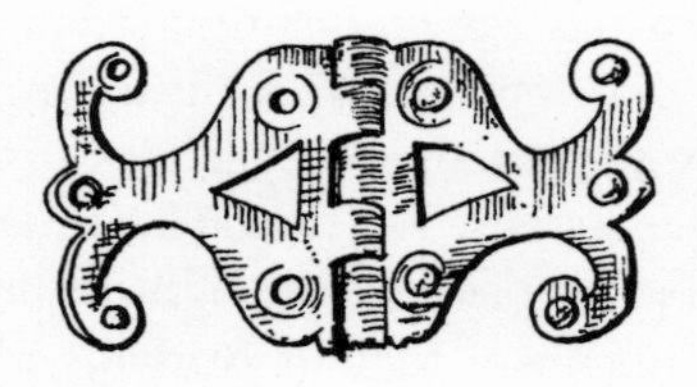

FIG. 17 *A hinge from the back of a* lorica segmentata *from Waddon Hill, Dorset (actual size)*

---

[1] It is clear from the Column that legionaries had to dig ditches and erect fortifications in full armour with shields, *pila* and helmets stacked near by. Corbulo in restoring discipline to slack troops in Lower Germany is on record as executing a soldier for digging without his side-arms (*non accinctus, Annals,* xi, 18).

[2] H. Russell Robinson, *Roman Armour*, 1969, an exhibition guide, National Museum of Wales.

buckles are frequently found on forts of this date.[1] It continued to be the standard legionary equipment up to the time of Trajan on the evidence from the Column. The legionaries on the Column of Marcus Aurelius are also shown in strip armour. Here, however, there are no front or back plates visible, the strips continuing up to the neck.[2] Variations in the pattern of the *lorica segmentata* are found when actual fragments are studied, and on the evidence

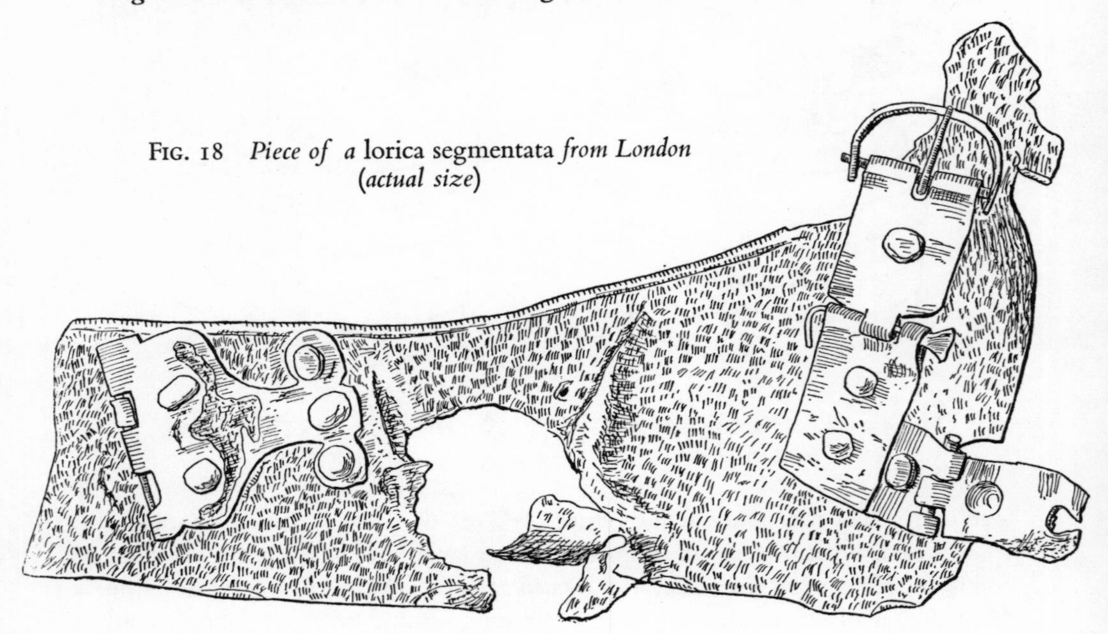

FIG. 18 *Piece of a* lorica segmentata *from London (actual size)*

from the Column it would seem that the plates were much reduced in size by the mid-second century. The strip *lorica* was replaced by scale and mail at a later date and this accounts for the examples of these other types of body armour frequently appearing in legionary fortresses. The reason for this change is not clear since it is hardly likely that these later types were lighter and more flexible. They are clearly shown on monuments of Marcus Aurelius, all three types are seen on the *allocutio* on the Arch of Constantine.[3] From the positions

[1] This is not, however, the view of J. Alfs who credits Trajan with the introduction of strip armour (*Zeitschrift für historische Waffen-und Kostümkunde,* Heft, 3–4, 1941).

[2] As they do on the bronze statuette in the B.M. (*A Guide to Greek and Roman Life,* 1929, Fig. 81).

[3] It appears in the attic; see Stuart Jones, *Pap. Brit. School at Rome,* 3 (1906), pp. 251–68.

of these soldiers it might be deduced that the one in scale armour nearest the Emperor is a praetorian, the middle one a legionary and the third an auxiliary. In this case the praetorians had already changed from strip to scale armour and presumably the legionaries followed suit at a later date. Round the neck was worn a scarf, knotted at the front, to prevent the metal plates from chafing the skin. The legionary had a wide belt, studded with decorated metal plates, which carried the dagger and an apron. The latter consisted of a number of leather thongs to which were riveted metal plates, and weighted with bronze terminals. It swung between the legs on the march and gave protection to the stomach and private parts. On the Column this apron is normally shown tucked into the belt, but its true length and nature appears on the tombstones.

For protection of the head there was a carefully designed, bronze helmet which had inside an iron skull-plate. At the back a projecting piece shielded the neck and a smaller ridge fastened at the front gave protection to the face. At the sides were large cheek-pieces hinged at the top. Two distinct variations of the helmet have been found and shown by H. Klumbach[1] to have a chronological significance, one replacing the other during the reign of Claudius. The earlier type was a sturdy solid bronze helmet with a horizontal projection at the back. Its disadvantage was the lack of protection it gave to the neck from an upward or horizontal sweep of the sword. The revised shape brought the back of the helmet over and round the neck and the projection was inclined to fit slightly over the shoulders, the extra coverage also permitted the fitting of ear-protectors; at the same time a thinner gauge of metal was used, and the skull-plate reinforced with iron (Fig. 19). Three examples of the earlier type have been found in Britain and must belong to the Invasion Army.[2] By A.D. 60 the new type was standard issue, since it is examples of these which were found in association with the Boudiccan Revolt on the Sheepen site at Colchester.[3] Another difference in the two types of helmet is in the plume-holder; the three early examples have a conical knob with a central slot,

---

[1] *J.R.G.Z.M.*, 8 (1961), pp. 96–105.

[2] They include one probably from the river Thames and now in the British Museum (*B.M. Quarterly*, 16, p. 17, and *B.M. Guide* (1951), Pl. xxv, No. 5); one from Bosham Harbour, Chichester, now in Lewes Museum (I. D. Margary, *Roman Sussex*, 1951, Pl. x); one found near Northcott Hill, Tring, Herts, now in Colchester Museum (*V.C.H., Herts*, 4, p. 158 and Pl. 1).

[3] *Camulodunum*, 1947, Fig. 62; these helmets presumably belonged to the veteran colonists and if so might have been standard issue some time previous to this event, but not necessarily as early as A.D. 50, the presumed date of the founding of the *colonia*, since Tacitus implies that land grants continued to be given after the foundation date (*Annals*, xiv, 31).

while the later ones consist of a flat strip raised across the central axis to form a long box-like space into which a metal tongue could be slipped. In the new version crests could be attached with greater ease.[1] It is clear from the Columns of Trajan and Marcus that plumes, in the second century, were only worn on parades and took the form of tufts of material fastened to the top of the helmet.[2] But Caesar, in describing the confusion which arose in a

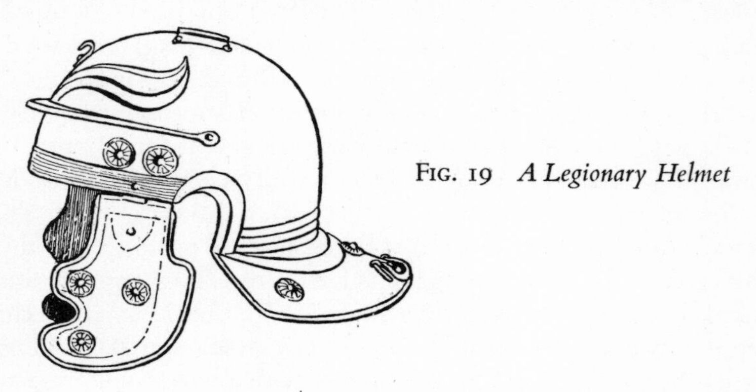

FIG. 19 *A Legionary Helmet*

battle against the Nervii in 57 B.C., says that his troops were 'so pushed for time by the enemy's eagerness to fight that they could not even take the covers off their shields or put on helmets—not to speak of fixing on crests or decorations'.[3]

The men's legs were bare, protection here being sacrificed for mobility. Each man carried a large shield (*scutum*) curved to fit the body. Examples from Dura-Europos show that they were made of a kind of plywood, thin sheets of wood, glued together so that the grain of each piece was at right-angles to its neighbour. The whole was bound round the edges with wrought iron or bronze and the centre was hollowed out on the inside for the hand-

[1] An example of this type of plume-holder has been found at Waddon Hill, Dorset, where pottery clearly indicates an occupation continuing until *c.* A.D. 60. There is another type of plume-holder which has a vertical stem 4.5 cm high with a Tee-piece at the top from which rise two vertical arms, clearly for a more elaborate type of plume (*Die Römischen Donau-Kastelle Aislingen und Burghöfe*, 1959, *Limesforschungen*, I, Taf. 20, No. 15, found at Aislingen).

[2] On the Column of Marcus there is an exception to this in the scene of legionaries on the march and auxiliary standard-bearers. It is curious that on both Columns and on the Arch of Constantine, helmets have a large ring attached to the top, but few examples have been found. One from Syria has been published by C. C. Vermeule (*J.R.S.*, 50 (1960), pp. 8–11), who considers that the ring was to enable the helmet to be hung up or carried on a lance.

[3] *Bell. G.*, ii, 21: Penguin translation.

(a)

(b)

Plate XIV

(a) Moorish horsemen on Trajan's Column (p. 143).

(b) Part of the tombstone of a standard-bearer of the *Ala Petriana*, at Hexham (p. 148).

Plate XV
Part of the tombstone of a trooper of a Thracian *ala*, from Cirencester (p. 148).

grip and protected by the metal boss.[1] On the outside the surface was covered in leather, on which was fastened gilded or silvered decoration, probably in bronze. Those of the legionaries represent Jupiter's thunderbolts. It is probable, as Vegetius implies, that each cohort had its shields coloured differently to aid recognition in the confusion of battle; they also carried the name of the soldier and that of his centurion. On the march the shield was hung by a strap round the left shoulder.

For taking the offensive the legionaries had two kinds of weapons, the *pilum* or javelin, of which each man had two, and the sword. The *pilum* was a disarming weapon and its early history has been discussed above (p. 28). Caesar clearly describes its function: 'The Gauls were much hampered in action because a single spear often pierced more than one of their overlapping shields and pinned them together; and, as the iron bent, they could not pull it out. With their left arms thus encumbered it was impossible for them to fight properly, and many, after repeated attempts to jerk their arms free, preferred to drop the shields and fight unprotected'.[2]

The *pilum* as seen on reliefs of Imperial times and actual excavated specimens, was seven feet long. The top three feet were of iron with a hardened point of pyramidal form which varied in length from about two to seven inches.[3] It is probable that more sturdy types of spear or pike were available for defence against cavalry.[4]

---

[1] A finely decorated bronze boss from the river Tyne, and now in the British Museum, bears a punched inscription stating that it belonged to Junius Dubitatus of the century of Julius Magnus of VIII *Augusta,* a vexillation of which was dispatched to Britain from the Rhine under Hadrian (*B.M. Guide to the Antiquities of Roman Britain,* 1951. Fig. 35 and p. 67). As indicated above (fn. 2, p. 22) this type of shield was not introduced until the time of Caesar.

[2] *Bell. G.,* i, 25: Penguin translation.

[3] A collection of over fifty is illustrated from Caerleon (Prysg Field Report, Pt. II, Figs. 20 and 21). It is seen here that very little of the soft shank has survived. The points from Claudian forts seem to be somewhat smaller than these. The two illustrated from Hod Hill are each three inches long, one of which has twenty-three inches of shank surviving (*Durden Coll.,* Pl. II B).

[4] There are interesting varieties of the *pilum* illustrated by L. Lindenschmit, *Tracht und Bewaffnung des römischen Heeres während der Kaiserzeit,* 1882. Those furnished with handguards could hardly have been designed for throwing and may be a form of pike or possibly for ceremonial occasions. Vegetius tells us (i, 17) that in the late third to early fourth centuries two Illyrican legions honoured by the titles Jovian and Herculean used a special loaded javelin known as the *martiobarbulus,* each legionary carrying five in 'the hollow of his shield'. These weapons enabled them to reach the enemy beyond the range of normal missiles.

The sword (*gladius*)[1] was a double-bladed weapon two feet long and two inches wide, often with a corrugated bone grip (Fig. 20), and it was used for thrusting at short range. It was carried in its scabbard high on the right-hand side so as to be clear of the legs and the shield arm and was withdrawn in an upward movement and used much like the modern bayonet for close fighting. On the left-hand side the dagger (*pugio*) was attached to the belt. This was slightly waisted in a leaf-shape and nine to ten inches long. Some highly ornamented scabbards have been found, but whether these represent normal issue or belonged to centurions is not clear.[2]

Apart from these weapons each man, according to Josephus,[3] carried a saw, a wicker basket much like a small modern wastepaper basket, for shifting earth, a piece of rope or leather for handling turfs (Pl. III) and a sickle. He also lists a pickaxe (*dolabrum*) with a broad cutting blade at one end and a tine at the other[4] (Pl. IIb, and Fig. 21), which was slung from the belt, its edge in a bronze sheath; it could be used for cutting down trees as well as digging ditches. The illustration on the Column of the legionaries marching across the

---

[1] A typical example of what must be of early Imperial date was found in the river Thames at Fulham and is now in the British Museum. The scabbard is decorated with embossed bronze panels with scroll-work, reminiscent of Augustan forms, and the wolf suckling the heavenly twins. Another notable example from Mainz in Germany, also in the B.M., has been known as the sword of Tiberius, as it bears a relief of that Emperor receiving Germanicus in A.D. 17 after his German campaign, during which the lost eagles of Varus were recovered and the bones of the Roman dead decently interred. It has been suggested that this was one of the special presentation swords made to officers at the time (*B.M. Guide to Greek and Roman Life,* 1929, Fig. 101, pp. 98–101). Scabbards more typically of the legionary have also been found at Mainz with niello decoration (*M.Z.,* 12–13 (1917–18), p. 175, Abb. 6), while from Britain there are fragments from Hod Hill (*Durden Coll.,* A1–A8) and bone handles from Dorchester, Dorset (*P. Soc. Ant.,* 2nd ser., 21 (1905–6), p. 153) and *Newstead* (Pl. xxxiv, No. 13).

[2] The *pugio* might have been regarded as a personal weapon and tool and its scabbard decoration subject to individual taste. Examples have been found at Hod Hill (*Durden Col.,* B5, Pl. iv) and Waddon Hill, Dorset (*Arch. J.,* 115 for 1958 (1960). Pl, xii). For examples on the Rhine see K. Exner, 'Römische Dolchscheiden mit Tauschierung und Emailverzierung' (*Germania,* 24 (1940), p. 22), and for Vindonissa, G. Ulbert, 'Silbertauschierte Dolchscheiden aus Vindonissa' (*G. pro Vindonissa,* 1961–2, pp. 5–18).

[3] *Bell. Jud.,* Bk., iii, 5.

[4] It is interesting to note that this tool was re-invented by the U.S. Army during the last war for providing soldiers with rapid means of digging themselves fox-holes against close-pattern bombing.

Danube shows them carrying some of these items of equipment on long poles or stakes (*pila muralia*), presumably for a kit inspection, since this is the only time they are shown with their gear. Marius is supposed to have introduced the forked pole for carrying equipment in this manner (p. 38), and Josephus

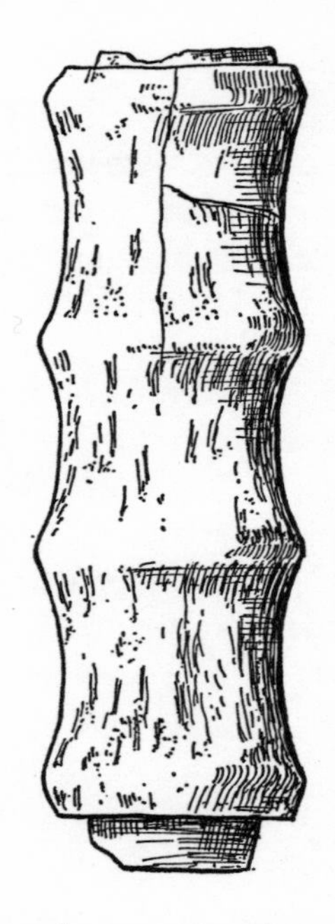

FIG. 20 *A bone hand-grip from a* gladius *from Dorchester, Dorset (actual size)*

implies that even in Flavian times the unfortunate soldier had to carry most of his gear. By the second century the transport system must have improved, for there are several scenes on the Column of baggage being unloaded from carts. The heaviest and most bulky piece of equipment was the leather tent (*papilio*) shared by eight men in the field. This was carried by a mule together with the

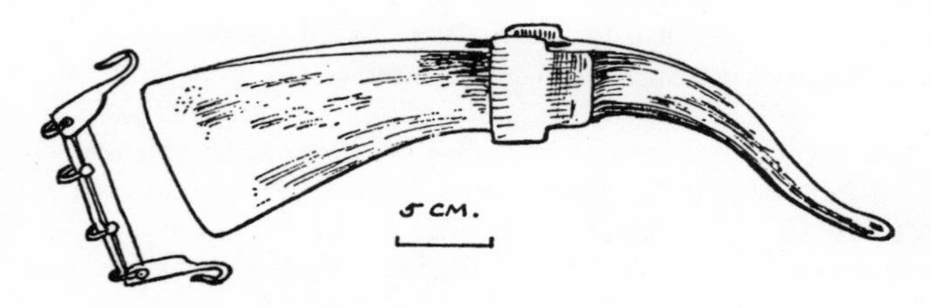

FIG. 21 Dolabra *and sheath*

pair of mill-stones for grinding the corn ration, which on the march was normally three days' supply. For the cold nights on sentry duty in northern Britain the legionary would have needed a heavy cloak (*sagum*), and a thick woollen one, probably with a hood, may have been the standard issue.

The Centurion

This officer was distinguished from the men by his uniform. A splendid first-century example can be seen on the tombstone of Facilis at Colchester (Pl. I). He wears a corselet of leather[1] with metal shoulder pieces and a beautifully ornamented belt. Below his corselet is a double-pleated kilt-like garment and on his shins are thin metal greaves. Unlike the legionary, he carried his sword in the orthodox position on the left swinging from a baldric, and from his left shoulder his cloak made of fine material hangs in elegant folds. In his right hand he carried his emblem of office, the twisted vine-stick (*vitis*).[2] Another tombstone illustrating a centurion's equipment is that of T. Calidius Severus of XV *Apollinaris* at Carnuntum[3] (Pl. IV). This legion was stationed here until A.D. 60, when it was moved to Syria, returning to Pannonia under Vespasian. Each item of equipment is shown separately. There is what appears to be a mail corselet, decorated greaves, and the helmet appears to have a transverse crest, which would distinguish him from the other ranks.[4] Below these items his servant is depicted holding his horse, since centurions would normally have ridden on the march. The saddle-cloth shows the two rolls and one of the stiff side-pieces which were weighted with decorated bronze plates.[5] A

---

[1] It is possible that this was fine mail and indicated only in the paint which picked out all such details originally on the stone.

[2] This may be analogous to the modern officer's swagger-cane, but some centurions were never loath to use it on the backs of lazy or dimwitted legionaries. Lucilius, lynched for this habit in A.D. 14, was nicknamed *cedo alteram* (Fetch me another), as he broke his staff over a soldier's back and bellowed for a second and even a third (*Annals*, i, 23).

[3] *C.I.L.*, iii, 11213 = *I.L.S.*, 2596; *R.L.Ö.*, iv (1903), p. 130; *Arch. Epig. Mittheilungen*, v, p. 206.

[4] Vegetius ii, 13, indicates that centurions wore their crests transversely across the helmet to distinguish them from the legionaires whose crests were fixed in the normal way. The tombstone of Marcus Petronius of VIII *Aug* also shows a front view of the helmet with a large transverse crest (A. Schober, *Die Römischen Grabsteine von Noricum und Pannonien*, 1923, Fig. 59, p. 60).

[5] J. Werner, *Beiträge zur älteren Europäischen Kulturgeschichte*, Band 1, p. 423.

centurion of about the same period is Q. Sertorius Festus of XI *Claudia pia fidelis*, whose tombstone is at Verona[1] (Pl. V). He also wears a scale jerkin and his decorations. His belt is, however, quite different from that of Facilis and his greaves are finely decorated. To judge from these four centurial tombstones, there was a difference between the dress worn by these officers as between the north-western provinces and those on the Danube. What is most probably another centurion appears on a fragment of a Trajanic relief of the Dacian Wars and now in the Louvre.[2]

The legate and his staff officers were distinguished by their fine cloaks, dyed according to rank, and doubtless they had their own armour and uniform suitable to personal taste. On the tombstones of officers it is common to find them shown wearing their decorations. These consisted of gold and silver necklaces (*torques*), armlets and sets of nine discs (*phalerae*) finely decorated with heads of deities in relief and worn on the chest. Different sets of decorations were accorded to each rank.[3] A fine example is worn by the effigy of M. Caelius (Pl. VI), centurion of XVIII, killed in the great disaster under P. Quinctilius Varus in A.D. 9,[4] when three legions were ambushed and annihilated. Caelius wears the civic crown of oak leaves (*corona civica*) awarded for saving the life of a fellow citizen under great danger.[5] Round his neck are two torques and on his wrists are thick bracelets (*armillae*), while he proudly bears on his chest six *phalerae*,[6] each having designs in relief of deities or their

---

[1] *C.I.L.*, v, 3374. The legion was stationed at Burnum in modern Yugoslavia until A.D. 70.

[2] Mrs. A. Strong, *Roman Sculpture*, 1907, Pl. XLIX. He wears a civic crown and a crest. His helmet is decorated with reliefs and he wears a corselet of large scales.

[3] Vegetius (ii, 7) says that these men were known as *torquati* and received a special allowance. Those who received the honour twice (*torquati duplares*) received a double sum.

[4] *C.I.L.*, xiii, 8648; [*ce*]*cidit bello Variano ossa* [*i*]*nferre licebit;* this fine stone is now in Bonn Museum. The scattered remains of the three legions were collected together and buried on the spot by a force under Germanicus in A.D. 15 (*Annals*, i, 61).

[5] This award was given to M. Ostorius Scapula, the son of the second governor of Britain, while serving on his father's staff here during the first revolt of the Iceni of A.D. 50 (*Annals*, xii, 31). Later this fine man of exceptionally large physique and distinguished military record fell a victim to Nero's enmity and was forced to take his own life (*Annals*, xvi, 15). Pliny says that in the early days victorious generals were awarded wreaths (*Nat. Hist.*, 16, iv, 4–5).

[6] A set of nine plain bronze discs was found at Newstead (Pl. XXXI) which may have been the backing plates for the decorated reliefs; the owner, Dometius Atticus, had scratched his name on each. It is hardly surprising that examples of the *phalerae* themselves are rare, since

attributes. An *aquilifer* of XIV *Gemina* at Mainz (Pl. VII) wears his decorations in a schematic arrangement on his chest; the two torques and nine discs appear to have a simple geometric pattern, but this may be due to artistic convention rather than a difference in rank. For being the first over the defences of an enemy stronghold, a centurion could win the *corona muralis* (a crown with battlements). For distinguished service legionary commanders and tribunes were awarded small replicas in silver of spears (*hasta pura*) and standards in numbers according to rank,[1] and these were presumably carried on ceremonial parades by an attendant and usually listed on official documents and tombstones where a career was given.

The Standards

There is nothing quite comparable in modern armies to the Roman standards, except perhaps the regimental colours. Both performed a dual function, a recognition signal and a rallying-point. Army units need a device to watch and follow in battle conditions and the soldiers also need to recognize their own at a glance. From this simple necessity arose the ceremony of Trooping the Colour which is still regularly observed and which forms part of the splendid annual spectacle on the Horse Guards Parade. Its original purpose was to show the new recruits their flag so that they would know which to follow in battle. It was thus closely identified with the unit and its loss to the enemy would be a permanent disgrace and stain on the history of the regiment. In this sense, the Roman standards were held in similar awe; one has only to remember the famous occasion on Caesar's first raid on Britain when, as the ship drew near the beach and his comrades held back, the *aquilifer* of X leapt into the sea with the eagle, crying, 'Jump down, comrades, unless you want to surrender our eagle to the enemy.'[2] There was also a special

---

they would have been highly prized possessions. The most famous set is that found at Lauersfort, near Krefeld, in 1858, in a copper chest lined with silver. They are made of bronze, but are covered with thin silver plating and have deities and mythological animals in high relief (Jahn, *Die Lauersforter Phalerae*, 1860; *Germania Romana*, 1922, Taf. 95, Fig. 1; two are illustrated by Helmut Schoppa, *Die Kunst der Römerzeit in Gallien, Germanien und Britannien*, Taf. 44).

[1] The centurion Petronius Fortunatus, whose career is mentioned above (p. 119), also recorded on his stone—*consecutus ob virtutem in expeditionem Parthicam coronam muralem vallarem torques et phaleras*. After the Jewish War, Titus presented small silver standards and gold spears to his officers (Josephus, *Bell. Jud.*, vii, 1, 3). See also *The Roman Soldier*, pp. 114–117.

[2] *Bell. G.*, iv, 25.

campaign launched against the Germans to avenge the tragic defeat of Varus and recover the lost standards.[1]

The standards also played an important part in pitching and striking a camp. The site being selected, the first act was to set up the standards by thrusting their pointed ends into the ground. When camp was struck the standards were plucked out by means of the large projecting handles.[2] It would have been a serious omen had they stuck fast and the men might have refused to move, saying that the gods meant them to stay there.

The standards played an important part in the many religous festivals which the Army scrupulously observed. On these occasions they were anointed with precious oils[3] and decorated with garlands; special battle honours and laurel wreaths may have been added. It is hardly surprising that a Christian writer has stated that the Army actually worshipped their standards.[4] On the Column they appear in full array at the forefront of all ceremonial occasions, at sacrifices and when Trajan addresses the troops.

In the line of battle the *signa* have key positions. This is clear from Caesar, who often refers to the *ante* and *post signani*, i.e. troops in front of or behind the standards. Orders were also given for movements in relation to the *signa*, as in the African war, when during one engagement the troops became disorganized and were commanded not to advance more than four feet beyond their standards. Later on, when surrounded, Caesar extended his line and gave orders for every other cohort to about turn and face to the rear of the standards, while the alternate cohorts faced the front.[5] Another important function was in the system of signals in the battlefield. Commands were relayed through the standard-bearers and *cornicines*, i.e. men with a large circular instrument. A blast on the *cornu* drew the soldiers' attention to their standard; where it was carried they would follow in formation.[6] A limited number of signals

---

[1] *Strat.*, ii, 8, gives a number of examples of the use of standards in raising morale. Augustus had an extensive coin issue to celebrate the return of the standards lost by Crassus in Parthia (*R.I.C.*, i. p. 46), while special mention is made of this episode in his *Res Gestae* (*Monumentum Ancyranum*, tab. v).

[2] The verb to strike camp is *signa tollere*.

[3] *Nat. Hist.*, xiii, 3, 23; Suetonius, *Claud.*, 13.

[4] Tertullian, *Apologia*, 16; oaths could be taken before the standards (*religio Romanorum tota castrensis signa veneratur, signa iurat, signa omnibus deis praeponit*), e.g. Paetus in Armenia, *Annals*, xv, 16.

[5] *Bell. Af*, 15 and 17.

[6] This chain of command is clearly seen on a Column scene where two lines of soldiers are marching to a prepared position (Pl. VIIIb). The centurion turns to the *cornicen*, who

by up and down or swaying movements were indicative of commands to close or extend ranks, etc.

When one comes to the standards themselves and their various types and patterns throughout Imperial times, there are some serious gaps in our knowledge. The use of effigies of an anthropomorphic nature by groups of warriors is a common feature of primitive society. There was doubtless some totemic influence at work in the obvious wish that the physical qualities of the chosen creatures be transferred to the warriors; that they should become as swift as the eagle, as strong as the bull, as cunning as the fox, as savage as the boar, nor must one forget mythical beings imbued with supernatural powers.

Celtic society can provide many examples, e.g. the array of standards on the Arch at Orange;[1] and there are historical references[2] back as far as Homer. Into this category might also be placed the dragon pennants carried by the Sarmatian horsemen and shown on the Column (Pl. VIIIa). Behind the dragon mask flowed a long tapered streamer, the shape of a modern windsock which gives the direction of the wind on an airfield. The pennant did not merely flutter in the breeze, but cavorted and wriggled like the tail of a dragon.[3] These Sarmatians were enrolled in the Roman Army during the second century and naturally brought their standards with them. There is a fine description of them by Ammianus Marcellinus in his description of the entry of Constantius into Rome in A.D. 357:[4] '. . . he was surrounded by dragons, woven out of purple thread and bound to the golden and jewelled tips of spears, with wide mouths open to the breeze and hence hissing as if roused by anger, and leaving their tails winding in the wind.' By the end of the fourth century dragon standards were common to the legions and according to Vegetius[5] one was carried by each cohort, the bearer being known as the *draconarius*.

---

would then sound a note, eyes would alert to the standard, now perhaps raised aloft, and on it being brought down all would smartly halt.

[1] *L'Arc d'Orange*, 1962.

[2] The Cimbri, according to Plutarch (*Marius*, xxiii), released Roman prisoners on parole after making them take an oath on their bronze bull.

[3] According to Arrian (*Tactica*), these standards were Scythian in origin and he says that they were 'made by sewing together scraps of dyed cloth and look like serpents from head to tail. . . . When the horses are urged forward the wind fills them and they swell out so that they look remarkably like live creatures and even hiss in the breeze which the brisk movement sends through them. These standards are not only a delight to the eye but are helpful in distinguishing one section from another and preventing confusion on parade.'

[4] xvi, 10, 7.

[5] ii, 13.

It can be assumed that animal standards were used by the Roman legions from earliest times and that they gradually became rationalized. The Republican legion is reputed by Pliny the Elder[1] to have had five standards, an eagle, a wolf, a minotaur, a horse and a boar; Marius made the eagle supreme because of its close associations with Jupiter, and the remainder were relegated or abolished. In late Republican times the eagle standard (*aquila*) was made of silver[2] and a golden thundeibolt was held between the talons, but later it was entirely of gold and carried by the senior standard-bearer, the *aquilifer* (Pl. IXa). There are numerous references and illustrations of this standard; it appears many times on the Column with wings displayed, and as late as the early fourth century on the Arch of Constantine, where the standard is shown as a miniature.[3]

While the eagle was common to all legions, each unit had several of its own symbols. These were often associated with the birthday of the unit or its founder or of a commander under whom it earned particular distinction, and took the form of the signs of the Zodiac. Thus the bull signifies the period 17th April to 18th May, which was sacred to Venus the goddess mother of the Julian family; similarly the Capricorn (Fig. 22) was the emblem of Augustus. Deities were also held in esteem, probably originating in an oath taken at a time of peril and it is hardly surprising to find Mars featuring in this context. Thus II *Augusta*, one of the British legions, displayed the Capricorn (for, as its name denotes, it was founded by Augustus), Pegasus and Mars. They probably appeared as ceremonial standards, but more often as decorative motifs on antefixes[4] and other legionary products. Only one example of this type of standard is seen on the Column, that of a ram, the Zodiacal Aries,[5] and it is

---

[1] *Nat. Hist.*, x, 5; it is clear from the context that Marius gave formal recognition to a practice which had been established for some years (*Romanis eam legionibus Gaius Marius in secundo consulatu suo proprie dicavit. erat et antea prima cum quattuor aliis: lupi, minotauri, equi aprique singulos ordines anteibant; paucis ante annis sola in aciem portari coepta erat, reliqua in castris relinquebantur*).

[2] *Nat. Hist.*, xxxiii, 19, 58. Appian, *Bell. Civ.*, 4, 101, mentions silver eagles at the Battle of Philippi. It was small enough in Caesar's day for him to have taken it from the staff and concealed it in his girdle (Florus, iv, 12).

[3] Many of the panels on this arch are from earlier monuments, but this, near the base, appears to be contemporary. It is possible that these standards depicting winged victories were awards granted by an emperor to officers (as in the case of Titus (*Bell. Jud.*, vii, 1, 3).

[4] This is an elaborate faced tile placed at eaves-level as a closing stop in the line of *imbrices* or curved tiles which covered the join between the *tegulae*, i.e. flat tiles.

[5] Cichorius, Taf. 35, No. 122. *Leg.* V was awarded an elephant standard (Appian, *Bell. Civ.*, 2, 14).

doubtful if these effigies would have played an important part in the organization and drill formations of the legion. The *imago* was of special importance in bringing the Emperor into a closer relationship to his troops, and this became of increasing significance with the rise and fall of dynasties. This standard bearing the image of the emperor was carried by the *imaginifer*. In later times it also had portraits of other members of the ruling house, as was the case of the *imagines* of the Praetorian Guard. That the standards of this body are far more elaborate

FIG. 22 *A capricorn standard from Wiesbaden*

than those of the legions is clearly seen from the Column and one can use this distinction to identify units of the Imperial bodyguard.[1] One of these standards of the Severan period appears on a small monument in Rome, the Arch of the Money-changers[2].

The *aquila* and *imago* were in the special care of the first cohort[3] but there were other standards for each century. The maniple was a very ancient division of the legion consisting of two centuries; that it survived into Imperial times may be indicated by the arrangement of the barrack blocks in pairs.[4] The Romans themselves seem to have no information about the

[1] This also contributed towards their weight; Caligula on a long march allowed the praetorians to use pack animals to carry their standards (Suetonius, *Gaius*, 43).

[2] *Arcus Argentariorum*, stands near the Arch of Janus.

[3] Vegetius, ii, 6; Val. Max., i, 6, 11.

[4] It is clear that Vegetius did not understand the nature of a maniple and confused it with

origins of this standard and it was supposed to have derived from a pole with a handful of straw tied to the top.[1] The hand (*manus*) had a significance, although it may not have been understood, since it appears on the top of those standards on the Column (Pl. IXa) outstretched with fingers and thumb together. Does it signify the hand of comradeship or depict a military salute with arm upraised, or would it be too fanciful to imagine it stretching out towards the gods, claiming divine protection? Below the hand is a cross-bar from which could be hung wreaths or fillets, and attached to the staff, in vertical array, are discs of varying numbers (four to six), the precise significance of which is not understood but they may have indicated the numbers of the cohort and/or century.[2]

There is a relief of three standards from Rome, in which the central one is an *aquila* bearing an embattled crown indicating special honours gained in a siege, and also the prow of a ship appears below the discs, probably denoting the origin of the legion from the marines of the Fleet (Pl. X). Evidence of simpler standards merely giving the names of units come from Dio[3] in his account of the crossing of the Euphrates by the legions under Crassus. A sudden desert storm blew up and swept into the river the boards bearing the names of the legions and their cohorts and the high purple flags carried in the van, a disastrous omen amply fulfilled in the deserts of Parthia.

The standard which most closely resembles the modern flag is the *vexillum*, a small square piece of cloth attached to a cross-bar carried on a pole. It is a type of standard more commonly borne by cavalry, the senior standard-bearer of an *ala* being known as the *vexillarius*. Its use in the legion was not, as some may suppose, solely for the small cavalry unit,[4] mainly scouts and messengers in the early Empire. The importance of the *vexillum* for the unit

a *contubernium* and also states that each century had its ensign (ii, 13); Caesar clearly implies that standards were associated with maniples (*Bell. G.*, vi, 40; *in signa manipulosque coniciunt*); Polybius, however, gives each maniple two standard bearers (p. 29 above).

[1] This probably originated in some way with the *sagmen*, the tuft of sacred herbs gathered by the consul in the Capitol, and the carrying of which signified the *fetiales*—a ceremony of treaty-making (*Nat. Hist.*, xxii, 2, 3; see C. Renal, *Cultes militaires de Rome; les enseignes*, Paris 1903).

[2] As there were only three maniples in a cohort the number of discs could not have indicated the maniple number, but it would support the suggestion that each century had its own standard.

[3] xl, 18.

[4] A rare example of an illustration of a legionary cavalry *vexillum* is seen on a stone at Pettau (*C.I.L.*, iii, 4061) and bears the inscription *vex eq*.

as a whole is shown by its appearance on stones commemorating building works completed by the legions. Some fine examples of these have been found in Scotland associated with the Antonine Wall. In panels at the side there are reliefs of the ceremony with its sacrifices, and legionary insignia appear together with a *vexillum* bearing the name of II *Augusta*,[1] Another illustration is seen on the Column where a legionary carrying this standard marches with the other standards in front of the troops over the bridge of boats across the Danube for kit inspection (Pl. IXa), and it is seen in a dominating position in Scene 24, where the standards are clearly pitched in a camp. It is difficult to explain the absence of this ensign in the several parades which are featured on the Column.[2]

An actual *vexillum*, the only example of a Roman standard to survive, is now in the Museum of Fine Arts, Moscow.[3] It was found in Egypt, perhaps in a grave, and consists of a piece of coarse linen fifty centimetres square with the remains of a fringe on the lower edge and a hem to take a transverse bar on the upper. The cloth has been dyed scarlet and bears an image in gold of a victory standing on a globe, but no lettering. The identification of this object as a *vexillum* has been made possible by the painting found in the temple of Bel at Dura. This shows a sacrifice being made by Julius Terentius, commander of *Cohors XX Palmyrenorum* with some of his officers and men, including a *vexillarius* holding his standard. The word *vexillum* was also used several times by Caesar and those who completed his memoirs[4] to denote a flag for signalling. It seems normally to have been customary to hoist a large red flag to denote that the battle was about to be joined.[5] Doubtless there were other kinds of flags and banners used for special occasions and by certain commanders. Further consideration might take us into aspects of signalling and communications over a distance, which belong in another context (p. 246). When one considers the special value the Army placed on their standards, it is not surprising that no others have survived, as would be the case with objects casually lost.[6]

---

[1] G. Macdonald, *The Roman Wall in Scotland*, 1934, Pl. lxi.

[2] There are two *vexilla* carried by dismounted horsemen (Taf. 9, Nos. 20 and 21) in a ceremonial parade, but these are presumably cavalry.

[3] M. Rostovtzeff, '*Vexillum* and Victory', *J.R.S.*, 32 (1942), pp. 92–106.

[4] In Gaul (*Bell. G.*, ii, 20, 1), and at Munda in Spain when he found that Pompeius had lined up for battle—*Hoc nuntio allato vexillum proposuit* (*Bell. Hisp.*, 28).

[5] Plutarch, *Fab. Max.*, 15.

[6] Bronze eagles are nevertheless often identified as *aquilae*; e.g. the fine example found at Silchester.

Under the heading of ensigns or insignia might be considered the official staves denoting the rank held. A *beneficiarius*, for example, on official business bore the authority he derived as the agent of the officer he served. His ensign is shown on a relief on an altar erected by Tertinius Severus of VIII *Augusta*, a *beneficiarius consularis*.[1] It consisted of a disc with two circular holes in it and a projecting spike, very similar to an actual example found in Germany.[2] The relief of an *optio* at Chester[3] shows him holding a long staff with a knob on the end (Pl. XI), presumably denoting his rank.

Finally it should be noted that the standard-bearers wore animal skins over their uniform. This follows Celtic practice; the Suebi, for instance, wore boar masks,[4] and these had the same kind of totemic influence as the animal standards. There are some fine illustrations of praetorian and legionary standard-bearers on the Column, the former with lion skins[5] and the latter with those of bears. The heads of the animals are carried over the men's helmets so that the teeth are actually seen on the forehead.

Musical Instruments

As noted above, the standards were used for relaying simple commands in close association with appropriate notes from instruments. There appear to have been at least four different ones used by the Army, the *tuba*, the *cornu*, the *bucina* and the *lituus*,[6] and most probably they had Etruscan origins. The *tuba* played by a *tubicen* was a type of trumpet over a yard long built in sections with a detachable mouthpiece. It was used to sound the advance and retreat and the watches for guards and small working parties.[7] The *cornu*,

[1] *C.I.L.*, xiii, 7731; the stone is in the Musée de Liège.

[2] *Germania Romana*, 1922, Taf. 94, No. 7. See also A. Alföldi, 'Vom Speerattribut der altrömischen Könige zu den Benefiziarierlanzen', in *Limes-Studien*, Basel, 1959, pp. 7–12.

[3] *R.I.B.*, 492; Wright and Richmond, 1955, No. 38.

[4] Tacitus, *Germania*, 45.

[5] Also as seen on the fragment now in the Boston Museum (Pl. IXb).

[6] Vegetius (ii, 22) lists only the first three.

[7] There are very few illustrations of this instrument; part of one appears on a relief from Remagen now in the Bonn Museum (*M.Z.*, vii (1912), p. 43) and the outline of another (*M.Z.*, xii, xiii (1917–18), Taf. x). Another example at St Germain (*C.I.L.*, xiii, 8275) shows a cone-shaped end which is perhaps a decorated cover put over the mouth of the instrument when not in use (Pl. XIII). Procopius states that the Army in its early days had only two calls (*Bell. Goth.*, vi., 23); Frontinus (*Strat.*, i, 5, 17) gives as an example of a tactic to deceive the enemy, the dispatch of a trumpet-player to a distant part of the field to blow calls to give the false impression of the presence of a body of troops.

played by the *cornicen*, was a large circular instrument, the expanded mouth curving over the player's shoulder something like a large version of the French horn. This instrument is well illustrated on the Column (Pl. VIIIb), with its decorated wooden bar fastened across its diameter; it also appears at the sacrifices and occasions when Trajan addresses his troops. The *cornu* is clearly associated with the standards[1] and its calls drew the soldiers' attention to the latter as indicated above. These two instruments must have given clear piercing sounds, making them suitable for tactical use in battle conditions, whereas the other two, the *bucina* (played by a *bucinator*) and *lituus* were purely for ceremonial occasions. Of the former little appears to be known, but it is thought to have been derived from an animal horn or shell,[2] and if so was presumably curved or twisted. According to Vegetius, it was used at executions and also for the *classicum*, a fanfare in which the other instruments joined, indicating the presence of the commander-in-chief. The *lituus*, an elongated J-shaped instrument with an enlarged mouthpiece, may have had religious connections, appearing only on such occasions; it receives very little mention.[3] The *aulos* or *tibia*, a small reed instrument with a shrill note, was used at sacrifices to help to prevent any inauspicious noises from being heard. The instruments were purified twice a year at a ceremony known as the *tubilustrium*.

THE AUXILIA

Introduction

The allies of Rome began very early in Republican history to play an effective part in the annual campaigns and large-scale wars. The citizens of Rome provided first-class heavy infantry in the form of legionaries, but in other types of fighting they were not so adept. In particular, they did not take so easily to the horse and their own cavalry troops were no match against nomadic peoples nurtured in the saddle. There were other notable differences. In some parts of the Mediterranean local conditions had evolved special methods of attack, originating, no doubt, in hunting game. Among these

---

[1] Also on the evidence of Vegetius, ii, 22.

[2] *The New Oxford History of Music,* i, 1957, p. 407.

[3] It is listed with the standards and the *cornu* in an incident in A.D. 272 in Aurelian's reign (*Hist. Aug., vita Aurelian.*, xxxi, 7).

were the archers from the eastern parts of the Mediterranean and the slingers from the Balearic Islands. Likewise, against nimble, light-footed mounted tribes, the legionaries were too slow and clumsy. The need for the Romans to equip themselves with these specialized arms and ways of fighting was felt as early as the third century B.C. It was not always possible to obtain the required skills from within the circle of accepted allies and so it became necessary to hire mercenaries. Livy, in describing the engagement of a thousand archers and slingers from Syracuse in 217 B.C. for use against Hannibal, implies that the practice was by no means new.[1] All the non-Roman forces, whatever their status, became known as *auxilia*—aids to the citizen legionaries. As Rome extended her influence over more and more countries, so she was able to make demands on their forces and call an increasing number of different kinds of *auxilia* into her armies. What may have been unusual in the third century B.C., to judge from the almost apologetic tone of the statement from Livy, soon became an accepted fact and many strange garbs and weapons were to be found side by side with the legionaries in most major wars.

In some of these conflicts the Romans came in contact with new forms of warfare and they were able to appraise their value and occasionally adopt them. They were not always quick, however, to appreciate this kind of lesson. In Spain, for example, the Romans put down repeated revolts, but usually judged the Spaniards to be too wild and unpredictable to make good soldiers. The Roman officer Sertorius, using Spain as his base for waging civil war against Rome, demonstrated that, when well led and disciplined, they made first-class troops, and the revolt was only crushed after the death of its leader. Caesar, during his conquest of Gaul, was given many opportunities of seeing the Gallic horsemen in action and it is hardly surprising that he was soon recruiting them, taking a large contingent with him to fight against Pompey. He may have regarded them as hostages, but he doubtless valued their fighting qualities more. Similarly the wars against Jugurtha demonstrated the value of the nimble Moorish horsemen whom Trajan later found so useful against the Dacians. They are to be seen with their characteristic African hairstyle on the Column (Pl. XIVa).

Augustus, upon assuming power, had the urgent and difficult task of rationalizing the chaos caused by the divided loyalties of the various armies which survived the civil wars. His practice, whenever possible, was to work to Republican precedent and although one might argue that he created for

---

[1] xxii, 37.

Rome the first fully professionalized standing army, this was only giving official recognition of what had been the actual state of affairs for many years. The auxiliary units were completely reorganized and given regular status. Instead of raising levies from the provinces as occasion required, the numbers of units and yearly intake of recruits were worked out according to a fixed annual scale, doubtless organized in close connection with the census of population, the initial purpose of which was the reorganization of taxation. Not every tribe was treated alike and there does not appear to have been a rigid, standardized system throughout the Empire. Tacitus tells us that the Batavi, on the Rhine, paid no taxes at all, but 'reserved for battle, they are like weapons and armour, only to be used in war'. [1] Conditions of service were also regularized and, most important, Roman citizenship was to be given on honourable discharge. This probably did not come into full effect until the time of Claudius. Spanish auxiliaries had received this privilege as early as 89 B.C., after the seige of Asculum, although at the time this was regarded as a special case.[2] This gave a real incentive in the first century to join the Army and serve it well. The cumulative effect of this steady extension of the franchise could hardly have been foreseen with at least 5,000 men ready for discharge each year from the *auxilia*.

Augustus was careful to avoid the posting of auxiliaries too far from their homeland, but with the development of the power of the principate any tender feelings of this nature were soon forgotten and there was occasional trouble. There was an example of this with the Thracians during the reign of Tiberius. These hardy warriors were very alarmed when they heard rumours that the system of service was to be altered and that units were to be drafted to distant parts of the Empire.[3] They were prepared to defend themselves against what they regarded as virtual enslavement and a short, but difficult, campaign ensued before they submitted. Another and more interesting example of natives rebelling against foreign service comes from the biography of Agricola. During his Caledonian campaign in the first century a cohort of Usipi from the Lower Rhine in Germany revolted. They were evidently raw recruits levied by Domitian and had been stationed for training in south-west Scotland.[4] They assassinated the centurion and soldiers in charge and seized three warships from the fleet, forcing their pilots to set course for their homeland. After suffering many privations and being reduced by starvation to cannibalism, for which they drew lots, they eventually sailed

---

[1] *Germania*, 29. [2] T. Ashby, *Class. R.* (1909). [3] *Annals*, iv., 46. [4] *Agricola*, 28.

round the north coast of Scotland and across the North Sea, were shipwrecked on the Frisian coast and captured as pirates. Finally they returned to the Rhine as slaves and gained much publicity from the narration of their adventure. Augustus, while allowing some of the eastern kings a limited independence, had permitted them to keep their own considerable armed forces. Gradually such kingdoms were absorbed into the Empire, their armies taken over, and broken down into auxiliary regiments and moved to other frontiers.

The movement of units, and their service in distant parts, raise the question as to whether auxiliary units continued to be manned by the countrymen from the province in which the unit was originally formed. Were British units, for example, always recruited from Britain wherever they were stationed? It is a matter which can only be studied with the evidence of the discharge certificates, the diplomas, and the tombstones on which the auxiliary's unit and birthplace were usually given. The answer is by no means straightforward and depends on the type of unit and the circumstances.[1] There was, for example, in the first century a stiffening of the cavalry recruited for the East with Germans, Gauls and Spaniards, in order to raise quality and sometimes provide more experienced troopers.[2] By the second century general recruitment had become the practice, and not even those units which contained highly specialized kinds of fighting men or equipment were exempt. The main recruiting areas tended to shift from the older provinces to the more barbarized areas on the frontiers. Nor does there seem to have been any reluctance in using the local levies in their own provinces on the Rhine and Danube. But a notable exception to this appears to have been Britain. To date, almost all known Britons in the Army, apart from those in the levies raised by Agricola for his Caledonian campaign, have been found serving in another province.[3] There may have been a special reason for this policy of distrust,

---

[1] For a detailed survey of the evidence see K. Kraft, *Zur Rekrutierung der Alen und Kohorten an Rhein und Donau*, Berne, 1951.

[2] This is probably the case of Genialis, a Frisian serving in a Thracian *ala* at Cirencester (*R.I.B.*, 109). In the incident of the Usipi, quoted above, these new levies were in process of training and experienced soldiers were attached to the unit for teaching discipline and thereafter continuing as its officers.

[3] Among the rare exceptions is a Brigantian serving in *Coh. II Thracum* (*R.I.B.*, 2142). The citizen from Gloucester whose diploma found at Colchester records his service in *Coh. 1 Vardullorum* (*J.R.S.*, 19 (1929), p. 216) may also have been of British origin. The earliest record of British cohorts is in 69 in a mixed force under Caccina (*et Britannorum cohortibus*, *Hist.*, i, 70).

but our information is very limited and a few new inscriptions may well correct this impression.

Organization

There were three kinds of units in the *auxilia* of the early Empire. In order of seniority they were: (1) the cavalry *alae*, (2) the infantry cohorts, and (3) the mixed infantry and horsemen, *cohortes equitatae*.

(1) *The Ala* was composed entirely of cavalry. The name is the Latin for a wing and derives from the use of horsemen on the flanks of an army, where they could give protection to the infantry centre. When necessary they could deliver flank attacks themselves and thereafter deploy against a retreating enemy or attempt to divert pressure on a withdrawal. The large-scale employment of cavalry serving as *auxilia* was in large measure due to Caesar following his experience of his Gallic allies. These units were originally led by their own chiefs and it seems probable that their internal organization was left to the commanders and local custom. Eventually the units became organized into troops (*turmae*). The *alae* were normally of *quingenaria* strength (i.e. five hundred), but there were a few of *milliaria* strength (i.e. a thousand), although, as in the case of the legion's century, it does not mean that there were exactly these numbers in the unit. The evidence on this point is by no means clear. Hyginus, a name given to an anonymous second-century author of a treatise on the arrangements for a military camp,[1] states that the *ala quingenaria* was divided into sixteen *turmae* and the *ala milliaria* into twenty-four *turmae*, from which it might appear that the size of the *turma* in these different sized units was not the same. Arrian who, as governor of Cappadocia, was concerned with troops in the eastern provinces in the time of Hadrian, says that an *ala quingenaria* was composed of 512[2] men which, on the above basis, gives thirty-two men to the troop, the same figure incidentally which another military writer, Vegetius, gives for the cavalry *turma*. The great German scholar, von Domaszewski, on the basis of an inscription from Coptos,[3] has argued that the troop of an *ala milliaria* had forty-two men, giving a total of 1,008 men, but it is impossible to establish this point

[1] *de munitionibus castrorum.*

[2] *Tactica*, 18.

[3] *Rang.*, p. 35; *C.I.L.*, iii, 6627; the *vexillatio* was drawn from three *alae* and consisted of 424 troopers and ten officers, i.e. five *decuriones*, one *duplicarius* and four *sesquiplicarii*, and it could be equally argued that these are five troop commanders and their subordinates rather than ten commanders.

beyond doubt and further evidence must be awaited from inscriptions or military documents.

The commander of the *ala* was a *praefectus*. At first he would have been a chief of his tribe taking his rightful place at the head of his troops. As the system became rationalized in the first century A.D. this command became a step in the ladder of promotion for young equestrians, i.e. men of the knightly class, entrance to which was based on a property qualification and had no more connection with horses than our knights of today. The military steps in the *cursus* were normally: *praefectus cohortis* (commander of an auxiliary infantry unit), *tribunus legionis* (military tribune in a legion) and *praefectus alae* (commander of an auxiliary cavalry unit). Often in the early Empire the men who aspired to these positions were not only young men from wealthy families but ex-centurions from the legions. The chief centurion of a legion (*primus pilus*), on attaining his rank, qualified automatically for equestrian status and he could then obtain an independent command in the *auxilia* if he so wished. But this system was changed by the end of the first century, when almost all the entrants into the auxiliary officer class were knights who had passed through the magistracies of the thriving municipalities of Italy or the provinces. This useful potential supply of educated provincials had become available as a result of the policy of Romanization and extension of the franchise deliberately pursued by the early emperors. It was a process developing in the *auxilia* itself. An uncouth barbarian from a frontier province or beyond could join an auxiliary unit and soon became accustomed to the ways of civilized life. After completing his service he would retire as a Roman citizen and his sons could become legionaries and rise to the centurionate or stay in civil life and as citizens join the town council by first being elected to a magistracy. One way or another the family could progress towards the equestrian class and this in turn could be used as a stepping-stone by their children, who could rise to the higher ranks of army command or the civil service.[1] Although abuse of wealth, graft and nepotism were rife, the life of the Empire was vigorous and new blood was constantly being pumped into its veins. There was little to stop the ambitious riding rough-shod over the weaker members of this strangely assorted society. Auxiliary officers can be traced from almost all parts of the Roman world, Britain being one of the few exceptions, and it is far from clear why there is no evidence of this province producing natives who rose to these positions.[2]

---

[1] A. Stein, *Der römische Ritterstand*, 1927.

[2] Dessau, 'British Centurions', *J.R.S.*, 2 (1912), p. 21.

The troop commander was the *decurio*, the same name given in civil life to a town councillor. A man promoted to this position could come from the lower ranks or from a legion, since this was an accepted step from the ranks into the legionary centurionate. The duties of an equestrian officer as set out early in the third century were 'to keep the troops in camp, to bring them out for training, to keep the keys of the gates, to go round the guards from time to time, to attend the soldiers' mealtimes and sample the food to prevent the quartermasters from cheating, to punish offences, to hear complaints and inspect sick quarters'.[1] It might almost be taken from a modern book of regulations.

The senior standard-bearer was the *vexillarius*, who carried a small square flag, decorated with tassels, on a pole. Each troop presumably had its standard carried by a *signifer*. The representations of these emblems vary greatly. The well-known example from Hexham Abbey of a *signifer* of the *Ala Petriana* (Pl. XIVb), a unit at one time stationed at Corbridge and later at Stanwix near Carlisle, shows a standard which looks like a large medallion having the appearance of an *imago* or image of the reigning emperor.[2] A similar emblem appears on the tombstone of Genialis, a trooper of a Thracian *ala* from Cirencester (Pl. XV). This example is fitted with what appear to be two streamers, while another standard-bearer, from Worms,[3] merely carried a pole with a cross-bar from which hang pear-shaped objects, similar to an example found at Zugmantel.[4] There was also an armourer (*custos armorum*) attached to each troop, and a headquarters staff with its clerks and keepers of records and the buglers and trumpeters. As with the legions, in the lower grades immunity from fatigues was given with seniority and special responsibilities.

(2) *Infantry Cohorts* were also organized on the basis of units of five hundred and a thousand strong and, like the legions, divided into centuries. From Hyginus we learn that the former consisted of six centuries and the latter ten. The excavation of the fort at Fendoch[5] in Scotland revealed ten centurial barrack blocks, each with ten pairs of rooms for sleeping and equipment. If all this space was used, it suggests centuries of eighty men. There are examples of the six barrack blocks of a quingenary fort at the Welsh fort of Gellygaer.

---

[1] Aemilius Macer, *Digest* 49, 16, 12, 2.

[2] Domaszewski, 'Die Fahnen im römischen Heere', *Abh. d. Arch. Epig. Seminars der Univers. Wien*, Heft 5 (1885), p. 70.

[3] *C.I.L.*, xiii, 6233; *B.J.*, 114/115 (1907), Taf. 1, 3; *Germania Romana* (1922), Taf. 30, No. 3.

[4] *Germania Romana* (1922) Taf., 95, No. 4.

[5] *P. Soc. Ant. Scot.*, 73 (1939), pp. 110–154.

Here the timber partitions between the men's rooms were not observed by the excavator, but the total length of these quarters is 92 ft as compared with 120 ft at Fendoch, suggesting sixty men to the century of a unit, 500 strong (see p. 217 below). The commander of the infantry cohorts was a *praefectus cohortis*, who, as seen above, was of a rank inferior to the cavalry commander. Under him, each century was led by a centurion and the other officers, *optio*, *signifer*, *tesserarius* and headquarters staff were probably very similar to those of the legion, but, of course, on a very much smaller scale.

(3) *The cohors equitata*[1] was a unit composed of both cavalry and infantry. According to Hyginus the milliary unit had a thousand men of whom 240 were mounted probably in eight *turmae*. The *cohors quingenaria*, he wrote, had six centuries and was divided in the same proportion as the larger unit. This would allow for 380 infantry and 120 mounted men. If the centuries of the *pedites* were legionary size, i.e. 80 men, there would have been ten in the milliary and six in the quingenary unit. There is at Birrens, Dumfriesshire, an example of a completely excavated fort of a *cohors milliaria equitata*, occupied by the Second Cohort of Tungrians.[2] Sixteen barrack blocks have been identified (Fig. 46), but there are six other buildings in the south end of the fort, which could account for two further barracks, leaving four as stables which would allow each to be shared by two *turmae*.

These troops were not of the high standard in physique and training as those of the *alae* and their pay was lower. It was possible for an infantryman after ten years service to become a trooper and the cavalry and infantry operated separately in the field. Mr R. W. Davies suggests a comparison with the cavalry and dragoons of a later age. The former (the *alae*) represented the striking force and kept for that purpose in a state of readiness. But the *cohors equitatae* like the dragoons were for general purpose work, skirmishing, patrolling, reconnaissance, escort duty and messengers. This becomes clear from a study of the Dura rosters.[3]

Numeri[4] and *cunei* were other kinds of infantry and cavalry units which

---

[1] The most recent study is that by R. W. Davies in *Historia*, xx (1971), 751–763.

[2] *P. Soc. Ant., Scot,*, 30 (1896), p. 81; 72 (1939), p. 275.

[3] *P. Dura*, 100 and 101; R. O. Fink, *Roman Military Records on Papyrus*, 1971, pp. 18–81.

[4] The word was used in the late Republic and early Empire merely to indicate a group of soldiers of any kind, from legionaries to irregular forces (*Bell. G.*, vii, 76; *Agricola*, 18; *Hist.*, i, 6 and 87). In the early second century it was given to distinguish a special group of soldiers, i.e. *numerus frumentariorum* (*Rang.*, p. 34).

seem to have been raised from the more barbarous provinces on the frontiers in the second century by Trajan and regularized by Hadrian. By the second century the process of Romanization had been carried so far that the recruits into the *auxilia* were reasonably civilized and lacking in the tough, warlike qualities of the tribes beyond the frontiers which they had to face in battle. These irregular formations were thus used in the frontier districts against similar barbarians, but of hostile intent. By this very sensible, practical policy the Romans were able to absorb the potential hostile tribes on the frontiers and use them as a screen between the more distant barbarians and the regular Army. One of the best examples is that of the *numeri* of Britons, of which ten at least are known, settled in Upper Germany, on the outer parts of the German frontier in the rugged Odenwald. They built watch-towers at irregular intervals. A German scholar, E. Fabricius, has suggested that these towers represent a controlled fence to keep the Britons in, rather than for protection from barbarians without.[1]

In Britain we find a *numerus* of bargemen from the Tigris at South Shields, presumably for ferrying troops and stores up the Tyne, and a similar unit of Moors at Burgh-by-Sands. Part of the garrison at Housesteads in the third century seems to be a *Cuneus Frisiorum* and a *Numerus Hnaudifridi*. These Teutonic units left their traces with altars dedicated to Mars Thincsus and the two goddesses Alaisiagae, called Beda and Fimmilena, who appear on another altar as Baudihillia and Friagabis. Among such barbarous units, inscriptions are much more rare than in the *auxilia* and therefore we know much less about their organization and officers, except that their commander was a *praepositus*. The districts from which most of this fighting strength was drawn, apart from Britain, were Germany, Syria, Africa and Dacia. The main differences between these units and the *auxilia* were that they did not receive citizenship on discharge and it seems that recruitment continued from the natives of the country of their origin. Words of command and battle cries were in their native tongue,[2] and the Palmyrenes in North Africa, for instance, kept up the worship of their own gods through the centuries.[3]

The Auxiliaries' Equipment

The fighting equipment and dress of the *auxilia* present a very complicated picture: not only were there changes from time to time, but most of the units

---

[1] *Ein Limesproblem*, 1902. [2] *Tactica*, 44. [3] *C.I.L.*, viii, 2505, 2515.

were differently equipped from the beginning. The only literary description which has survived is that by Arrian writing at the time of Hadrian about the Army in the East. He says 'Cavalry may or may not be provided with armour. In the armoured cavalry (*cataphractarii*), both horse and man are protected, the horse with protection for the sides and front, the rider with body-armour of mail, canvas or horn and with thigh-guards. The other kinds of cavalry have no protective armour and of these, some carry spears, some pikes, some lances, while others use only missiles. The "lancers" are those men who fight at close quarters with pikes or spears, driving back the enemy in a charge like the Alani and the Sauromatae; the "sharp shooters" are trained to use missiles at long range, like the Armenians and some of the Parthians. Of the former, some carry oblong shields, others fight without shields, merely with spears and pikes . . . some with missiles, some use javelins, others bows and arrows . . . Their long flat-bladed sword hangs from their shoulder and they carry oblong shields and iron helmets, breastplates and small greaves . . . they also carry small axes with a circular edge.' This last point is confirmed by a tombstone from Housesteads on Hadrian's Wall where an archer carried a bow in his left hand and an axe in his right (P. XVI).[1]

Arrian's description also tallies with the details on Trajan's Column. There are many examples here of the light horse, probably Gallic units, wearing mail cuirass[2] and leather jerkins with a zigzag finish at the sleeves and lower edge. They wear knotted scarves round their necks like Boy Scouts, tight trousers, probably of leather, and helmets. They carry oval shields with raised decoration, which varies considerably, quite different from the stylized thunderbolt motif on those of the legions (Pl. XXIVb).

The horse, a smallish animal little more than pony size (probably over-reduced on the Column like all other features), is beautifully groomed and carries a saddle-cloth. The military saddle had no seat like its modern equivalent but was a substantial leather construction stiffened with bronze plates and with projections to help the man maintain his seat.[3] There are no stirrups, equipment introduced much later, but the leather trappings are richly decorated with bronze medallions and pendants (Fig. 23), often with a silvery

[1] This archer may have belonged to another and as yet unknown unit (*A.A.*, 4th. ser. 46 (1968), pp. 284–291).

[2] The indications of mail are faint on the Column and almost invisible on the casts, but

[3] There is a reconstruction from leather fragments from Valkenburg by W. Groenman van Waateringe (*Romeins lederwerk uit Valkenburg Z.H.*, 1967, Fig. 41).

tinning and gleaming with bright enamel. Apart from the horse's bit and the soldier's sword and lance, these metal objects are the only part of the equipment likely to survive. They are fairly common in excavations of military sites, but the finest collection comes from Holland, where a magnificent hoard of them was found at Doorwerth from which a reconstruction of the trappings has been attempted in Leiden Museum (Pl. XVIII).[1] Another series, now to be seen in the National Museum of Antiquities, Edinburgh, comes from the Tweedside fort at Newstead near Melrose.[2] Those medallions which are decorated with bold relief have often been confused with military awards

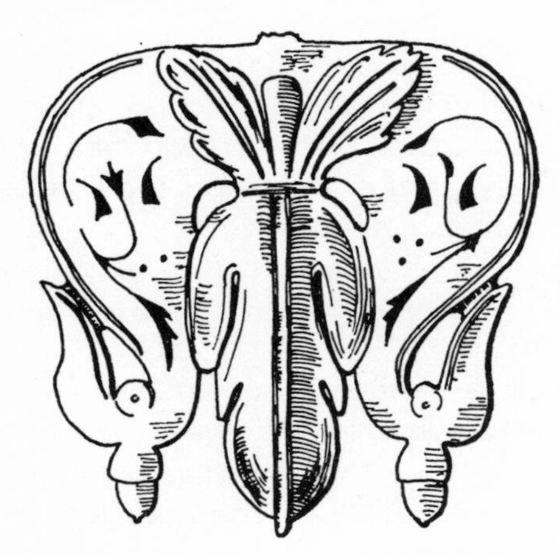

FIG. 23 *A pendant from a horse trapping from Newstead (actual size)*

and both seem to be covered by the general term *phalerae*. Like other forms of equipment, the Claudian-Neronian examples are tinned and decorated in scroll patterns with black, niello inlay, but the high quality of this work was not maintained.

Medium and heavy cavalry were often equipped with mail shirts. Of these the most striking are the Sarmatians, who are seen on the Column (Pl. XIX) fighting against the Romans. As shown here, both they and their horses are equipped with remarkable scale armour from top to toe. This is clearly an artist's convention since no horse could possibly be covered in this way and remain mobile. Actual fragments of the horse-armour found in the excavations at Dura-Europos (Pl. XX), show that the horse-armour was a rectangular piece placed over the back of the horse with a hole in the centre where the saddle was attached. Examples of special bronze eyepieces for the horse (Fig.

[1] Holwerda, *Oudheidkundige Mededeelingen uit het Rijksmuseum, Leiden*, xii, 1931.

[2] *Newstead*, Pl. lxxii–lxxiv.

24) have been found at Chesters[1] and elsewhere, but are more likely to belong to parade armour.[2] Soon after the Dacian Wars, Sarmatians were recruited into the Roman Army and in 175 a force of 5,500 was sent to Britain by Marcus Aurelius.[3] These units were a very important arm in the late Army, when the value of heavy cavalry became more fully appreciated and it could almost be said that here was to be found the precursor of the medieval knight with his mail and hauberk.

On Trajan's Column there are interesting details of the distinctive Eastern archer's uniform. These Levantine-looking auxiliaries are using a short bow,

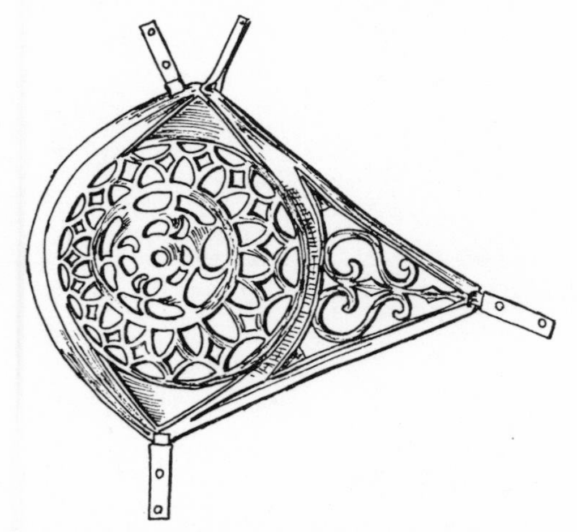

FIG. 24 *An eyepiece from a horse's parade frontlet (half-size)*

a composite weapon made of bone and steel, of which fragments have been found at Caerleon (Fig. 25).[4] They have a conical helmet, very reminiscent of the Norman *spangenhelm* and probably having an iron framework. Their bodies are protected by a shirt of mail made of large scales, below which is a flowing skirt-like garment. The quiver is carried high on the back so that the arrow can be plucked out from over the shoulder (Pl. XXI). The *auxilia* never became so standardized in their equipment as the legions and it is probable that every regiment had its own distinctive features, which once established were

[1] G. Simpson, *Britons and the Roman Army,* 1964, Pl. iii, p. 44; for a complete example see *Germania Romana,* 1922, Taf. 95 No. 3, from Mainz.

[2] J. Keim and H. Klumbach, *Der Römische Schatzfund von Straubing*, 1951, Taf. 20 and 21.

[3] Dio, lxxi, 16; I. A. Richmond, *J.R.S.*, 35 (1945), pp. 15–29.

[4] Prysg Field, pt. ii, Fig. 42; also *The Roman Forts on the Bar Hill*, 1906, Fig. 44, and *R.L.Ö.*, 2, Taf. xxiv, for examples from Carnuntum. See also Faris, *Arab Archery*.

jealously guarded. They must have added that gaiety, colour and eccentricity which today seems the prerogative of some of our older regiments.

On Parade

In one respect the cavalry of the *auxilia* differed notably from the legionaries in that they were equipped with special parade armour. The most remarkable feature of this was the helmet. When these fine objects have been found they have always excited wonder. In this country an early discovery was the Ribchester helmet, (Pl. XXIIa), now in the British Museum, and considered

FIG. 25 *A bone stiffener from a composite bow from Bar Hill (half-size)*

worthy of inclusion in *Vetusta Monumenta*, an illustrated catalogue of antiquities published by the Society of Antiquaries in 1747. Since then, other helmets have been found in Newstead and in Norfolk,[1] but the greatest find of all has been at Straubing in Bavaria, where a whole hoard of these and other pieces of equipment have been recovered[2] (Pl. XXIIb).

Again we turn to Arrian, who gives us a brilliant picture of cavalry on parade: 'The horsemen enter (the exercise ground) fully armed, and those of high rank or superior in horsemanship wear gilded helmets of iron or bronze to draw the attention of the spectators. Unlike the helmets made for active service, these do not cover the head and cheeks only but are made to fit all round the faces of the riders with apertures for the eyes . . . From the hel-

[1] *J.R.S.*, 38 (1948), p. 20. [2] See fn. 2, p. 153 above.

mets hang yellow plumes, a matter of décor as much as of utility. As the horses move forward, the slightest breeze adds to the beauty of these plumes. They carry oblong shields of a lighter type than those used in action, since both agility and smart turnout are the objects of the exercise and they improve the appearance of their shields by embellishment. Instead of breastplates the horsemen wear tight leather jerkins embroidered with scarlet, red or blue and other colours. On their legs they wear tight trousers, not loosely fitting like those of the Parthians and Armenians. The horses have frontlets carefully made to measure and have also side-armour.'[1]

It is difficult for us in this day and age to understand the motives behind the expense and trouble of providing this special equipment merely for the parade. But ceremony has always played a very important part in army life. In giving pride in appearance, it deepens the feeling of superiority. The flashing splendour of these arms and the elegant and skilful drill movement must have produced a deep effect on both Roman and barbarian alike. Nor must we forget the deeper religious significance. This was no mere show to impress an audience, like the Royal Tournament, but an affirmative identification of the Army with the gods. Perhaps we may see this in the presence of the gilded parade masks. The swarthy barbarian faces were hidden behind this uniform, classical façade. This must have been either to make them all look like Romans or like gods. It may be significant that there was a similar practice in the classical theatre. Not all the masks are classical; some are definitely barbarian or have eastern features with flowing locks, and the suggestion has been made that the cavalry enacted the battles of the Trojan War in the annual festival celebrating the founding of Rome.[2]

THE NAVY

The Roman Navy was always considered an inferior arm and strictly under Army control.[3] During the Punic Wars unexpected successes had been obtained by a logical Roman idea that a warship was little more than a floating platform on which the soldiers could be brought into close contact

---

[1] *Tactica*, 34. For cavalry training grounds see *Arch. J.*, 125 for 1968 (1969), pp. 73–100.

[2] H. Petrikovits, 'Trojaritt und Geranostanz', in the *Egger Festschrift*, Band 1 (1952), pp. 126 ff.

[3] The best general study is that by Chester G. Starr, *The Roman Imperial Navy 31 B.C.—A.D. 324*, 2nd ed., 1960.

with enemy ships. Rome almost lost all she had gained at sea by lack of seamanship and ignorance of navigation. It was possibly these factors which prevented the development of a navy and made her rely entirely on the Greek cities to provide ships when they were required. But as Rome gained control of the lands of the eastern Mediterranean, so the sea power of the Greek cities declined, and in the years 70–68 B.C. the pirates of Cilicia were able to carry on their trade with impunity right up to the Italian coastline.[1] The threat to the corn supply was such that the Senate was stung into action and gave Pompey an extraordinary command to clear the seas. He achieved this in the remarkably short time of three months,[2] far too short a period in which to have built any ships. His fleet was largely composed of vessels pressed into service from the Greek cities. After this there is evidence of fleets kept in being in the Aegean, although they may not always have been in a high state of efficiency. It was the Civil War between Caesar and Pompey which so clearly demonstrated the true significance of sea-power and at one time there may have been as many as a thousand ships engaged in the Mediterranean. As the struggle continued, Pompey's son, Sextus, acquired a fleet sufficient to keep Octavian at bay and endanger the grain supply to Rome. Octavian and Agrippa set to work to construct a large fleet at Forum Iulii (Fréjus), and train the crews. In 36 B.C. Sextus was finally defeated at Naulochus and Rome became, once more, mistress of the western Mediterranean. The final event of the Civil War was the Battle of Actium, which destroyed Antony. Octavian was left with some seven hundred ships of various sizes ranging from heavy transports to the light galleys (*liburnae*).[3]

Some of the best ships and crews were picked out to form the first permanent squadron of the Roman Navy and established at Forum Iulii,[4] the

---

[1] A Roman fleet was even attacked as it lay in the port of Ostia (Plutarch, *Pompey*, 24; Dio, xxxvi, 20–23). The pirates swept the seas with their extravagantly decorated ships, their gilded sails, purple awnings and silvered oars. At the height of their power they had over a thousand ships and held four hundred cities to ransom.

[2] This remarkable success was due to careful planning; the Mediterranean was divided into thirteen areas each with a fleet under a selected legate. The co-ordinated sweeps scattered the pirate bands in forty days; Pompey then converged on the main Cilician stronghold, which soon fell to his direct attack.

[3] Augustus claimed to have captured 600 ships 'besides those which were smaller than triremes' (*Monumentum Ancyranum*, C.3, 19–20). The oft-quoted idea that Octavian's victory was due to his use of these swift galleys against Antony's heavier craft is open to doubt (Starr, pp. 7–8).

[4] For summary and bibliography, see Paul-Albert Février: *Forum Iulii* (Fréjus), 1963.

base constructed for the campaign against Sextus. The *princeps* saw, as with the Army itself, the need for a permanent arrangement for maintaining the peace, but the most strategic and economical situations for the main bases had yet to be evolved. Forum Iulii controlled the north-western Mediterranean and the Rhône, but soon further bases were needed to protect Italy herself and the corn supply to Rome, and the Adriatic. An obvious choice was Misenum on the Bay of Naples equidistant from Sicily, Sardinia and Corsica, and considerable harbour works and buildings were started by Augustus, the port remaining the most important naval base throughout Imperial times. Detachments of the Misenum Fleet served at Ostia, Puteoli, Centumcellae and probably at ports elsewhere. Sardinia and Corsica had yet to be pacified and there is epigraphic evidence of naval units operating at Aleria and Carales (Cagliari) from time to time. It is likely that once established Misenum soon dwarfed Forum Iulii, which lost its independent status, becoming merely a detachment of the main fleet. Sailors of the Misenum Fleet were also stationed in Rome itself, at one time in the praetorian camp and later under the Flavians in barracks near the Colosseum. Their tombstones suggest these were young soldiers and they may from an early date have been responsible for the large awnings in the amphitheatres and theatres where active men skilled in rope and canvas would be needed, in addition to the naval spectacles occasionally enacted. The same qualities would have been suitable for the fire brigade and it may not be so surprising to find them in association with the *vigiles*. They could also have policed the harbour, acted as couriers and escorts and been a useful military reserve, as both Nero and Vespasian realized when they each raised a legion, I *Adiutrix* and II *Adiutrix* from the fleet.

Augustus also constructed a new naval harbour at Ravenna at the head of the Adriatic.[1] At this time trouble might still be expected from the coast of Dalmatia and the Illyrian hinterland. Another important area which Augustus felt needed special care and protection was Egypt, and it is probable that he founded the Alexandrine Fleet, which was mainly recruited from the Greeks of the Fayûm. For services to Vespasian in the Civil War, it was rewarded with the title *Classis Augusta Alexandrina*. The squadron had a detachment along the African coast at Caesarea when Mauretania became a province and may have been responsible for supplying the armies sent there under Claudius.

---

[1] This was much later to be completely silted up and it is all now dry land. A mosaic at San Apollinare Nuovo included a view of the harbour in late Imperial times.

The *Classis Syriaca* is attested in the reign of Hadrian, but it was probably created much earlier. The important trading area of the Aegean and the great harbour in the Piraeus had no local fleet, but sailors from Misenum and Ravenna served there.

Along the northern frontiers, squadrons were created to meet the needs along the coasts, rivers and inland seas as the Empire expanded. The conquest of Britain involved massive naval preparations which Caesar was quick to appreciate.[1] For the Claudian invasion of A.D. 43 the ships were assembled at Gesoriacum (Boulogne) and this harbour remained the main base for the *Classis Britannica*. The fleet played an important part in the conquest of Britain, in bringing supplies to the troops advancing into terrain where there were only tracks, but they were never far from the sea or one of the deep estuaries. One of the finest recorded achievements is the circumnavigation of Scotland under Agricola.[2] The squadron may have played a part in the campaign of Suetonius Paulinus against the Druids on Anglesey in A.D. 60, and doubtless operated from Chester. There is little epigraphic evidence for the British squadron. Tile stamps bearing the mark CLBR are common in Kent, especially at Lympne, where also a *praefectus* is recorded.[3] A late mosaic in the temple of Nodens at Lydney which overlooks the Severn estuary bears an inscription which may indicate that its dedication was by an official in charge of a supply depot of the fleet.[4] Units may have been stationed at various points along the British coast as the distribution of tile stamps suggests. Those from London may indicate barracks or stores and certainly ships would have been needed here for the courier service. The other harbours

---

[1] For the second expedition he ordered the construction of special shallow-draught, but wider, transports which could be easily beached (*Bell. G.*, v, 2).

[2] Tacitus, *Agricola*, 10 and 38.

[3] L. Aufidius Pantera (*C.I.L.*, vii, 18; *R.I.B.*, 66).

[4] This interesting mosaic, now lost, was carefully drawn by the Rev. William Bathurst (*Roman Antiquities at Lydney Park, Gloucestershire*, 1879, Pl. viii). The inscription is not complete, but appears to be a dedication by ITILAVIVS SENILIS (thought by Wheeler to be a mistaken drawing for T. FLAVIVS SENILIS) PR REL which Mommsen (*C.I.L.*, vii, 137) restored as PR(AEPOSITVS) REL(IQVATIONI CLASSIS), the work having been carried out by a man who from the drawing is described as INTERETINE which Wheeler read as INTERPRETE (an interpreter) in which case the mosaicist or the artist must have fallen into serious error. The Bristol Channel would seem to be an obvious place for a supply depot at this late period when there were serious threats of piracy from Ireland.

were Richborough (Rutupiae), Dover (Dubris), Lympne (Lemanis) and Pevensey (Anderita).[1] The Weald would have provided supplies of iron and timber.

In the campaign against the German tribes the Rhine, both as a barrier and a highway, played a crucial role. Squadrons of the fleet were operating along the lower stretches as early as 12 B.C. under Drusus the Elder,[2] but with as yet little understanding of the tides, since his ships were left high and dry in the Zuyder Zee and his forces only saved by the Frisian allies. He also constructed a canal to shorten the distance from the Rhine to the North Sea. This was used by his son Germanicus in A.D. 15, in whose campaigns the fleet was, once more, much in evidence. Four legions were taken by river to the mouth of the Ems, but they must have been too crowded, since on the return half the force was ordered to march along the bank. But the autumn gales struck and there is a dramatic description by Tacitus of their difficulties as the troops struggled with rising waters until they could all be rescued by the fleet.[3] Nor did the Romans seem to learn by this, since after the great victory over Arminius, Germanicus sent some of his army back to winter quarters by land, but the main force embarked on a thousand ships and sailed down the Ems to the sea. Once more caught in a strong gale the fleet was scattered, some ships blown out into the open sea, some driven on to rocks, some cast on to shoals or beaches. When the storm subsided ships began limping back and on being repaired were sent round all the islands searching for survivors, some of whom had been captured and had to be ransomed.[4]

The fleet continued to be effective along the lower Rhine, although Frisia was given up by Tiberius. The invasion of Britain in A.D. 43 made it necessary for the strengthening of the forces on the lower Rhine and the first fort of

---

[1] Stamped tiles have been found at Dover, Lympne and Pevensey and also at a villa at Folkestone, which has led to the suggestion that this was an official residence (S. E. Winbolt, *Roman Folkestone*, 1925), but this is far from conclusive. There is also a *gubernator* of Leg VI at York (*R.I.B.*, 653). He would presumably be a river pilot with duties along the lower Ouse and in the Humber. Tacitus implies a rank senior or equivalent to a centurion (*Hist.*, IV, 16). For further information about the British Fleet see D. Atkinson '*Classis Britannica*' in *Historical Essays in Honour of James Tait*, 1933.

[2] Dio, liv, 32.

[3] Tacitus, *Annals*, i, 60–62.

[4] One boat at least had been driven across the North Sea to Britain (*Annals*, ii, 24) and returned, presumably by Cunobelinus, who at this early stage of his reign might not have wished to cause Rome offence.

Valkenburg can be dated to this period.[1] This was, however, destroyed only five years later in the rise of the Chauci under Gannascus, who had deserted from the Roman *auxilia*. The rebels then took to piracy with a fleet of small ships, preying on the coast of Gaul. The new governor of Lower Germany was Cn. Domitius Corbulo, the famous general and disciplinarian, who soon restored order, but his efforts to regain control over the east bank of the Rhine were frustrated by Claudius, who did not wish for further involvements. Corbulo, reluctant to have idle soldiers under his command, set them to work digging a new twenty-three-mile canal between the Maas and Rhine.[2] In the confusion of the revolt of Civilis auxiliary units deserted and so did twenty-four ships of the fleet of the lower Rhine. Many of their rowers were Batavians and they managed to turn their ships towards the east bank, putting to death the centurions and pilots who opposed them.[3] In the long and bitter struggle that followed the complete control of the lower Rhine was a great advantage to Civilis. The British Fleet was ordered to ferry over XIV *Gemina* to distract the Batavians. The legion successfully disembarked and invaded the territories of the Tungrians and Nervians. Another tribe, the Canninefates, however, attacked the fleet and destroyed much of it.[4] Nor did Cerialis fare much better; a daring but impetuous commander, he neglected to maintain the discipline and watchfulness which keeps an army alert. The Germans quietly sailed up the river one dark night and took his fleet and forces encamped on the bank by surprise. Cerialis narrowly escaped; had he been on his warship he might have been killed or captured, but he was on shore pursuing personal pleasures. Most of the Roman warships were captured[5] and Civilis now once more had command of the rivers, but his vanity caused him to organize a great naval display at the mouth of the Maas, giving even small vessels the appearance of light Roman

---

[1] *Jaarverslag van de Vereeniging voor Terpenonderzoek*, 25–28 (1944); 33–37 (1953); but more recently it has been suggested that the foundation was earlier: J. E. Bogaers, 'Praetorium Agrippinae', *Bull. van de Koninklijke Nederlandsche Oudheidkundige Bond*, 17 (1964), pp. 210–39.

[2] The Rhine at this time flowed by Utrecht, where there was a fort, which is now partly under the cathedral. It then flowed into the North Sea at Katwijk, this point being protected by Valkenburg. Corbulo's canal must thus have anticipated, in part, the course of the modern Rhine below Nijmegen.

[3] *Hist.*, iv, 16.

[4] *Hist.*, iv, 79.

[5] The flagship was given to Veleda, a prophetess who had foretold the destruction of the Roman legions (*Hist.*, iv, 61).

Plate XVI
The tombstone of a Hamian archer from Housesteads, now in the Museum of Antiquities, Newcastle upon Tyne (p. 151)

Plate XVII
Two auxiliaries with oval shields from Trajan's Column (p. 151).

galleys. Although Cerialis had fewer vessels, he could not allow this to go unchallenged. He sailed downstream to engage in what might have been an interesting river battle, but after an exchange of missiles, the Batavians thought it prudent to withdraw.[1] Although its activities could hardly be called distinguished, the fleet of the Rhine did receive the title of *Augusta* from Vespasian and later shared with the other Lower German units the title *pia fidelis Domitiana*, following the suppression of Antonius Saturninus.

The headquarters of the *Classis Germanica* remained at Alteburg near Köln (Colonia Claudia Agrippinensis), where remains have been found of its depot and cemetery.[2] There were probably other stations lower down, especially near the mouth, where navigation became hazardous. This may account for the presence of stamped tiles at Katwijk and Arentsburg.[3]

The Danube has a natural division into two at the Iron Gates in the Kazan Gorge and it was probably difficult in times of low water, as it is today, for ships to pass safely through this stretch. The river thus came to have two fleets, *Classis Pannonica* to the west and *Classis Moesica* to the east. The Pannonian fleet owed its inception to the campaign of Augustus in 35 B.C. The natives attempted naval warfare on the Save with dug-out canoes[4] but with short-lived successes. Hostile patrols and supply routes along the rivers Save and Drave must have been factors in this campaign and the later ones under Tiberius. As soon as the Danube became the frontier the fleet was moved there, although patrols may have continued along the other main rivers. The main base appears to have been at Taurunum, near the junction of the Save and Danube, while there is evidence from the tombstones of sailors stationed at Brigetio and Aquincum, and tiles stamped CL(ASSIS) F(LAVIAE) P(ANNONICAE) have also been found at Brigetio and Carnuntum.

The conquest and control of the lower Danube came more slowly and was not achieved until A.D. 6 by Caecina Severus. Peace and settled conditions did not, however, materialize until A.D. 15, by which time the province of Moesia

---

[1] *Hist.*, v, 23.

[2] The defences appear to be of two periods, an earthwork of Tiberian and stonework of Flavian dates (A. Grenier, *Quatre villes romaines de rhénanie*, 1925, pp. 154–5). Identification has been by numerous tiles stamped C.A.G. (*Classis Augusta Germanica*) and C.G.P.F. (*Classis Germanica pia felix*) and tombstones (*C.I.L.*, xiii, 8166, 8168, 8198, 8322 and 8323).

[3] *C.I.L.*, xiii, 12565 and 6, but it is thin evidence on which to claim a naval base (J. H. Holwerda, *Arentsburg, een Romeinsch Militair Vlootstation bij Voorburg*, 1923).

[4] Dio, xlix, 37.

had been formed and presumably a fleet established. Little more is heard until the time of Claudius in A.D. 49, when Mithridates, a King of Bosporus who had been deposed by his half-brother Cotys I, raised an army and attempted to regain his throne. A mixed force of Romans and local levies completed a successful campaign, leading to the capture of the King. But this must have involved a difficult operation along the north coast of the Black Sea, and Tacitus implies that it was mainly from ships.[1] While this may have been the task of the Moesian Fleet, the need for a separate Black Sea unit was soon apparent. But the Danubian forces had other and more serious troubles as the Dacians began their inroads into Moesia. Domitian strengthened the garrison and split the province into two, placing the fleet under the command of the governor of Moesia Inferior. In the subsequent campaigns under Trajan, the evidence from the Column clearly shows the demands put on the fleet for transporting supplies and troops, no doubt making full use of the navigable northern tributaries of the Danube. The spiral of reliefs begins with a view of the south bank with its watch-towers and signal-stations,[2] then comes a permanent legionary fortress with a small harbour and store base[3] surrounded by a stockade. Three transport vessels are anchored and receiving tents and barrels probably filled with grain. There are three other harbour scenes[4] (Pl. XXIVa) and a fortified enclosure in which naval shipwrights[5] are building or repairing boats. This event must be deep in hostile territory and in the pre-

---

[1] *Annals*, xii, 15–21. Some of the ships went aground on the return voyage and casualties were suffered from hostile natives.

[2] Cichorius scenes 1–6.

[3] Cichorius scenes 7–12, the column of legionaries and praetorians emerge from a gate and cross a bridge of boats. As on all the Column scenes, against the size of the men all buildings and other features are scaled down.

[4] Cichorius scenes 80–87, where seven vessels are seen, three of which are biremes. The view in the background appears to indicate a legionary fortress with an amphitheatre outside and harbour works; 118–20 depicts a more modest harbour similar to the first one and in the same manner the troops march out of a gateway, crossing a bridge of boats; 207–13 is a more imposing place and the five biremes are shown in finer detail; it is followed almost immediately by another harbour in which is moored the only large transport ship shown on the Column (228–9).

[5] Cichorius scene 356. The site appears to be an island or bend in the river and there is a strong log fence protected by timber fences and ditches, unless the arrangement of posts indicates an embankment. Richmond suggested that the two soldiers (equipped as legionaries) are carving out log canoes for use as pontoons (*Pap. Brit. School at Rome*, 13 (1935), p. 28), and cited Vegetius for this practice (iii, 7). Although the men are working vigorously with hammers and chisels, the finely shaped prows may imply another type of boat.

vious scene auxiliaries are seen guarding a kind of trestle bridge across a small river.[1] The annexation of Dacia and its development as a province necessitated a military reorganization, and while the main bases of the fleet may have remained on the Danube, the northern tributaries can hardly have been neglected. To the area controlled by this fleet was also added the north-western coast of the Black Sea.

The Pontus Euxinus is a vast sea, larger in area than the whole of Italy, and its shores witnessed the mingling of a great diversity of cultures. Extensively colonized by the Greeks in the eighth to sixth centuries B.C., it did not attract serious attention from Rome until the reign of Claudius; until then power was invested in friendly or client kings. Little attempt had been made to control piracy.[2] It was the annexation of Thrace in A.D. 46 that brought part of the shoreline under direct Roman control and there is a hint of a Thracian Fleet, the *Classis Perinthia,* which may have been of native origin. The Armenian campaigns under Nero led to the taking over of Pontus (see p. 63), and the royal fleet became the *Classis Pontica.* During the Civil War the Black Sea became a battleground. A freedman Anicetus, commander of the fleet, raised the standard of Vitellius, destroyed the Roman ships and the town of Trapezus and then turned to piracy assisted by tribes from the eastern shore who used a type of boat known as *camera.*[3] A new fleet had to be fitted out and this, with the help of a legionary vexillation, drove Anicetus into his stronghold at the mouth of the river Khopi on the east shore, but the local king clearly saw that it was in his best interests to surrender the rebel. After this there was no further trouble for a long period and the peaceful conditions are well illustrated by Arrian, Hadrian's governor of Cappadocia, who, in an official report, describes the coastal conditions from Trapezus to Sebastopolis.[4] The *Classis Pontica* was responsible only for the southern and eastern parts of the Black Sea; the mouth of the Danube and the coastline to the north as far

---

[1] This flimsy-looking bridge is very reminiscent of Vegetius—'piles driven into the bottom of the river and floored with planks' (iii, 7).

[2] Ovid, exiled in one of the old Greek colonies at Tomi, records the extent of this activity, *Ex Pont.*, 4, 10, 25–30.

[3] Tacitus, *Hist.*, iii, 47; these boats appear to have been plank-built without any metal fasteners and so arranged that they could add depth to the sides with planks and build an arched roof. They had prows at each end and the oars could readily be used in either direction.

[4] This remarkable document, *The Periplus of the Euxine Sea,* written in A.D. 131 is a supplement to a full report of an extensive tour of inspection written in Latin and which has unhappily not survived (for a useful general account of the *Periplus* see 'Arrian as Legate of Cappadocia', *Pelham's Essays,* 1911).

as the Crimea was the responsibility of *Classis Moesica*.[1] A naval base may have been established at Chersonesus (Krim) on the tip of the Crimea.[2] Roman control on this coastline was firm enough as late as the early third century, when an auxiliary records a march along it as far as this harbour.[3]

Organization of the Fleet

The commanders of the fleet were *praefecti* recruited from the equestrian order like those of the auxiliaries. Their status in the military-civil hierarchy underwent changes in the first century. At first there was a tendency to use army officers, tribunes and *primipilares*, but under Claudius it became linked with civil careers and some commands were given to Imperial freedmen. The Civil War proved that this was very unsatisfactory—witness the example of Anicetus in the Black Sea. There was a reorganization under Vespasian, who raised the status of the praefecture, and that of the Misene Fleet became one of the most important equestrian posts and the crown of a long and distinguished career.[4] This, with the prefecture of Ravenna, became a purely administrative position with active service a very unlikely event. The praefectures of the provincial fleets ranked with auxiliary commands.

The lower commands present a complex pattern. In the first place many of these positions were Greek and the names of their offices persisted, their precise meanings becoming a little obscure, but, as Chester G. Starr has shown,[5] the navarch must have been a squadron commander and the trierarch a ship captain, but just how many ships constituted a squadron is not known, although there are indications that it might have been ten.[6] The basic difference between the Army and the Navy was that these officers could never hope for promotion into another arm, until the system was changed

---

[1] It appears to have been the practice of the governor of Moesia to appoint a tribune of the I *Italica*, stationed at Novae, as *praepositus vexillationibus Ponticis aput Scythia(m) et Tauricam* (*C.I.L.*, viii, 619).

[2] A trierarch set up a dedication here (*C.I.L.*, iii, 14214 and 14215).

[3] This is a fragmentary inscription in Greek on a shield found at Dura-Europos; the decoration includes a ship which may represent the one needed to sail from Chersonesus to Trapezus, since these two names follow one another preceded by Tomi, Istropolis (Istrus), Tyras and Borysthenes (Olbia) (F. Cumont, 'Fragment du bouclier portant une liste d'étapes', *Syria*, 6 (1925), pp. 1–15.)

[4] This command was held at his death by Pliny the Elder.

[5] Starr, p. 39.

[6] The *duoviri navales* of the early republic commanded ten ships each (Livy, xl, 18.8; xli, 1.2).

by Antoninus Pius. The highest rank any sailor could achieve until then was the navarchy and he then returned to take up an honoured position in his local municipality. In the lower levels the Romans took over the Greek organization of the crew, but superimposed its own military commands, making the whole a complex and obscure pattern. Each ship had a small administrative staff under a *beneficiarius* and the whole crew was considered as a century under a centurion assisted by his *optio*. Presumably the centurion was responsible for the military aspects and had under his direct command a small force of trained infantry who acted as a spearhead in an assault party. The rowers and other crew members would have had some arms training and would have been expected to fight when called upon. The exact relationship between the centurion and trierarch may have been difficult at times, but custom must have established precise spheres of authority.

The sailors themselves were normally recruited from the lower ranks of society, but were free men or *peregrini*. The Roman had never readily taken to the sea and we find few of Italian origin; most of the sailors appear to have originated from amongst the sea-faring peoples of the eastern Mediterranean.[1]

Service was for twenty-six years, a year longer than auxiliaries, marking the fleet as a slightly inferior service, and citizenship was the reward on discharge. Very occasionally whole crews might for a special piece of gallantry be fortunate enough to receive immediate discharge and there are also the cases where they were enrolled into the legion.[2]

---

[1] This is clear from Starr's Table I (p. 75), where he has listed the known sources of members of the Misene and Ravennate Fleets; of a total of 317 only thirteen are from Italy. One of the largest groups (thirty-eight) came from Egypt, one of the few provinces which did not otherwise contribute towards the military levy. Two interesting letters from Egyptian recruits to their parents have been preserved and published (*Class. P.*, 22, p. 243, and, *Ägyptische Urkunden aus den Museen zu Berlin: Griechische Urkunden.*, 423; also in Hunt and Edgar, *Select Papyri*, i (Loeb), 1932, Nos. 111 and 112). Both have gone to Misenum; the latter gives the name of his ship and adds significantly that his name is now Antonius Maximus, the first step towards his Romanization.

[2] I and II *Adiutrix*; also Hadrian brought X *Fretensis* up to strength with men from the Misene Fleet.

Chapter 4

CAMPS AND FORTS OF THE FIRST AND SECOND CENTURIES

The terms 'camp' and 'fort' have often been used very casually for Roman sites. For example, on the Ordnance Survey maps of prewar vintage the word 'camp' has been applied not only to all kinds of military works, but also to towns and settlements and occasionally to defensive earthworks of other periods. One ought to be more precise in these definitions and confine the word 'camp' to a marching camp and 'fort' to the more permanent establishment, normally housing a single unit, while 'fortress' should be reserved for legionary establishments. These differences are clear on sites of the late first century and later, but in earlier periods there is some confusion. In the early Empire the Army had not become consolidated along its frontier zones in permanent stone-built forts. Units were grouped together, living under tents even in the winter. The process of giving the Army permanent comfortable quarters came gradually. Thus we find that some of the characteristics of the marching camp are carried forward into the more permanent sites and this has given rise to the use of the term 'semi-permanent'. There is a further complication, in that occasionally one finds timber buildings, presumably for stores, inside what is otherwise an enclosure for tents. It is unfortunate, too, that most of the literary authorities, Caesar, Polybius, Josephus, Tacitus and the so-called Hyginus, were describing only marching camps and temporary winter quarters. These descriptions have often been freely used by modern writers when dealing with the details of the permanent forts. There are also defensive earthworks which have been shown to have been constructed purely for practice and training. While they often give much valuable information, these sites often have many peculiarities like those on Llandridod Wells Common, which appear to be a series of entrances and corners formed into small squares.[1] The classic example of this type of site is

---

[1] *Arch Camb.*, 91 (1936), p. 69; 118 (1969), pp. 124–134; see also 117 (1968), pp. 103–120 and *Arch. J.*, 125 for 1968 (1969), pp. 73–100.

Cawthorn in the North Riding of Yorkshire, which was excavated by Sir Ian Richmond.[1] The two Cawthorn camps and two practice forts have most of the usual features, but were arranged in an abnormal pattern; the excavation and interpretation has, however, told us much about them and the methods of construction. There must be other similar sites waiting discovery.

The Marching Camp

When the Roman Army went on campaign the soldiers slept in leather tents. These were quite different from the round canvas tents of modern times, their measurements exclusive of guy-ropes being ten Roman feet square. They each housed eight men (*a contubernium*) and were made of best-quality leather[2] with access back and front (Pl. XXIVb and Fig. 26). They could be rolled up into a long sausage-shape and in this form were carried by the mule or

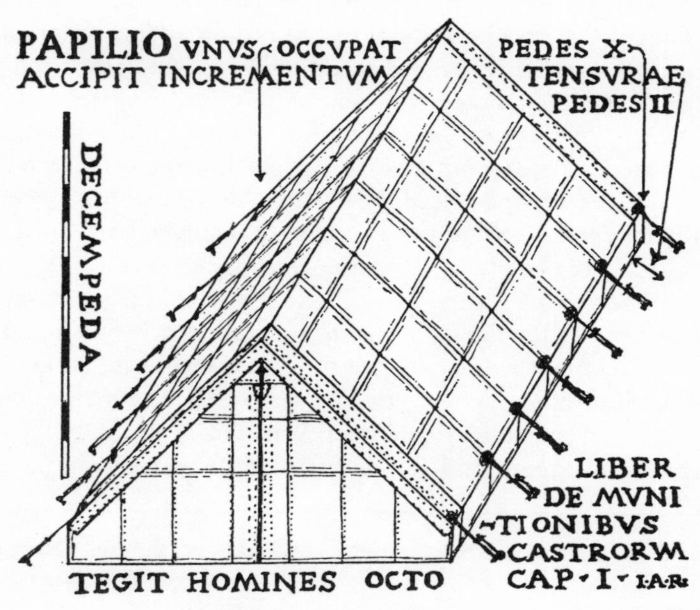

FIG. 26 *A leather tent*
(as reconstructed by I. A. Richmond)

[1] *Arch. J.*, 89 (1933), p. 17.

[2] Pieces from Newstead and Birdoswald have been identified as calf but the Valkenburg fragments proved, on microscopic examination, to be goat leather (*Romeins lederwerk uit Valkenburg Z.H.*, 1967, pp. 17–18).

pony. This shape may have given rise to the Roman name *papilio*, a butterfly, as it rolled up like a grub and with its wings may have reminded the soldiers of the insect emerging from the chrysalis.[1]

The tent lines were carefully laid out by the surveyors and the areas allocated for each unit were arranged and marked out in advance in order of seniority. One of the most important detailed descriptions of a marching camp is that with the title *Hygini Gromatici Liber de Munitionibus Castrorum*,[2] written, probably, in the late second century. It is clearly intended to be part of a handbook for Roman military surveyors and includes in the surviving fragment details of defences and the order of precedence for the units inside the camp. The author brings together, for the latter purpose, a hypothetical army. There are praetorians, three legions, auxiliaries, marines and even a camel corps. One can imagine the surveyors[3] going ahead of the main body and selecting a site, one fairly level, well drained and clear of trees and undergrowth. They first marked out the site of the commander's tent (*praetorium*), which contained not only the senior officer's living quarters but also the headquarters (*principia*). From this central position they set out the two principal streets, one across the front of the *principia* (*via principalis*) and one at right-angles to it (*via praetoria*) forming a T-shape and so arranged that the central site, the *principia*, faced the enemy. All the tent lines were then set out in the three areas, behind the *principia* (the *retentura*) and in front of it, one on each side of the *via praetoria* (the *praetentura*), packing the men in at an average of 220 to the acre.[4] According to Hyginus, the legionary century had only eight tents for the eighty men, since sixteen men were always on guard duty. The main streets were sixty feet wide and the lesser ones fifty feet, while a third category were twenty feet wide. The inter-vallum space between the tail of the rampart and the first tents was also sixty feet across to allow ease of movement and full access to the defences. In the annex, a

---

[1] The classic description of the tent is that by I. A. Richmond and James McIntyre, 'Tents of the Roman Army and leather from Birdoswald' (*T. Cumb. and West.*, 34 (1934), p. 62).

[2] The best edition remains that of von Domaszewski, Leipzig, 1887.

[3] In the Republican Army it would appear that these officers were known as *metatores* (Lucan, i, 382) as distinct from the *mensores*, the ordinary surveyors.

[4] Based on the calculations of Roy (*Military Antiquities*); however, no allowance was made by the General for changes the Romans may have made in these camping arrangements from Republican to Imperial times. Richmond has shown, by the arrangement of the gates, that Reycross held a legion (*T. Cumb. and West.*, 34 (1934), p. 55 and Fig. 1). The area of this camp is 20 acres and this gives a slightly higher number of men per acre than Roy.

wagon park would be established and here, too, no doubt, would have been special ablution tents where cauldrons of hot water would be available for the men to wash off the grime and sweat of the march, or they may have been content to use the nearest stream. The camp was usually pitched near a river. Along its banks places would be allocated for drawing off drinking water for men and lower downstream, for horses. Spaced out inside the camp would be latrines and rubbish pits and cooking ovens, while surrounding the whole would have been the defensive ditch and bank. On Trajan's Column there are a number of scenes showing these activities. Perhaps one of the most instructive is that with two advancing columns of praetorians arriving at the camp (Pl. VIIIb); at the head of each column are their officers giving the command to the horn-players to sound the halt, and soon when all eyes are on the standard they will be raised up and brought smartly down and each column will halt as one man. A corner of the camp is seen and within are orderlies unloading the tents and baggage from the carts. Outside, down by the stream, a watering-place has been established with boarding laid down to prevent the banks from being trodden down and muddying the water; here an orderly is filling a kettle.

Hyginus describes two types of defensive ditches, the normal V-shaped (*fastigata*), five feet broad and three feet deep and with a vertical outer face (*punica*); the rampart is given as eight feet broad and six feet high, surmounted with a breastwork (*lorica*). Vegetius, on the other hand, gives larger dimensions for the ditch—nine feet broad and seven deep in a normal situation, but twelve feet broad and nine deep when danger threatened—though these measurements are more akin to the defences of a permanent fort.

Archaeologically the remains of these marching camps are very slight and often all that survives is the surrounding ditch. Under exceptional conditions one may see more, and the most outstanding examples are the siege-works and camps at Masàda in Palestine.[1] On this high arid plateau the Roman Army carried out in A.D. 72 the final stages of the subjugation of the Jewish Revolt. Here, surrounded by precipitous cliffs, a group of fanatics had ensconced itself in the old fortress of Herod. The site and the existing remains are of particular importance, as we have, from Josephus, a description of the actual events.[2] The Romans completely encircled the site with a wall 4,700 yards long and built eight camps to command the approaches and exits. In

[1] *Antiquity*, 3 (1929), p. 195; *Israel Exploration J.*, 7 (1957), p. 1; for a popular account of the recent excavations on the Herodian fortress see Yigael Yadin, *Masàda*, 1966.

[2] *Bell. Jud.*, vii, 8 and 9.

these camps the troops built stone walls two feet thick and probably four feet high on which their tents could be fixed.[1] This would have afforded more capacity and some protection against the great midday heat and fierce, cold winds at night. Inside the tents of the legionaries (Camp B) were traces of three-sided couches of earth and stone on which the soldiers slept and dined. Even the cooking hearths outside the tents are preserved.[2] On a less spectacular scale are some of the remains in Numantia,[3] and in Germany, where under wet winter conditions the soldiers dug little drainage channels round their tents to prevent themselves from being flooded out.[4]

Here in Britain many marching camps are known from aerial photographs which show the single ditch, but any other features are rarely discovered. The camps vary greatly in size and shape, but most of them have pairs of sides in parallel and are usually rectangular in plan although some are rhomboidal.[5] A departure from this shape is seen in the notable series of camps in Scotland,[6] the most northerly of which is at Raedykes, where the defences are more formidable than usual. These polygonal enclosures, about 120 acres in extent, were at one time thought to mark the line of Agricola's advance in the season A.D. 84. Lately, however, opinion has favoured Severus as their originator in his great campaign of A.D. 208–10.[7] The ditch of the normal marching camp is usually quite small, not more than a yard deep and across. On the upcast mound, or bank of turfs, were planted the palisade stakes (*pila muralia*). Each soldier had two of these seven-foot wooden stakes sharpened at either end and having in the centre what has been termed a 'hand

---

[1] As suggested by Sir Ian Richmond (*J.R.S.*, 52 (1962), p. 146), who finds a parallel in the turf-walled structures in Cawthorn B.

[2] Richmond here noted the distinction between the baking ovens in the intervallum space operated on a century basis and these hearths where each *contubernium* could cook its meals.

[3] A. Schulten, *Numantia; die Ergebnisse der Ausgrabungen* 1905–12, vols. 3 and 4, 1927–9.

[4] As at Hofheim.

[5] One of the best series is that of the fourteen camps in Redesdale, I. A. Richmond, 'The Romans in Redesdale', *Hist. of Northumberland*, 15, 1940, pp. 116–29.

[6] Crawford, *Topography of Roman Scotland*, 1949, p. 108 and Pl. xix.

[7] Aerial reconnaissance by Dr J. K. St Joseph has added much to our knowledge of marching camps in Scotland. He now distinguishes three different series, the first of varying size, but all having a typical double clavicular gate as at Stracathro, shown by excavation to be Flavian; in the second series the camps are all about sixty-three acres and have six gates protected by a *titulum*, of unknown date, but probably Flavian or Antonine; the third series consists of the very large ones of about 120 acres with many irregularities and probably the work of Severus (*J.R.S.*, 48 (1958), pp. 93–94; 55 (1965), pp. 81–82).

grip', the main function of which was actually to facilitate the tying together of the stakes with pliable withies or thongs, to form a serviceable fence (Fig. 27). This was never considered as a defensive structure, to be manned by the Army in an attack, but was merely a fence to keep out stray natives and wild animals and, no doubt, also to prevent the Roman soldiers from wandering beyond the camp. Only at the gates were there any special precautions

FIG. 27 *A palisade stake from Castleshaw*

designed to prevent a group of attackers from forcing the position with a rush. These took two different forms; one was the short ditch (*titulum*)[1] in front of the gate spanning its width, and the other the ditch and fence which swung outwards in a curve forming a quadrant (*clavicula*),[2] forcing an oblique approach towards the gate, usually so that an attacker's sword arm faced the rampart, denying him the protective use of his shield (Pl. XXIVb and Fig. 28).

---

[1] This was formerly and erroneously called a *tutulus*, following Richmond (*Arch., J.*, 89 (1932), p. 23) and now corrected by Dr. J. P. Wild (*Arch. Camb.*, 117 (1969), pp. 133–134).

[2] The *clavicula* type seems to be confined to the first century, but the *tutulus*, although also of that period (as at Hod Hill), continued probably to Severan times (as in Scotland, if Dr St Joseph's third series is of this date). There is a distinctive type of entrance in a group of camps in Scotland, of which Stracathro is a typical example (*J.R.S.*, 48 (1958), Fig. 7, p. 92).

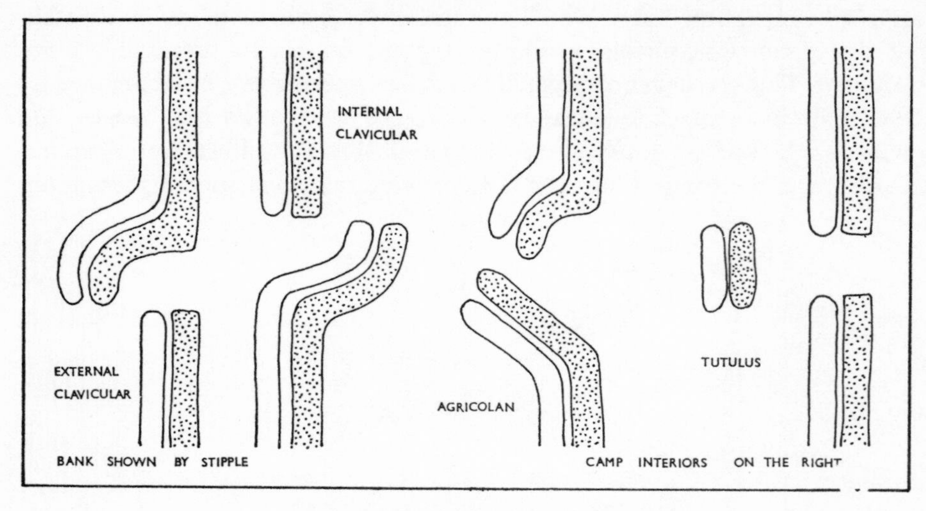

FIG. 28 *Plans of camp entrances*

The best study of these types of gate has been that of Sir Ian Richmond in his report of his investigations of the practice camps at Cawthorn. He showed here how these slight defences could be improved if troops were required to stay on and dig in, while inside it was possible to identify some of the cooking ovens, latrines, a small platform from which the commander could take the auspices and address his troops, and, most interesting of all, bracken-lined slit trenches for the officers to ensure warm and comfortable beds.

Permanent Forts

The arrangements for permanent forts were quite different. In the first place the site was carefully selected to hold a position of strategic importance, such as a river crossing or mountain pass, and was one of a network of similar positions, spaced out to control the movements of the inhabitants and prevent a hostile concentration from gathering. The most favoured site was a small plateau at a river junction which afforded views along both valleys. These forts were not, in the first century, regarded by the Romans as strongholds, but as bases from which the troops could emerge to deal with trouble in the open, where their superior equipment and discipline could be used to advantage. When looking for fort sites one must therefore usually disregard the question of tactical advantage. On the other hand, the Romans did not put themselves at any serious disadvantage when it could be avoided, as, for example, the presence of higher ground within arrow-shot. A small well-drained plateau with falling ground on every side was ideal for their purpose and the difficulties facing any would-be attacker were designed and constructed by the Army.

The first impression the student receives is of a diversity of defensive systems, but there are a few basic types with variations dictated by circumstances. Two factors influenced the design of ditches and rampart; the situation of the site itself and the potential degree of hostility to be expected from the natives. A study of the defences of forts of different periods can often reveal the Roman state of mind towards the enemy and this has obvious historical importance.

The function of a ditch system was to check a hostile force in any attempt to break into the fort. Any obstacles placed in the way of native warriors would hinder their progress and, while they were being held up, missiles could be discharged from the ramparts; accurate fire could cause casualties and, if persistent, force the attackers to withdraw. The first necessity, therefore, was to plan the defences with the range and strength of available fire-power foremost in mind. It would have been absurd, for example, to provide elaborate obstacles beyond the effective range of any missiles possessed by the unit. Most of the auxiliaries had only javelins, which have a killing range of about forty yards, but there were also archers, slingers and stone-throwers, while the legions were equipped with spring guns. Only rarely does the system extend to fifty or more yards beyond the rampart front. Another important factor was that of avoiding dead ground. It should in theory have been possible to cover from the rampart top all the bottoms of all the ditches, otherwise an attacker could crouch there in safety and recover his breath before the next rush, Here, the elevated platform of the rampart was a great advantage, and from it the interval and corner-towers rose another ten or twelve feet, enabling missiles to be discharged to greater range and effect (Fig. 29).

The normal ditch profile was the V-shape with a rectangular slot in the bottom. Sir Ian Richmond showed at Cawthorn that the Roman method of excavation was first to dig a narrow trench along the centre line of the ditch and then to take out the sides to the required slope. While doing the latter it was most important to estimate in advance the angle at which to work. This was dependent on the kind of subsoil; a stiff clay could, for example, be cut to a much sharper angle than soft sand or loose gravel. No doubt the officer in charge of defence construction dug trial holes at various parts of the site to enable him to judge this and issue precise instructions. The slot at the bottom

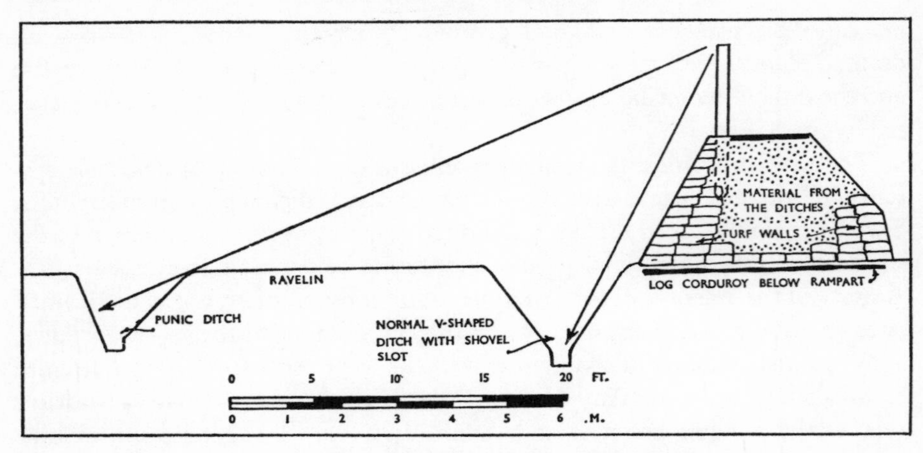

FIG. 29 *Fire control on fort defences*

was a shovel width and working parties would be sent round periodically to clear out the ditch bottom and prevent the growth of vegetation.[1] There are indications on some sites that the ditch sides have been deliberately plastered over with clay. This would not only help to maintain the slope cut in loose subsoil but also make it slippery in wet weather. In size the ditch was about ten to twelve feet wide at the top and six to eight feet deep, wide enough to prevent anyone from jumping it in one leap, but an attacker landing on the ditch side would then have to scramble out against the slope. There may have been additional obstacles which have left little archaeological trace, and these include such devices as a quick-set thorn hedge planted at the bottom or side of the ditch. At the legionary fortress at Inchtuthil, the depth of the ditch was increased by three feet by building a flat-topped mound in which a series of holes revealed the presence in the original of an obstruction formed by branches.[2] This acted as the modern barbed-wire entanglement and would have been more effective if so placed that an attacker would not see it until he had landed in it[3] (Fig. 30).

The usual pattern was three V-shaped ditches and any additions to this may reflect fear of attack; Birrens and Ardoch each have five. Occasionally the

[1] This slot is occasionally narrower than this and has the function of breaking or twisting the ankle should an attacker fall awkwardly into it.

[2] *J.R.S.*, 43 (1953), p. 104; they were known as *cervi* (*Bell. G.*, vii, 72) from their resemblance to a stag's horns.

[3] See a suggested reconstruction from Hofheim (Abb. 2.).

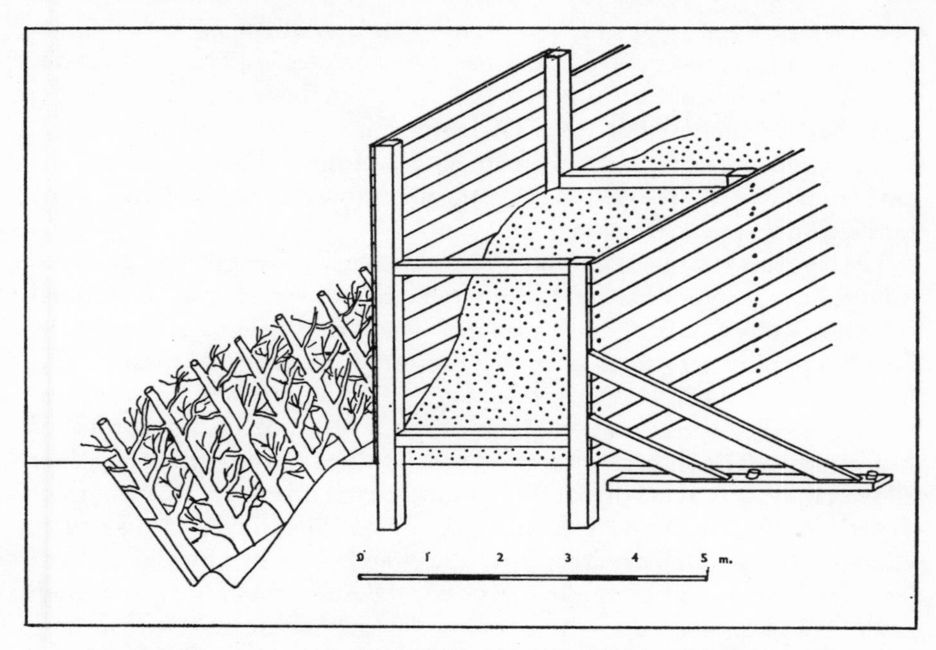

FIG. 30 *A box rampart (after Valkenburg) and thorn hedge in the ditch (after Hofheim)*

system is strengthened, not by multiplying ditches but by providing a much larger single one. A splendid example is the late Domitianic ditch at Newstead, in Scotland, which was found to have been twenty-five feet across and nine feet deep, a formidable obstacle immediately below the rampart, correspondingly heightened with the additional spoil. There was, however, another system which appears to show confidence, for it almost invited attack. It consisted of two ditches separated by a twenty to thirty foot platform, but while the inner is of the normal type, the outer one, known as a Punic type, had on the enemy's side a vertical face which must have required a timber revetment in most British subsoils (Fig. 29). To the attacker it would have presented little difficulty in reaching the platform between the two ditches. Fire would have been held until this moment and then a withering volley discharged and maintained. The shape of the outer ditch now revealed its true purpose. For in retreat the vertical face presented a hazard which had to be climbed, this check doubtless providing a further opportunity for the Romans. In the third and fourth centuries when the Army engineers had to turn their attention more and more to the question of defensive tactics, the problems of fire-power and artillery became of greater importance. Josephus records the use of stone missiles of about 55 lb with the heavy siege catapults in the first century, but at High Rochester (*Bremenium*) actual stones were found weighing a hundredweight each. Sir Ian Richmond has shown how the gun platforms were arranged to cover the main approaches[1] and two inscriptions record their construction in A.D. 220.

The defences in front of the gates were occasionally strengthened by means of ingenious, if somewhat fearsome, means. Holes were dug in the ground forming a quincunx pattern, so that they were not in alignment along the line of approach. Into these holes were securely fixed spears or stakes with their points upwards, but not projecting above the ground. Branches and leaves were then spread over the tops of the holes to obliterate all trace of their presence. The demoralizing effect on a horde of attackers on seeing their fellows falling into these dreadful traps can be well imagined. The troops of Caesar's army gave them the ironic nickname of lilies (*lilia*), as they were reminiscent of the stamens of that flower. Although described by Caesar,[2] and Dio,[3] the only known example in Britain is at the Antonine Wall fort at Rough Castle[4] but with modern scientific prospecting methods it is likely that more will be found (Pl. XXVa).

The Rampart

While the ditches and their attendant obstacles were carefully designed to hinder the progress of the attackers and bring them under fire, the rampart was the ultimate barrier which had to be both difficult to scale and break through. The Roman Army was considered, as will be seen below, extremely proficient at dealing with enemy fortifications. The normal method of breaking through the defences was first to build a causeway across the ditches and then to bring up to the wall itself a massive iron-shod battering ram, the rhythmical swinging of which gradually pierced the barrier and created a breach.

---

[1] *Hist. of Northumberland*, 15 (1940), Fig. 21.

[2] *Bell. G.*, vii, 73; Caesar describes other defensive measures he took at the Battle of Alesia to tighten the blockade (see pp. 236–43). Dr St Joseph has noted a single line of pits between two ditches at Glenlochar, which were probably some extra defensive structures in the Antonine period (*J.R.S.*, 41 (1951), p. 60, and Pl. vii, 1; 43 (1953), pp. 107–9).

[3] lxxv, 6; used by the troops of Albinus against Severus at the Battle of Lugdunum in 197.

[4] *P. Soc. Ant. Scot.*, 39 for 1904–5, p. 456; *The Roman Wall in Scotland*, 1934, p. 235–8 and Pl. xliii.

Plate XVIII
Horse-trappings and saddle displayed in the Rijksmuseum van Oudheden, Leiden
(p. 152).

Plate XIX
Sarmatian cataphracts on Trajan's Column (p. 152).

Against the sun-dried brick walls found in the East the ram was very effective, but against the stone and timber defences of north-west Europe it made little impression and efforts were normally concentrated on the gates. Here the timbers might first be set alight by rushing up loads of burning brushwood under a favourable wind.

In considering their own defences, the Romans would be thinking of these methods, although few barbarian armies were so well equipped with siege machinery or disciplined in its use. When it was available, the chief material used in rampart construction was turf, which made a resilient and fireproof barrier. The turfs were cut to regulation size (in Roman feet, one by one and a half and half thick) (Fig. 31) and could be used like bricks, stacked grass to

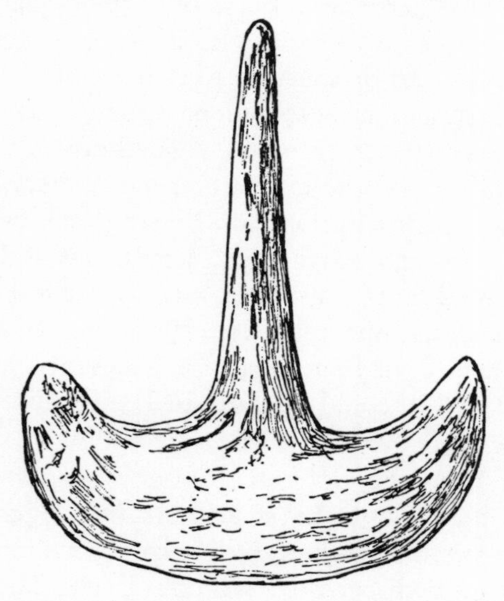

FIG. 31 *A turf cutter from Great Casterton, Rutland (half-size)*

grass, to form, in the front, an almost vertical face. A battering ram would merely have sunk into it without any shattering effect. The rampart was ten to eighteen feet thick and the turf work was concentrated on the steep front and the lower part of the back, the middle being filled up with the spoil from the ditch excavation. On Trajan's Column there is a fine illustration of this construction in progress, although the sculptors here give the turf work the appearance of masonry (Pl. XXVb).[1] One sees also the soldiers with turfs on

[1] This is exactly as they would appear in a drawing.

their backs held in position by a short length of rope, a very practical way of handling these awkward, heavy turfs, so liable to break when lifted.

The rampart was stiffened by timber in various ways. It was common practice to lay at the base a complete layer of logs at right-angles to the line of the rampart, as a foundation platform. In the extremely wet site of Valkenburg, near Leiden in Holland, the actual logs have survived, but usually one finds only brown streaks in the clean subsoil to indicate their presence (Pl. XXVI). Logs and brushwood were also laid in the body of the rampart, especially at the rounded corners of legionary fortresses, where special precautions were necessary to take the recoil of the heavy catapults mounted on them. On the Column, the ends of these logs can be seen about half-way up the rampart. When the site produced only poor-quality turf more use had to be made of timbering to form a solid structural unit, but this introduced a serious fire risk and precautions were necessary to ensure that some form of cladding in turf or clay was added to any timber revetment. Van Giffen was able to reconstruct, from his Valkenburg excavations, the box-like structures which formed the rampart structure there, but this is a type not at present known in Britain (Fig. 30). Possibly lack of survival elements has made it difficult to reconstruct these details, but at Lincoln[1] the main feature was a substantial front revetment held by vertical timbers housed in a trench, while at the back was only slight timbering. It was not possible to see if these two structures had ever been held together. Where timber was used at the front like this, it could be continued vertically to form the frame-work of the parapet and merlons to give protection to the soldiers on the rampart. It is clear from the Column that the merlons were placed fairly close together and they would be at least six feet high to give complete cover, while the breastwork between would have been four to five feet high to allow the troops to discharge their weapons freely. These structures were probably made of saplings or strong withies forming a tough bratticing or hedge, strong enough to stop enemy missiles. The rampart top, usually with a rammed gravel surface to give a secure foothold, was no mere patrol track, but a fighting platform eight to twelve feet wide to ensure free movement of troops without interfering with those already in action.

By the end of the first century maintenance of those timber forts must have become a serious problem; the steep turf fronts would have gradually weathered and been difficult to repair effectively, while the timbers would soon

---

[1] *J.R.S.*, *39* (1949), Fig. 8.

deteriorate in the British climate. Sir Ian Richmond was able to show that even the buried sleeper beams of the barracks had been attacked by wet-rot and then replaced. In times of peace this basic repair work could be carried out by the troops and a commander might indeed be grateful for these work jobs to keep his men busy, but with trouble brewing and the Army engaged in ceaseless patrol and fighting, the structures in the old forts would soon begin to show serious decay. We know that in northern Britain there had been a number of setbacks and general disorder at the very end of the first century but details are lacking. It is hardly surprising to find the soldier-emperor Trajan initiating a change of policy which altered fundamentally the appearance of the forts, by replacing turf and timber with stone. It was not an entirely new idea, for in exceptional cases where turf and timber were scarce, forts were provided with stone walls; a notable example is the fortress of the Twentieth Legion at Inchtuthil[1] in Perthshire built by Agricola *c.* A.D. 84. The added stone walls were not independent structures, but merely a replacement of the turf front with a more durable material. At Chester the wall is only four feet thick and with no inner face; the rampart was retained to provide solidity and the required width at the top for the fighting platform.

The Gates

Roman forts had four entrances, one on each side of the fort. One of the best examples of a timber gateway is that at Fendoch in Scotland.[2] It consisted of two towers each 12 ft by 17 ft, flanking the cobbled roadway, 10 ft wide. The towers were both built on six massive upright posts, each 12 ins square in section. Carefully cut, rectangular pits had been excavated in the subsoil and when the posts had been placed in position, the back filling had been thoroughly rammed and consolidated round them. At Lincoln the bottoms of the posts had been placed in a box made by setting stones on edge. Sir Ian Richmond calculated that the towers would have been about twenty-eight feet high allowing for a ten foot upper storey and a six foot crenellated top. Through this gate ran two water or drainage channels, below ground-level. There were double gates, the outer one being hung at the mid-point between the towers, which would have given complete command over a group of attackers. Above the opening would have been fixed an inscription cut on a wooden board, with letters picked out in red, commemorating the emperor under whose rule the work was carried out and indicating the name of the

---

[1] *J.R.S.*, 43 (1953), p. 104 and Pl. xix, 2. [2] *P. Soc. Ant. Scot.*, 73 for 1938–9, p. 115.

unit and that of the provincial governor. These timber inscriptions rarely survive, but by a fortunate chance a small chip from one from a turf wall milecastle on Hadrian's Wall has been found and identified.[1]

The conversion to stone would have given an opportunity for the improving of the gateway in appearance and dignity. Fragments of masonry often survive, and whereas one has to attempt to reconstruct the timber structure from a pattern of post-holes, there are examples like the fine west gate at High Rochester which still stands to the springing of the arch. Much more is, therefore, known about these gates, not only from actual remains in position but also from a study of the shapes of individual stones found in collapsed debris. The gateways of the forts on Hadrian's Wall have been subject to a detailed investigation resulting in some fine isometric drawings by Mr F. A. Child (Fig. 32)[2] One of the basic differences between timber and stone construction is the use of the arch and vault in the latter cases. There are no unnecessary decorative features on these structures; it is simple but bold functional architecture. There are two portals faced with large blocks and flanked with guard chambers. The upper storey, carried over the openings by vaulting was provided with small openings and a flat roof with crenallated surround above. The doors themselves, at least four inches thick, would have been of wood, but were probably covered on the outside with iron plates as a fireproof cladding. Stout cross-bars were inserted behind the door, as with medieval gates, as an additional strengthening and the holes in the masonry of the jambs have occasionally been found.[3]

The corner- and interval-towers would have been built in the same manner. In the first and second centuries they did not project in front of the wall, but were flush with it and extended to the rear. The towers were spaced out at 150–200-foot intervals along the stretches of wall and usually fifteen to eighteen feet by twelve to fifteen feet. The main function of these towers, as indicated above, was to make more effective use of missiles and, in the case of the legions, to house the small spring guns (*carro-ballistae*) in the interval-towers[4] and the heavy catapults (Pl. XXVIIIa) (*onagri*) on those at the corners.

---

[1] *T. Cumb. and West.*, 35 (1935), p. 220.

[2] *A.A.*, 4th ser, 20 (1942), p. 134.

[3] *A.A.*, 4th ser., 11 (1934), Pl. xvi, 2, and p. 108; 13 (1936), p. 270.

[4] These would be kept under cover and fired through the windows from an upper storey (see Richmond, *The City Walls of Imperial Rome*, 1930, Fig. 14). There was also the need to keep the cords dry, so as not to impair the torsion.

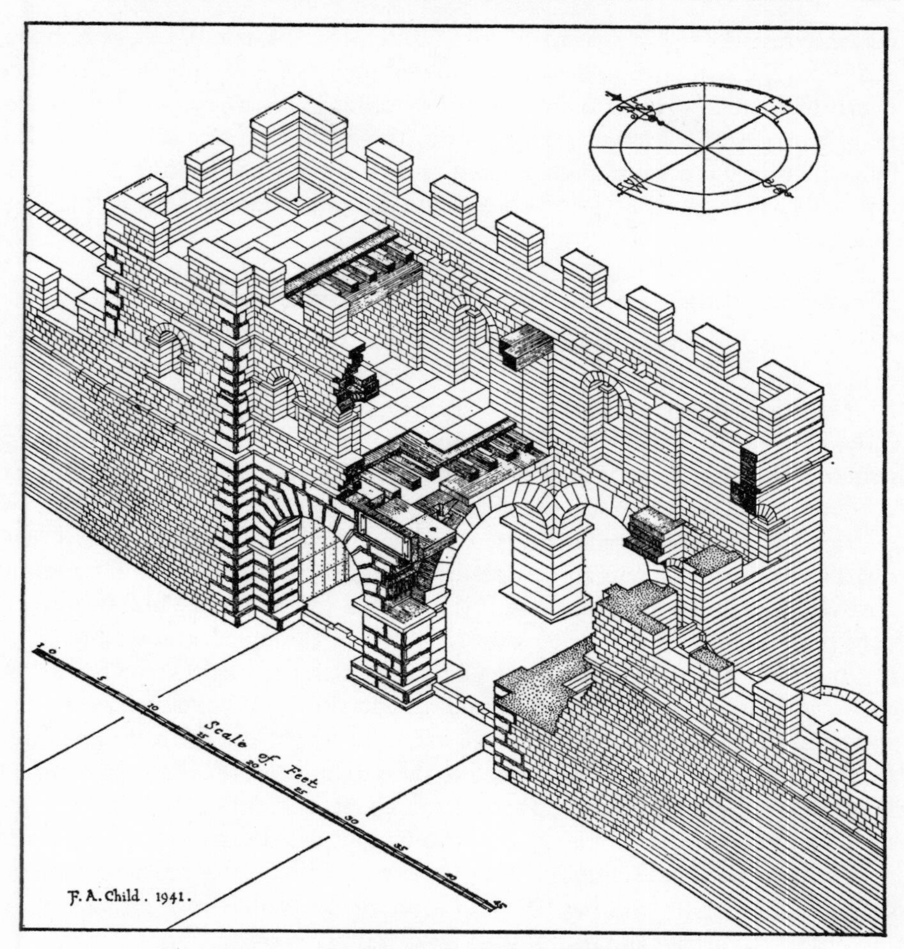

FIG. 32 *The west gate at Housesteads restored*

The Plan of the Fort

All forts had the same basic street pattern derived from the arrangements of the marching camp. Occupying the central position was the *principia* or headquarters building fronting the *via principalis*, and another street, the *via praetoria*, at right-angles, formed a T-junction; all other streets were secondary to these. One can see this layout very well today in the modern streets of Chester, where Eastgate and Watergate Streets are on the same line as the *via principalis* and Bridge Street on the *via praetoria*; hence the serious traffic delays when there is a stream of cars heading for North Wales at holiday times, for they all have to do a right-angled turn in the middle of Chester to conform to the Roman military street system. The *via praetoria* led to the main gate (*porta praetoria*), which was in the middle of this side of the defences, while there were gates at each end of the *via principalis*.[1] There was normally a rear gateway (*porta decumana*) in the same relative position, making four in all.

The Legionary Fortress (see Bibliography for detailed references)

The areas of legionary fortresses were fifty to sixty acres and, as one might expect, their buildings were on a massive scale. In Britain two of the permanent stone fortresses at York and Chester have been covered over and disturbed by medieval and later buildings and only fragments of the Roman structures have been, or will ever be, available for study. At Caerleon there is a better opportunity, since the site is covered only by a village; about a quarter has already been excavated and eventually something like three-quarters of the plan of the fortress should be recoverable, although the central area will remain beyond reach, as it is occupied by the church and its graveyard (Fig. 33)[2] With the earlier timber fortresses the situation is no better. It is now known that the fortress at Gloucester is buried below the medieval town, as is likewise the case at Lincoln. The site of Wroxeter is, happily, still open fields, but the military levels are eight feet below the remains of the Roman town and can only be studied to the detriment of the upper structures. Fortunately there remains the fortress of XX at Inchtuthil, Perthshire, constructed under Agricola. The plan of this uncompleted fortress has now been revealed by Sir Ian Richmond (Fig. 34), and will be the most important contribution Britain will have to offer towards the problems of the layout of the legionary fortress and the logistics involved in its construction.

Fortunately the meagre information from Britain can be supplemented with the results of work carried out elsewhere. The fortress at Neuss (Novaesium[3] (Fig. 35), on the Rhine, has been completely excavated, and another

[1] The *porta principalis dextra* and *porta principalis sinistra*, to right and left as one stood with one's back to the *principia*.

[2] A folding plan of the fortress showing discoveries made up to December 1966 by G. C. Boon and C. Williams, has been published by the National Museum of Wales in 1967. It includes, on the reverse, comparative plans of other legionary fortresses and notes on the buildings.

[3] *B.J.*, 111–12 (1904).

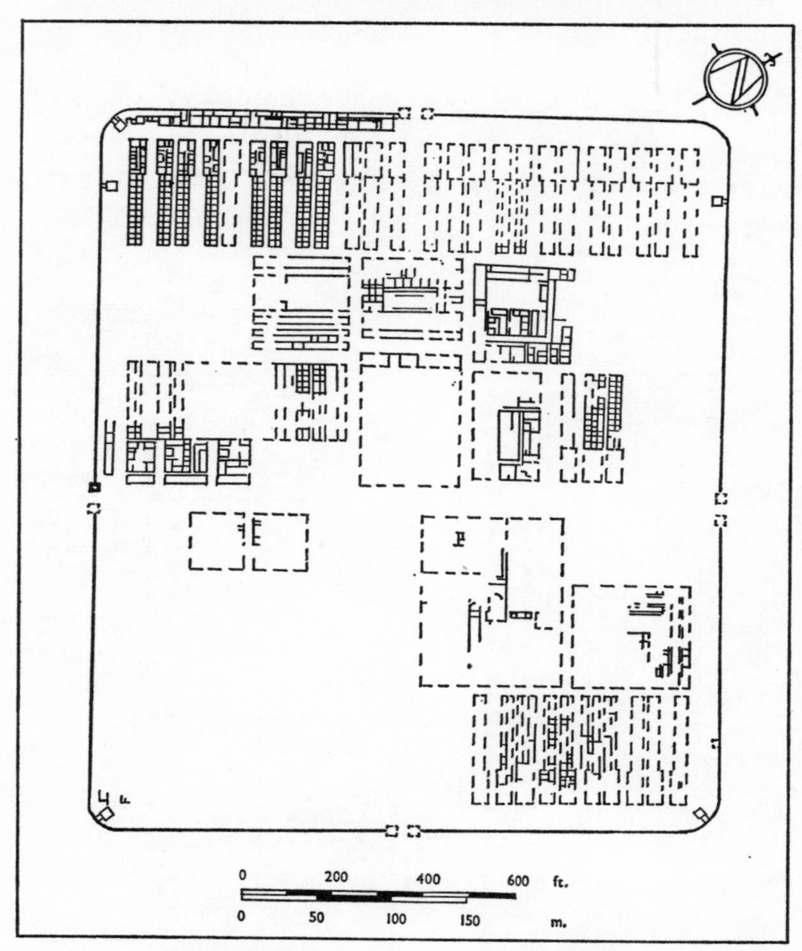

FIG. 33 *The Legionary fortress at Caerleon*

which offers important comparisons is Lambaesis in North Africa.[1] The Danubian fortresses Aquincum, Carnuntum and Lauriacum still present serious difficulties, as the excavators have not always been able to distinguish between the different periods and the plans are thereby confused. The excavations at Vindonissa have been of more recent date and over half the plan is now known; similarly more is gradually coming to light at Nijmegen on the lower Rhine.

[1] R. Cagnat, *Les deux camps de la Légion IIIe Auguste à Lambèse.*

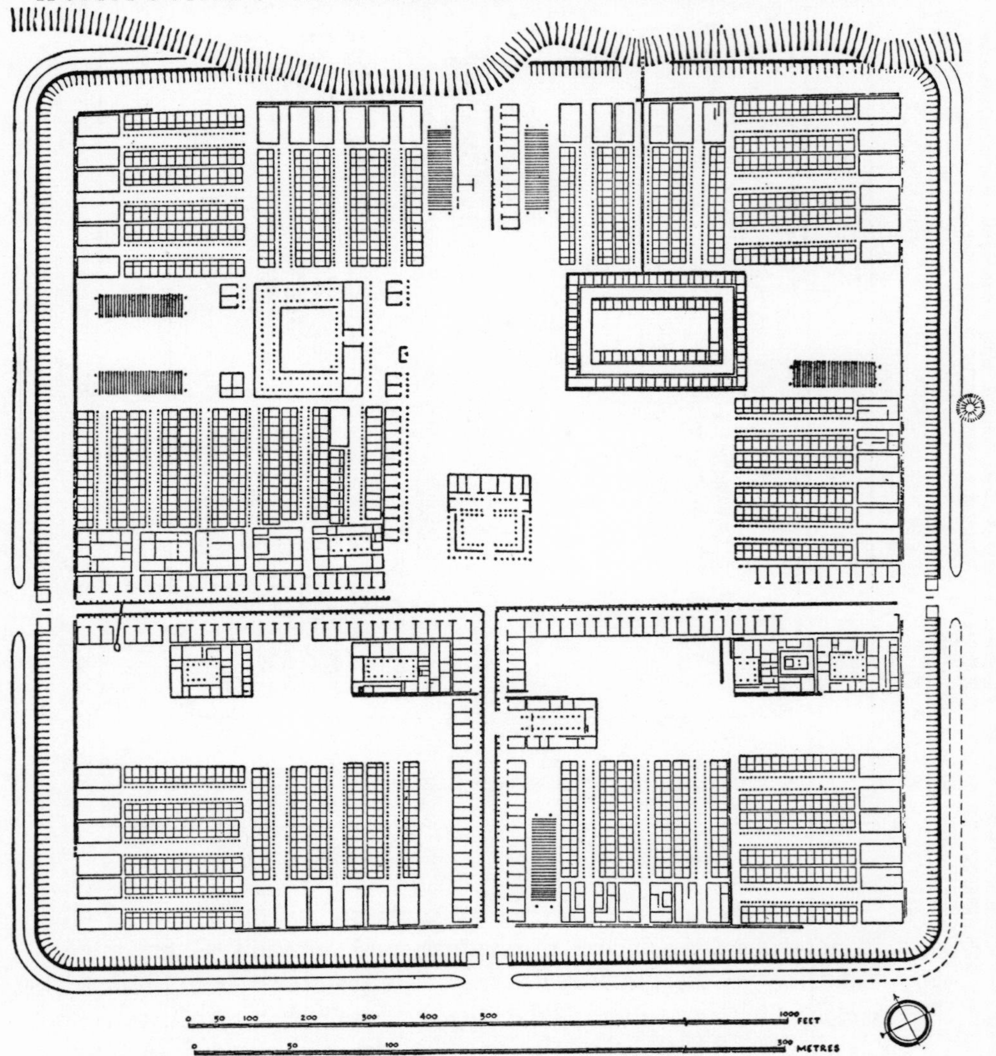
INCHTUTHIL ; GENERAL PLAN OF THE LEGIONARY FORTRESS
0 50 100 200 300 400 500 1000 FEET
0 50 100 300 METRES

FIG. 34

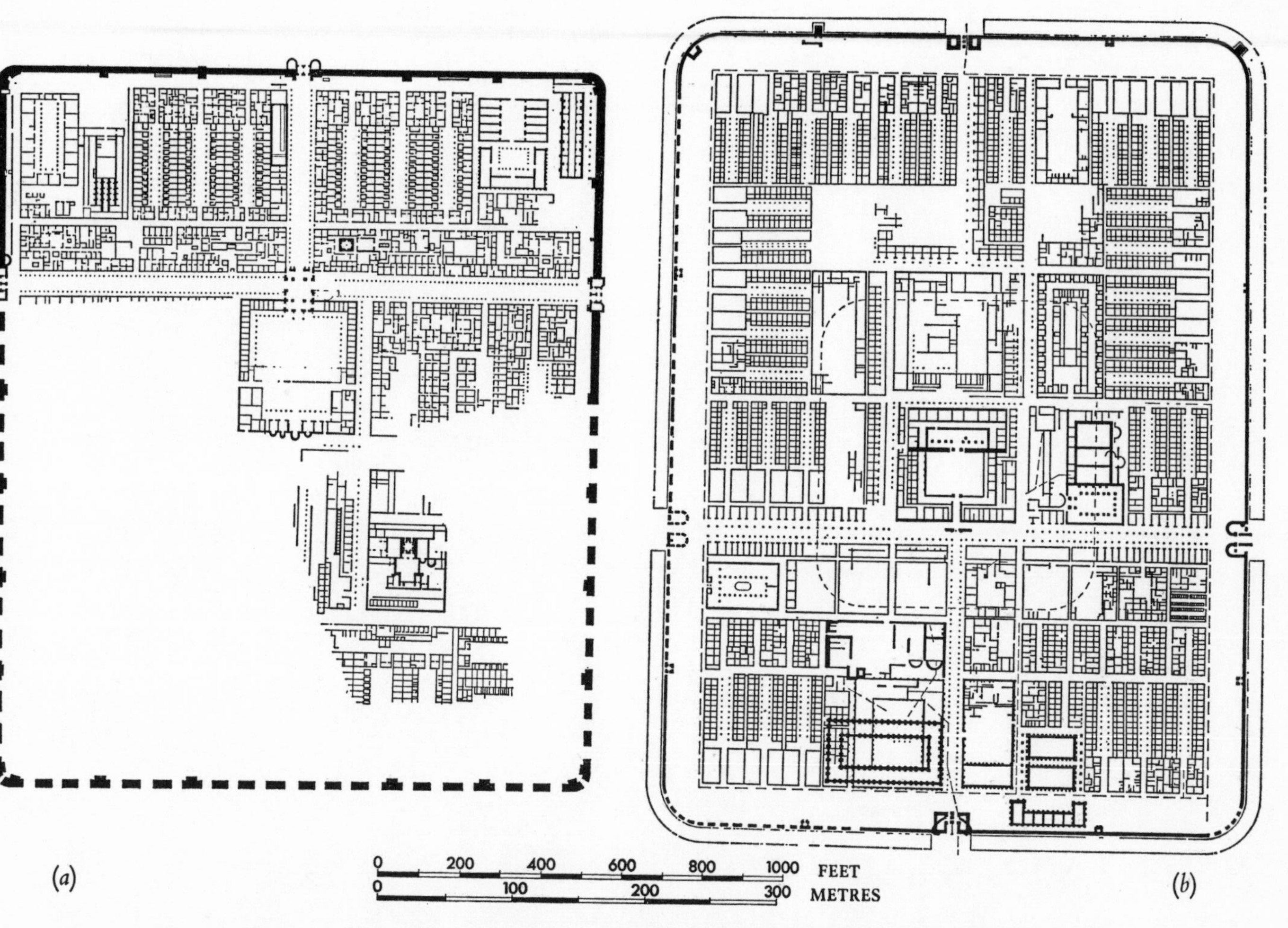

FIG. 35 *Plans of the Legionary fortresses (a) at Lambaesis and (b) at Novaesium*

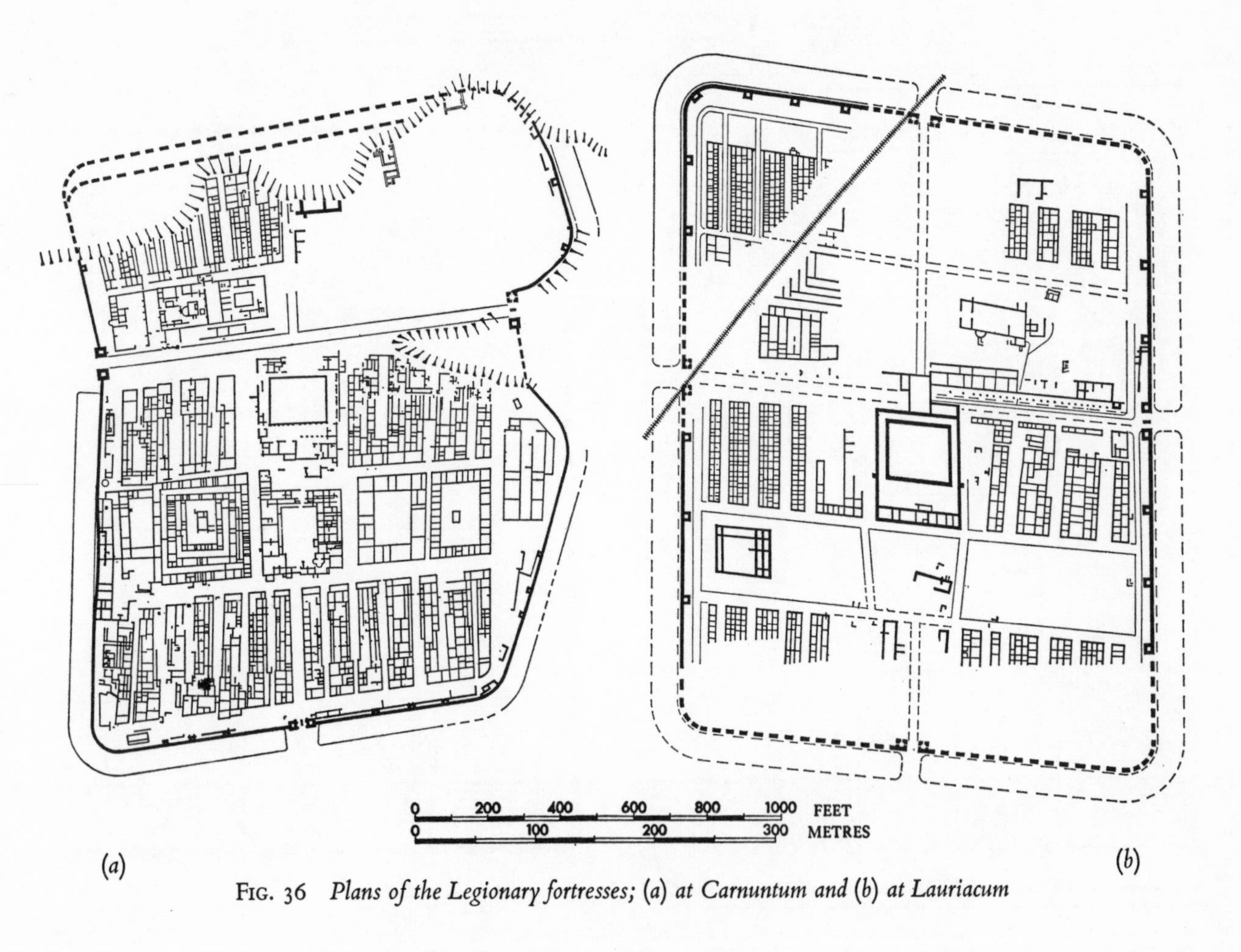

FIG. 36 *Plans of the Legionary fortresses; (a) at Carnuntum and (b) at Lauriacum*

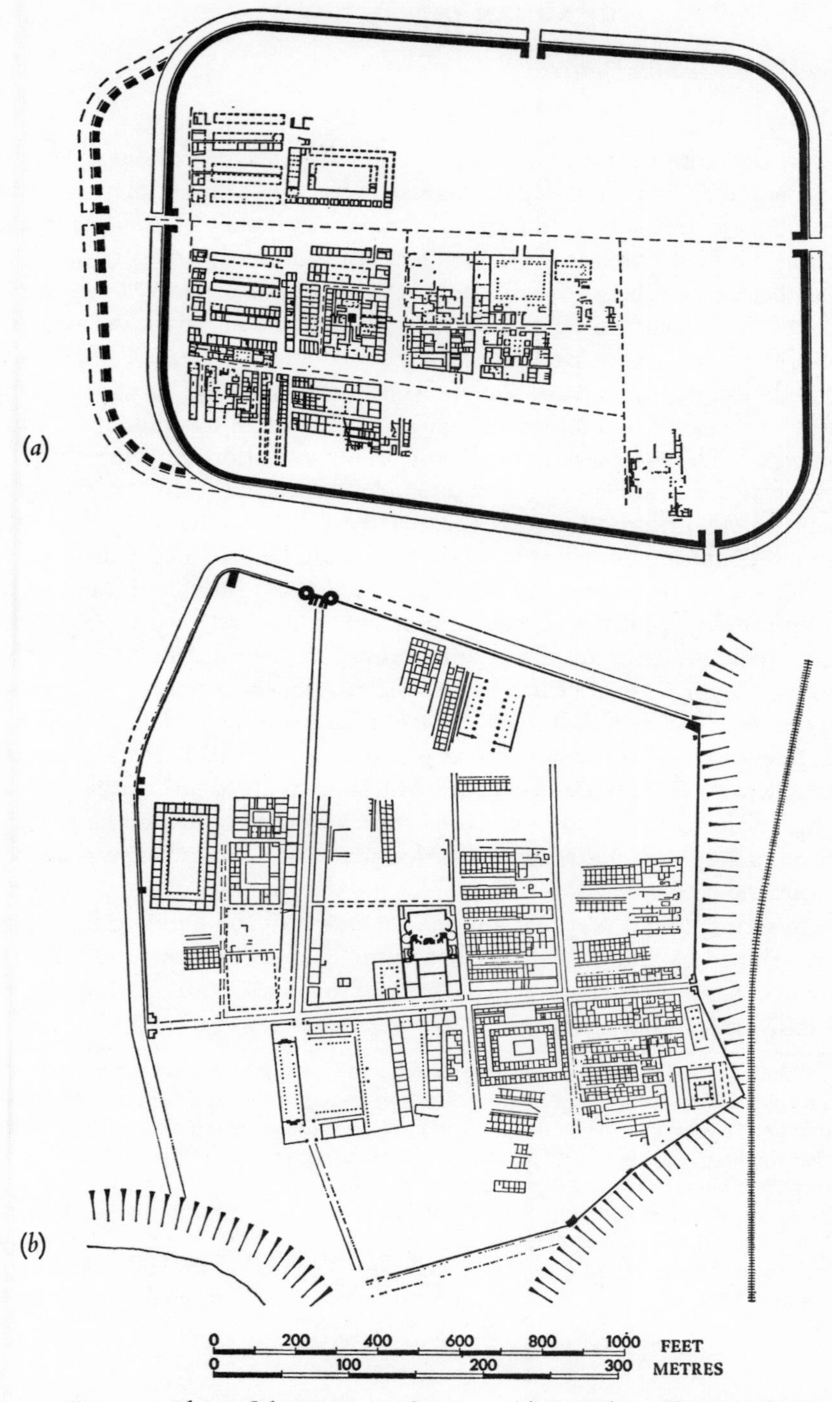

FIG. 37 *Plans of the Legionary fortresses; (a) at Haltern (b) at Vindonissa*

The Buildings Inside the Fortress

The Principia

This was the administrative centre, a large complex of buildings with a central courtyard. The building at Chester probably occupied an area about 250 feet by 330 feet, almost the size of a football pitch. The *principia* in the marching camps can be seen as an open space in front of the standards and the commander's tent. There would be no room for the whole army to assemble here, but sufficient for a small parade when the commander took the auspices and addressed the troops. Here, too, the guard could be changed and details assembled. Round the square the vital armaments could be stored. Out of the several distinct and important functions the plan of the building gradually emerged. The most important change in the evolution from the camp to the fort was the removal of the commander's quarters from the *principia*. Under field conditions he needed to be at the centre of the camp, but in a permanent establishment a separate building could be provided at his house, giving him more room and privacy. This has led to a little confusion, since Latin authorities, speaking of camps, give the term *praetorium* to the commander's tent and space around it and antiquaries have naturally given the central building in a fort the same name. The problem has been resolved by an inscription of Diocletian (A.D. 296–305) from Birdoswald on Hadrian's Wall[1] which refers to both a *praetorium* and *principia* which had been restored. A dedication from Rough Castle on the Antonine Wall found in the central building refers to it as the *principia*,[2] while altars from Chesterholm[3] are dedicated to the *genius praetorii* and were found in a building which was clearly the commander's house.

There seems little doubt that although the plan of the building had been fully developed by the end of the first century, it had a more complicated history than has hitherto been appreciated and that a closer study of examples of the mid-first century would be most rewarding.[4]

---

[1] *T. Cumb. and West.*, 30 (1930), p. 200; *R.I.B.*, 1912, the *praetorium* is mentioned first since it had been allowed to collapse and become overgrown, presumably because it had not been in use for a long time, while the *principia* and *balneum* were merely restored. The large *praetorium* might at this date have become an anachronism (see J. Wilkes, 'Rebuilding in Hadrian's Wall Forts', *Britain and Rome*, 1966, pp. 122–6).

[2] *The Roman Wall in Scotland*, 1934, p. 228; *R.I.B.*, 2145.

[3] *C.I.L.*, vii, 703 and 704; *R.I.B.*, 1685 and 1686.

[4] The best up-to-date study is by Rudolf Fellman, 'Die Principia des Legionlagers Vindonissa und das Zentralgebäude der römischen Lager und Kastelle', *G. pro Vindonissa*, 1958.

The broad façade of the *principia* along the *via principalis* would have been carefully planned to present to the visitor an impressive appearance, even in the case of the timber structure at Inchtuthil. There would probably have been an external colonnade and a massive central gate facing the *via praetoria*. One passed through this opening into the square with its paved or gravelled surface. On three sides behind a colonnade were the ranges of store rooms and the offices of the quartermaster and his clerks. Facing one was the great cross-hall, dwarfing the surrounding buildings. This enormous building stretching the full width of the *principia* was of basilican plan in having down its centre two rows of columns supporting a wall with clerestory windows and a great roof above. The nave would have a thirty to forty-foot span and on either side, aisles, each of half that width. A modern visitor would immediately have been struck by the similarity in appearance to a great Norman cathedral, but without the tombs and religious embellishments. This would be hardly surprising, as the Normans copied the Roman form without always understanding the method of construction and in consequence many of their buildings developed serious structural faults. At one end of the cross-hall stood the *tribunal*, a platform on which the commander could stand to address the troops. No one has yet satisfactorily explained the purpose of this enormous building. It would have been large enough to pack in shoulder to shoulder, the greater part of the legion, on those occasions when the commander addressed them, but this would never allow the dignity and formality required for the celebration of religious festivities which needed a large parade ground. It was undoubtedly used as a court of justice and probably for swearing in new recruits and other formal occasions, but these hardly seem sufficient reason for this large hall.

The hall was approached through a central doorway in the square and facing one across its width was the *sacellum* or shrine of the standards. Like the altar in a church, this was the focus of religious attention.[1] Stone screens decorated its doorway, through which the standards could be seen and their presence dominated the formal occasions in the hall. The standards would be ranged round the shrine with its stone vaulted roof, and it is noteworthy that in some of the timber forts the only stone structure has been this shrine, giving fireproof protection to the sacred emblems.[2]

The shrine had a secondary function, for below the floor (from the time of

[1] See Sir Ian Richmond's comments in *P. Soc. Ant. Scot.*, 73 for 1938–9, p. 125.

[2] As in Period 2 at Kastell Echzell (*Saal. J.*, 22 (1965), p. 140 and plan); and Caerhun (*Arch. Camb.*, 82 (1927), pp. 312 ff.).

Severus) was a small cellar[1] in which was kept the great iron-bound box which constituted the soldiers' bank, since the standard-bearers acted as treasurers of these funds. Thus any attempt at robbery was overshadowed by the greater crime of sacrilege.

Flanking the shrine was a range of offices where the headquarters staff did their daily routine business, filling in and filing away returns and reports. On the rare occasions when such fragile material as wood and papyrus survive, as at Dura-Europos and in Egypt, the evidence suggests that the Roman Army was as 'paper'-ridden as any modern army.

The only legionary *principia* about which we know anything in Britain, apart from Inchtuthil which was surprisingly small (160 feet square) is that at Chester. Excavations in 1948–9 revealed part of the western side and enabled a tentative reconstruction to be made of the plan.[2] One of the astonishing factors here is the survival of some of the column bases of the great cross-hall in what is now the basement of a fashionable gown shop. It says much for the interest and public spirit of the owner of the property in 1897 that when found these columns were left standing on rock platforms while the floor of the basement was taken down around them. But that is by no means all the story, for this strange situation provided a section of the rock below the level of the columns and here the eagle eye of Sir Ian Richmond saw the slight but significant evidence of filled-in post-holes and trenches of the timber headquarters block which preceded the stone one.

Perhaps the finest and most intriguing *principia* of them all is that at Lambaesis in North Africa. There are two striking differences to the normal plan; firstly, instead of a great cross-hall there is, more appropriate to the local weather, an open space with a colonnade in front of the shrine and offices; secondly there is also, a most elaborate gatehouse which spans the street and stands today almost to its full original height. There are also the strange apsidal ends not only to the shrine itself but to three of the offices, as if they had been turned into guild chapels, to which inscriptions found on the site may refer.[3]

---

[1] A good example is seen at Caernarvon (*Y Cymmrodor*, 33 (1923), pp. 54–57). It measured 10 ft 9 in by 5 ft deep and no less than 114 coins were found in the filling. In the Flavian *principia* at Inchtuthil, the strongroom was at ground-level but that part of the *sacellum* 'was underpinned by a grid of timber' (*J.R.S.*, 44 (1954), p. 85).

[2] *Chester Arch. Soc. J.*, 38 (1950), Fig. 9.

[3] *C.I.L.*, viii, 2554; *Rang.*, p. 43. The formation of guilds of various ranks appears to have been first allowed by Severus.

The Praetorium

The house of the commander of a legionary fortress would seem to merit the term used by the Germans—legate's palace (*Legatenpalast*). It was a complex of buildings covering as large an area as the *principia* itself and apart from the substantial private apartments would need to cater for a large staff and for entertainment on a considerable scale. The legionary commander was a senator, inevitably from a wealthy family, and the enormous size of this establishment reflects his social position. It enabled him to transport his whole family and household with table-ware and furnishings, and continue to live in a style not far removed from his customary one, even on a remote frontier. Here, too, a commander could try to impress the local chiefs and potentates with the dazzling brilliance of Roman civilization in the building, its decoration and furnishing. Julius Caesar had this idea in mind when he transported, in sections, a mosaic pavement to lay on the ground inside his tent on his Gallic campaigns.[1]

The best example of a legionary *praetorium* is that at Xanten (Vetera) reconstructed by Mylius[2] (Fig. 38). The building (250 feet by 310 feet) was planned round a central open court with a wide opening from the street on one side and the large dining-room on the other, while on the other sides were suites of rooms presumably for receiving guests. At the rear was a remarkable colonnaded 'garden' with rounded ends, ninety yards long and twenty yards wide. The main interest of this and other military houses lies in the adaptation of the Mediterranean styles to this rather special purpose, far removed from the normal houses which developed in town and country in Britain. Basically this legate's house consists of a number of open courts surrounded by ranges of rooms, but the whole carefully integrated into a simple but pleasing architectural pattern. It is doubtful if this kind of building, so open to the elements, was suitable to the British climate, but it certainly made the best use of natural light, so necessary in the absence of large windows. A better understanding of the needs of the commander can be gained from a study of the simpler plan to be found in an auxiliary fort, as will be seen below.

The Tribunes' Houses

The six military tribunes required separate houses and presumably the senior (a senator designate) had a rather more elaborate one than the others (eques-

---

[1] Suetonius, *Caesar*, 46.

[2] *B.J.*, 126, (1921), pp. 22 ff.

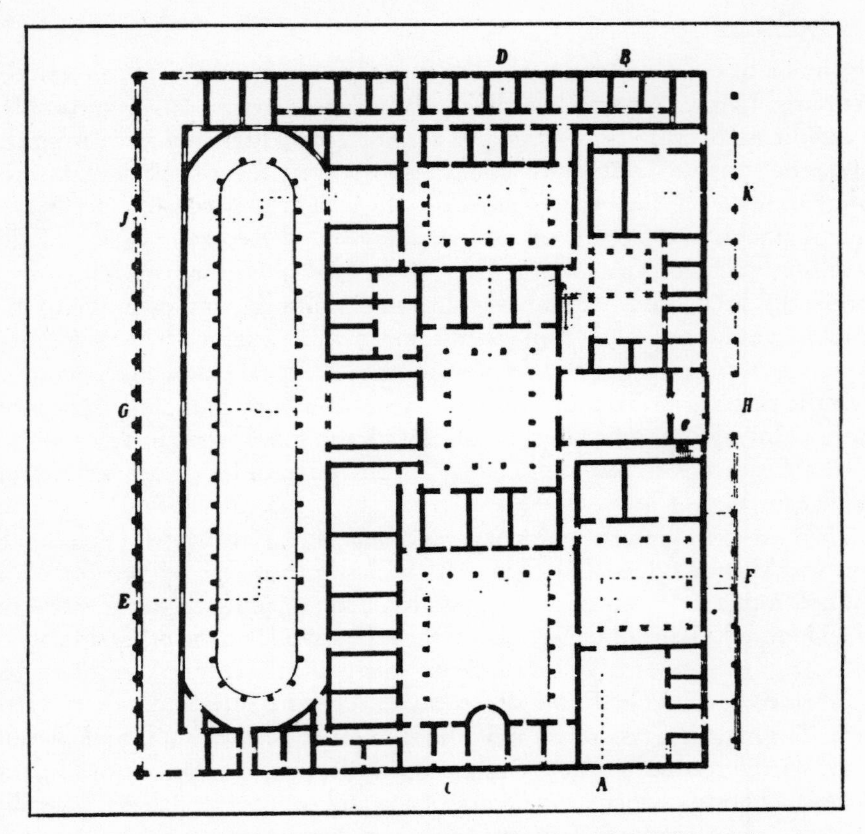

FIG. 38 *The Legate's Palace (praetorium) at Xanten* (1:1000)

trians). There are interesting differences in the four houses excavated at Inchtuthil (Fig. 34) which may reflect the variety of social status and function of the tribunes. Other evidence can be studied from the plans of Neuss, Lambaesis and Vetera. They are all planned round a central court with a large dining-room and an entrance from the street, with suites of rooms for staff and the transaction of military business.

Barrack Blocks

The greater part of the fortress would have been taken up by the soldiers' barracks. These were thin rectangular blocks arranged in facing pairs and clearly derived from the rows of tents in the marching camp (Fig. 39). The

Plate XX
A horse-armour from Dura-Europos (p. 152).

Plate XXI
Levantine archer on Trajan's Column (p. 153).

original tent-party of eight men (*contubernium*) now had a pair of rooms, one for storing equipment (about twelve feet by fifteen feet) and the other for sleeping (about fifteen feet square),[1] while along the front ran the veranda with its colonnade and wicker rubbish baskets for food waste, sunk into timber-lined holes in the ground and fitted with lids. Although strictly only ten pairs of rooms would be needed for each century, one finds at least eleven and sometimes twelve; the extra rooms were presumably for storage and

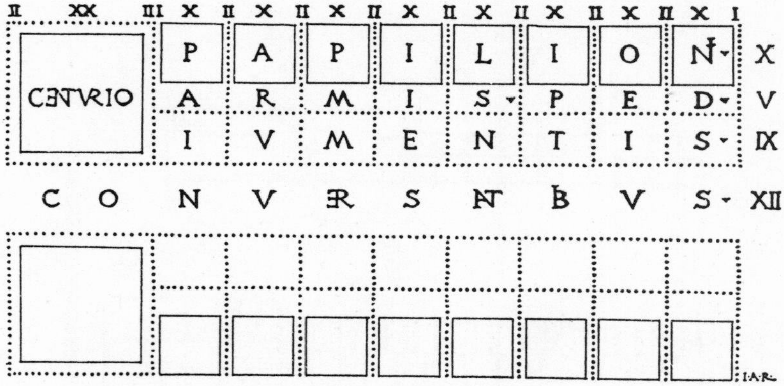

FIG. 39 *The tents of a maniple in a marching camp*

junior officer accommodation.[2] The arrangements vary very little from one fortress to another (Fig. 40). At Chester, in deference to the northern winter, each dormitory had a tile hearth, but the most notable distinction was seen at Lambaesis, where the blocks are very much larger and a passage separates the equipment room from the dormitory. These refinements may be due to the need for more air space per man in a hotter climate. The quarters of the centurion attached to each block were far more elaborate and he had eight or nine rooms, usually with a central corridor. Some of these were for his personal use and included a latrine and wash-room, but there would also be the company office and probably store-rooms and his *optio* presumably lived here, too. The five centurial blocks of the first cohort are much larger than those of

---

[1] This compares very favourably with recent British Army barrack accommodation at 300 cu ft per man.

[2] There is no such extra accommodation in the auxiliary fort at Housesteads and it has been suggested that the junior officers, i.e. *signifer* and *optio*, lived in the centurion's block (*A.A.*, 4th ser. 39 (1961), p. 282).

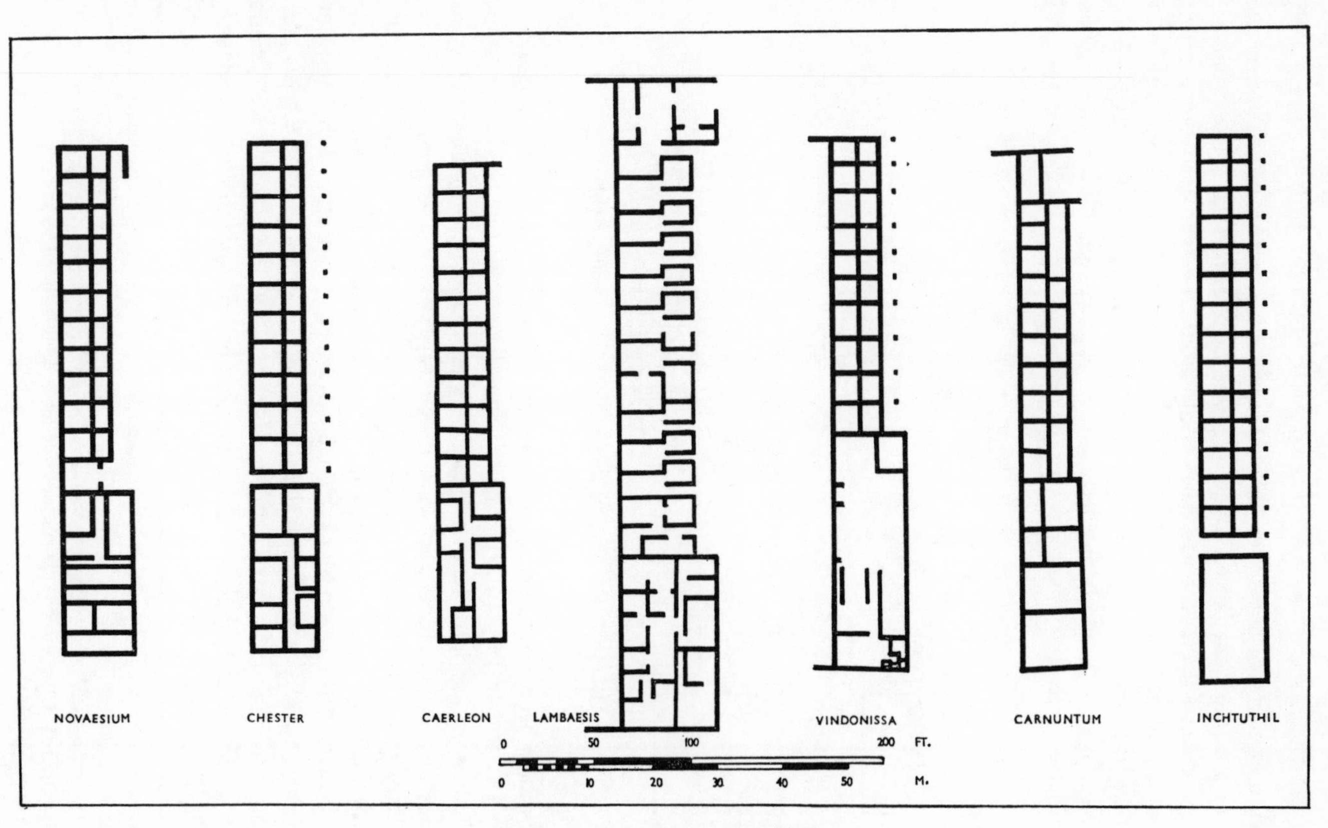

FIG. 40 *Legionary barrack blocks*

the other centurions and they are in effect small courtyard houses. That of the *primus pilus* is half as large again at Inchtuthil and is also provided with a room heated with a hypocaust, an unusual feature to find in a timber building.[1] The arrangements for roofs and windows in the men's blocks are not known, but as the pairs of rooms are back to back, yet with space between for an eaves drip, the most economical design would be for the roofs falling both ways from the centre with a break to allow for clerestory lighting for the dormitory.

The Hospital (Valetudinarium)

As will be seen below (p. 248), the medical service in the Army was remarkable for its apparently modern standards and ideas. One of the ways in which this aspect can be studied is through the planning of the hospital. In each of the legionary fortresses there would have been one of these buildings.[2] At Neuss it is half the size of the *principia*, but at Carnuntum it appears to be as large, while the only British example in a legionary fortress so far fully investigated, at Inchtuthil, is 298 feet by 193 feet[3]. In plan (Fig. 41) it bears a close resemblance to that of Neuss. The shape is rectangular, built round an open central space; there are sixty wards, in pairs, flanking a corridor which allows circulation round the whole block, an arrangement clearly evolved from a grouping of tents in a camp. The wards are not directly open to the corridor, but through a passage from which there was access to a latrine, so that each pair has a single dividing wall. This would effectively isolate the wards, each fifteen feet square, and ensure quietness. Other rooms having doors on the corridor would be for administration, staff and storage. At Neuss and Vetera there appears also to be a large hall at the entrance which could have been the casualty reception centre. Opening off this at Vetera is a room equipped with small hearths which, it has been suggested, were for sterilizing instruments before an operation; this hospital is also provided with its own bath-suite. These refinements are not present at Inchtuthil, but there is here a remarkable feature of construction in the form of double timber walls between the corridor and the wards. This meant that the corridor was an independent structure

[1] *J.R.S.*, 48 (1958), p. 132; three similar centurial houses are known at Caerleon from the Broadway excavations (*Arch. Camb.*, 87 (1932), pp. 394–5; 88 (1933), pp. 113–14.

[2] For legionary hospitals in general see R. Schulze in *B.J.*, 139 (1934), pp. 54–63.

[3] The *valetudinarium* at Caerleon has been located and partly excavated (*J.R.S.*, 55 (1965), p. 199 and Fig. 10).

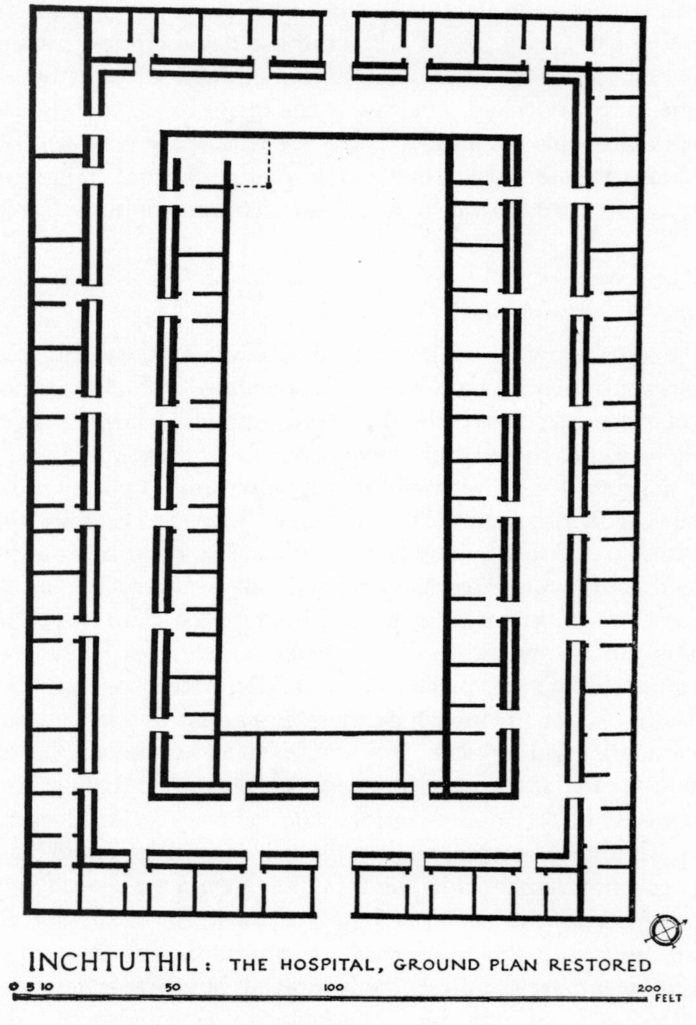

FIG. 41

carried higher than the wards and having clerestory lighting. The gap between the walls would have been an eavesdrip, giving very effective insulation from heat and sound. The wards at this fortress would hold five beds, giving about five per cent accommodation, which could readily be expanded into the other rooms and along the wide corridor (Fig. 42).

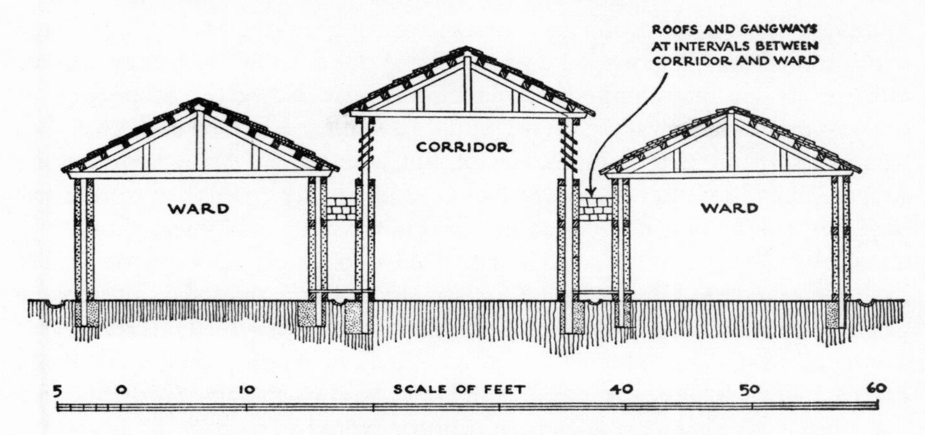

FIG. 42 *A section through the hospital at Inchtuthil*

The Granaries (Horrea)

The troops' food supplies had to be safe from theft and deterioration and the granaries which gave this protection were carefully planned and built. They were long but narrow in plan and the floor was raised above ground-level and supported by a series of sleeper walls with ventilators in the main walls to allow the free circulation of air. It would also be advantageous for off-loading the stores from carts on to the platform at the end of the building. The main walls were very massive and supported at intervals by buttresses. These were not, however, so much to take the extra weight as to allow the roof to project at least a yard clear of the building so that the rain water did not run down the walls and cause dampness inside. There would doubtless be louvred ventilators in the walls allowing air, but not small animals and birds, to get into the granary.

The best-preserved granaries in Britain are those at Corbridge, the great Severan store-base behind Hadrian's Wall, and many of these structural details can be studied there today by the visitor. Only at Chester have stone-built legionary granaries been found in Britain. Three of these massive buildings have been excavated by Mr F. H. Thompson near the Watergate. They were each 159 feet long and forty-five feet wide, and a fourth is known in the *retentura*.[1] Inchtuthil has a total of six scattered in different parts of the fortress, but these are each about 136 feet by forty-two feet. Five granaries of the Chester size would have given about the same capacity as the six Inchtuthil

[1] *Chester Arch. Soc. J.*, 34 (1939), pp. 7 and 8 and Pl. II.

buildings. One of the problems is that of working out the amount of storage space needed for a legionary garrison. This is not as simple as it may appear, as there are too many unknown factors, not least that of the proportion of grain to other kinds of food. Grain would have been delivered each year after harvest and parched to help preserve it, but whether the Army kept a year's supply in hand to offset disaster or bad harvest, as may be implied from Tacitus,[1] is not clear. Nor do we know if the grain was stored loose or in barrels, nor the height of the bunkers; the space needed for other types of food such as fresh and salted meat, oil and wine, etc, has also to be considered. Internally the granary had a central passage flanked by the bins, one of which it is reasonable to suppose was allocated to each century. Mr F. H. Thompson has calculated that each of the Chester granaries could have held 1,000 cubic yards, and had there been five such granaries each century would have had about eighty-three cubic yards.[2] R. G. Collingwood estimated a year's corn ration per man to occupy half a cubic yard and thus a century would need at least forty cubic yards.[3] One must assume that space was probably needed for the households of the commander and his officers and the servants and other non-fighting personnel. Consideration may even have been given to the civil population outside the fortress. The general conclusion seems to be that the fortress granaries were designed to hold two year's supply, but only on rare occasions would the whole of this capacity have been needed.

The Bath-house

Nothing is known of the arrangements for washing in the marching camps, but as hygiene would have played an important part and the soldiers bathed regularly, there must have been some method of heating water on a large scale. Possibly the problem of fire risk or water supply and drainage dictated the siting near a river, away from the camp, but protected by it. It is clear that whenever a Roman unit was permanently established in its quarters a stone bath-house was erected in such a position. The Roman method of bathing was totally different from ours. Not for them the brisk shower or private wallow in the tub—these modest facilities of our modern age would have appalled them. Here, perhaps, the machinery of civilization has slipped back a cog, for there is much to be said for their methods, which have survived in very modified form in the Turkish bath and can be seen better, perhaps, in the Finnish

[1] *Agricola*, 22.

[2] *Chester Arch. Soc. J.*, 46 (1959), p. 37.

[3] *T. Cumb. and West.*, 20 (1920), pp. 138–42.

steam bath. The Roman method demanded a series of rooms of varying temperatures and humidity which induced a perspiration subsequently sluiced off by warm or cold water, followed by massage and oils rubbed into the body. It must have been an exhilarating experience and its effect on the morale of the troops very considerable. The bath-house was also what has now become the club or NAAFI. In the towns much of the afternoon was spent in the bath gossiping and gambling with friends or watching the sports in the exercise court. How much of this kind of lounging about would have been permitted in the military bath-house is not certain. To enable a whole legion to indulge in such pleasures regularly would have needed a very large bath-house or a whole series of them.

In the auxiliary forts this building was usually in the annex, but in some of the legionary fortresses it formed an important central building. For many years a strange building at Chester with a basilican plan and mosaic pavements and hypocausts puzzled antiquaries. It was in the southern part of the fortress fronting the east side of the *via praetoria*. The mystery was solved in 1964 in large-scale clearance for commercial development when a large and very complex bath-house was revealed.[1] The colonnaded hall and its adjacent hypocausts revealed as long ago as 1725 are now seen as the *palaestra* of the bath-house to the south of it. Curiously enough a parallel situation exists at Caerleon, where in 1877 a large heated room with tessellated pavements was found in an unexpected area, but very similar in relation to the plan at Chester;[2] then in 1964, at the same time as the Chester bath-house was revealed, so was that at Caerleon. Here the *frigidarium* was fifty feet wide with a large plunge-bath with apsidal ends.[3] The bath-house at Neuss fronts the *via principalis*; behind the large colonnaded *basilica* there are three large rooms significantly placed to form a block. The building at Vindonissa occupies about the same position and is of the same type, but with greater architectural elaboration.

In such a large and complex organization as a legion there were status differentials which demanded special facilities for this important social occasion. It is not surprising to find bath-houses outside the fortress. These may have been for officers only, above a fixed rank or their guilds. Others may have been for the underprivileged, as at Chester, where the site of a bath-house is known

[1] It is most regrettable that the pace of modern development did not allow a full-scale investigation of this imposing building, some of the walls of which still stand to a height of six feet.

[2] *P. Soc. Ant. London*, 2nd ser., 7, for 1876–8; pp. 219–23.

[3] *J.R.S.*, 55 (1965), p. 199; 56 (1966), p. 198; 57 (1967), p. 175.

near the Watergate, now buried under eighteenth-century houses.[1] The building work in 1779 which revealed this structure also produced a fine altar now in the British Museum and dedicated to Fortune, Aesculapius and Salus by the household staff of a legionary commander who was graced by the names Titus Pomponius Mamilianus Rufus Antistianus Funisulanus Vettonianus.[2] It is possible, as Collingwood has indicated,[3] that it was the same man who was a friend of the younger Pliny and who became consul in A.D. 103. It is interesting to note the association of the bath-house with health and sickness, for the dedication to Fortuna Redux rather implies a safe return journey, and the other gods invoked wounds or illness. Here also can be identified a bath-house for freedmen and household slaves who, presumably, would not be allowed in the legionary bath-house in the fortress. At Caerleon, parts of two extensive bath buildings have been found, one as the result of an excavation in 1894[4] and in the case of the other only the fringe was skirted during the amphitheatre excavation—a fine opportunity still remains of stripping this extensive building, which had to be slightly rebuilt to accommodate the great amphitheatre.[5] One of the bath-houses at York was destroyed without adequate record, when the railway station was built in 1839–40, but it was clearly an extensive building and thought to have had the largest *caldarium* in Britain;[6] it also produced an altar to Fortuna.

Other Buildings

There would have been other kinds of buildings in the fortress, such as stables for the horses and workshops (*fabricae*), a very fine example of which was found at Inchtuthil.[7] At Neuss there was the little prison (*carcer*), with its cells, by one of the gates. Here also, and at Lambaesis, there is one building which is worthy of comment. This is the so-called *schola* which has been

---

[1] *Chester Arch. Soc. J.*, 27 (1928), pp. 110–12. A large bath-house is now known inside the fortress.

[2] *C.I.L.*, vii, 164; *R.I.B.*, 445.

[3] *Roman Britain and the English Settlements*, 1937, p. 120.

[4] *Isca Silurum*, 1862, p. 85, Pl. xxxviii, *Arch. Camb.*, 85 (1930), p. 147.

[5] *Arch.*, 78 (1928), p. 144. Two fragments of this bath-house were found and both showed evidence of the building going out of use during the Roman period; the pottery from the filling is mostly rubbish survival, but there are Antonine pieces.

[6] *Arch. J.*, 103 (1947), p. 76; the room is 30 ft wide and at least 50 ft long and may have been smaller than that at Wroxeter, which is 42 ft by 48 ft with additional side tanks (*Ant. J.*, 46 (1966), p. 232).

[7] *J.R.S.*, 51 (1961), Fig. 10.

considered a kind of officers' club. At Neuss a building so ascribed (No. 61) is 108 feet by 180 feet with an internal colonnade and a stone-lined tank in the centre, but with no internal walls at all. But this could equally be interpreted as a large stable with wooden loose boxes, all traces of which have gone, and a central manure pit. However, the building in the north-east corner of the Lambaesis fortress is rather different, as it has a series of rooms on each side of the central court with a veranda, and a similar building at Vindonissa has been identified as a store or magazine.[1] At Chester part of an odd building has been excavated north-west of the *principia*.[2] It has a curved wall fitted into a corner with traces of radial walls and was described as a theatre-like building, but its identification is far from clear. It could possibly be a drill-hall for arms practice and demonstrations with a small arena. A building has been identified with more certainty as a drill-hall (*basilica exercitatoria*) at Inchtuthil.[3] At Chester and Caerleon a number of rectangular buildings have been found built into the tail of the rampart, presumably weapon and ammunition stores. Those at Chester were deliberately demolished at the end of the second century. In this area one also expects to find latrines, and the ovens and cook-houses, put here to minimize the risk of fire. At Neuss, Vetera and Inchtuthil there were rows of buildings divided into store-houses fronting the two main streets.

Buildings Outside the Fortress

A building which rouses interest by its very size is the amphitheatre.[4] Those brought up to think of the Roman Empire in terms of gladiators and wild beast shows are naturally intrigued by these splendid structures which imagination readily fills with the wild orgies of popular belief. However, the shows which became so extravagant in Rome could not be repeated all over the Empire and in any case the emperors frowned heavily on any possible competition in the courting of favours from the populace. What is also forgotten is that these shows were organized for special festivals and that they were preceded by ceremonies and sacrifices. It is more than likely that the legionary amphitheatres were primarily for this purpose. Here the whole

---

[1] *Jb. G. pro Vindonissa*, 1959–60, p. 16.

[2] *Chester Arch. Soc. J.*, 34 (1939), pp. 5–45.

[3] *J.R.S.*, 50 (1960), p. 213.

[4] The standard work on amphitheatres generally is L. Friedländer, *Darstellungen aus der Sittengeschichte Roms*, 1922, ii, 318 ff.

legion could be easily assembled and could take part in celebrating the many festivals in the official calendar. When the rites had been performed the soldiers could relax and enjoy the modest games and spectacles afforded by provincial resources. Good gladiators, like our professional boxers and footballers, were very expensive and could demonstrate their prowess only with dummy weapons. Occasionally, perhaps, a few decrepit prisoners who were not physically up to the slave market or the mines might be slaughtered in a mock battle. Wild animals could also be used only on a modest scale. Britain could certainly produce wild boars, and bear and bull baiting was a possibility, but it is doubtful if the amphitheatres of Britain could ever achieve a shadow of the enterprise and scale of the Imperial extravaganzas in Rome itself. The amphitheatre could also have been used for demonstration of tactics, use of weapons, arms drill, etc.

In Britain, the amphitheatre at Caerleon was excavated by Sir Mortimer Wheeler in 1926–7 (mainly through the generosity of the *Daily Mail*). The accumulation of centuries was removed, some 30,000 tons of it, and the whole structure cleared and conserved, so that it can be seen today as one of the most spectacular buildings in the whole of Roman Britain.[1] This amphitheatre is altogether 267 ft by 222 ft and the arena precisely 140 Roman ft along its shorter axis and laid out, with the latter as a base line, from four centres. The eight entrances were placed symmetrically about the arena. The seating was built on an earth bank retained by a massive external wall, five and a half feet thick, reinforced by buttresses. It was possible from a study of the vaulting of the entrances to calculate fairly accurately the original height of the structure. The arena wall rose at least twelve feet above the floor and the seats went up beyond this in fifteen rows to the back wall, another thirty-two feet, giving accommodation for about 6,000, rather more than the full legionary complement.

The amphitheatre at Chester occupies precisely the same position relative to the fortress defences as at Caerleon. It was found during preliminary trenching for a new road scheme which was later amended to avoid the site. Although the surviving remains are not so extensive as at Caerleon, the building is larger. Excavation was started in 1960 on half of the area and the stone structure was found to have a timber predecessor.[2]

---

[1] *Arch.*, 78 (1928), p. 111–218.

[2] The discovery of this amphitheatre is recorded in *Chester Arch. Soc. J.*, 29 (1932), p. 5; see also *J.R.S.*, 51 (1961), p. 165 and Fig. 14.

Of the other military amphitheatres, that receiving the most detailed study is Carnuntum,[1] rather larger than Caerleon and thought to accommodate about 8,000 people. Built against one of the main entrances was found a shrine to Nemesis, the goddess of fate, which seems appropriate to the arena.[2]

Drainage and Water Supply

By no means least in the essential structures of the fortresses were the drains and water-pipes. Here the Romans, like the Russians of today, believed in letting water flow freely and scorned the use of taps and plugs. For drinking water, the engineers sought out a good clear source which they could lead into the fortress by gravity. Doubtless a piped supply would go into the principal buildings, especially the bath-houses, but for the rest, there would be a fountain somewhere near the centre from which supplies could be drawn by buckets. From here the water flowed away in drains and helped to flush latrines. Rainwater poured off the roofs into stone channels along the sides of the streets and added a cleansing force. The site of the spring used at Chester produced an altar charmingly dedicated to the Nymphs and fountains.[3] The water was fed by gravity into a main of lead piping which has been found lying along the edge of the main street and there are two lengths which bear a cast inscription giving the name of Agricola and dated to A.D. 79, when he was governor of Britain.[4]

The Canabae and Civil Settlements

Camps and forts acted like magnets on the civil population, drawing traders and purveyors of the modest facilities which soldiers need the world over and throughout all time. These are mainly food in greater quantity and quality than the Army normally supplies, drink, women and entertainments and the conviviality which goes with it all. Even on the torrid wind-swept plateau of Masàda some thirty little *tabernae* have been noted scattered on the bleak hill-

---

[1] W Kubitschek and S. Frankfurter, 'Führer durch Carnuntum', *J. Öest.*, 1923, p. 127, and L. Klima and H. Vetters, 'Das Lageramphitheater von Carnuntum', *R.L.Ö.*, 20 (1953).

[2] A similar altar dedicated by a centurion *ex visu*, i.e. 'as a result of a vision', was found in the Chester amphitheatre in 1966.

[3] *R.I.B.*, 460. Water deities and cults were very important in the Celtic world (*Arch J.*, 122 for 1965 (1966), pp. 1–12).

[4] Wright and Richmond, 1955, No. 199. A lead water-main has been found at York, 6 ft long and with a 4½-in bore, set in mass concrete (*Eburacum*, 1962, p. 38).

side near Camp F.[1] Traders followed the Army on campaigns, often at great peril,[2] and occasionally got in the way, as at Vetera in A.D. 69, when the civil buildings had to be destroyed to prevent the enemy from using them as cover in an attack.[3] The settlement round a legionary fortress was known as the *canabae,* a word which means simply 'the booths'.[4] During a campaign the commander would regulate the placing of the huts so as not to interfere with the camp defences. Once the fort was permanent, the traders held five-year leases ending at each *lustrum* and controlled by the *primus pilus.*[5] In some cases settlements outside the fortresses grew into rich communities and while the *canabae* remained under legionary control, these other communities often obtained self-government through a charter giving them the status of a *colonia.*[6] York is the only example in Britain, but only fragments of what must have been large and imposing buildings are known. It probably reached its peak of splendour when Severus held his court here in the early third century. At Caerleon there is the best opportunity for future investigation, since the land between the fortress and the original course of the Usk is clear of modern buildings and a start was made in 1954, but there is nothing so far to suggest a very large or wealthy community; the same applies to Chester where remains outside the fortress have been singularly scanty.

Along the Rhine and Danube there were extensive *canabae* and they must have had a considerable civilizing influence on these barbarian areas, and their markets would have brought a steady stream of trade across the great rivers. A good example is at Petronell, near Vienna, the *municipium* of Aelium Carnuntum, the area of which has been estimated at two by one and a half kilometres, and much of it excavated.[7]

---

[1] *J.R.S.*, 52 (1962), p. 151, these modest structures are clearly seen in the centre of Pl. xviii.

[2] As in Caesar's Gallic Wars. when the Germans attacked his camp so unexpectedly that the traders who had pitched their tents at the foot of the rampart had no time to escape (*Bell. G.*, vi, 37).

[3] *Hist.*, iv, 22.

[4] *Thesaurus Linguae Latinae*; for epigraphic evidence, for the use of the word and that applied to the inhabitants—*canabenses*—see P. Salway, *The Frontier People of Roman Britain*, 1965, p. 10, fn. 1.

[5] *I.L.S.*, 4222, 9103, and 9104; L. Barkóczi, 'Brigetio', *Dissertationes Pannonicae*, No. 22 (1951) and J. Szilágyi, *Acta. Arch. Acad. Scient. Hungaricae*, 1 (1951), p. 196.

[6] The distinction between the *canabae* and *colonia* is discussed by Sir Ian Richmond in *Eburacum*, xxxiv– xxxix.

[7] For a summary, see *Carnuntum, Römische Forschungen in Niederösterreich*, Band 1, 1964, pp. 83–181. Several of the town houses have been permanently exposed (Taf. xl and xliv).

Auxiliary Forts

The size of the fort and provision of specialist buildings depended entirely on the type of unit for which it was built and to some extent to its period and situation in the frontier area. Comparison between the sizes of forts in Wales with those on Hadrian's Wall shows that the latter are slightly smaller. This may have been due to different arrangements for provisioning the garrisons. The isolated situation of the Welsh forts spread in a network over difficult country would necessitate considerable reserves of food and fodder in case the units were cut off from one another in the winter. Hadrian's Wall, on the other hand, presents a different problem, since here the forts are close set along a permanent frontier line and they could all be supplied without difficulty from rearward depots and store bases. The forts of the Antonine Wall are smaller still, but here it is clear that the Romans were in difficulties over man-power and had to divide the auxiliary regiments and legions into small units to be able to spread them out effectively.

It is rash to make any general conclusions on sizes, since, when one gets down to detail, the situation is seen to be even more complicated. In planning the fort consideration had to be given to the type of unit. Cavalry, for example, would require more space than infantry, for not only would stables be needed but space for fodder considerably in excess of that for the men's food. The streets would also have to be wider to allow for the assembling and movement of the mounted men. It is unfortunate that in Britain we know so little about the internal arrangements of these different kinds of units, as so few forts have been completely excavated. Evidence of the type of unit from inscriptions is also rare and sometimes difficult to date, for many forts had a long life and different regiments would come and go according to circumstances. Consider, for example, the fort at South Shields, where there is evidence of the presence of the *Ala Sabiniana, Ala I Asturum, Cohors V Gallorum* and *Numerus Barcariorum Tigrisiensium,* probably in that order. The task of associating the structural remains with particular units is no easy one. More success is usually gained where the fort had a short life and there were no changes.

In Wales it is known that Brecon was the fort of the *Ala Hispanorum Vettonum,* a Spanish cavalry regiment, 500 strong, and its size is 7.8 acres. Forden Gaer and Caersws with 7.58 and 7.7 acres respectively must have had similar units. There is a group of 5-acre forts like Coelbren, Castel Collen I, Caernarvon I and Caerhun, and these may represent the mixed cavalry and infantry (*cohors equitata*). There follows a number of 4-acre forts varying from Caer Gai, 4.2 acres, to Tomen-y-Mur II, 3.3 acres, clearly for infantry regiments, and this is confirmed by an inscription from Caer Gai of the First Cohort of Nervians. On Hadrian's Wall the Second *Ala* of Asturians occupied Chesters in a 5.75-acre fort, but Stanwix is 9.32 acres and was held by a milliary cavalry regiment, the *Ala Petriana*. Housesteads, 5 acres, had the First Cohort of Tungrians, a milliary infantry unit[1] occupying less space than 500 cavalry at Chesters. The smaller wall forts are 3.5 or 3 acres and were clearly designed for infantry regiments 500 strong, except Birdoswald at 5.3 acres, the third-century garrison of which was the milliary First Aelian Cohort of Dacians.

The actual size of forts may not be very significant; more important is the identification of the internal buildings, but unfortunately all too few forts have been completely excavated. The following are the only auxiliary fort plans in Britain which can help greatly towards an understanding of the needs of different units:

1. *Gellygaer* (Fig. 43),[2] which although an early investigation, was excavated with a care and understanding much in advance of its day. No inscription has survived to give evidence of the name of its unit, but a study of the buildings has suggested that it was probably buit for a *cohors peditata*. It is 3.67 acres in area and there are six barrack blocks, the men's sections being almost 100 Roman ft long, suggesting ten *contubernia* with eight men each. Some of the other long, narrow buildings could be interpreted as stores.
2. *Housesteads* (Fig. 44),[3] perhaps the best-known fort, as the remains have been kept open, is now under the guardianship of the Ministry of Works, but the 1898 excavation presents problems of chronology which have yet to be resolved. There are at least ten recognizable barrack blocks which would accord with the ten centuries of a *cohors milliaria*.
3. *Caerhun*[4] (Fig. 45) was completely excavated except for the north-east

[1] In the early third century irregular Germanic units, a *Cuneus Frisiorum* and a *Numerus Hnaudifridi*, could hardly have had permanent quarters and may have been there to take part in a campaign.

[2] J. Ward, *The Roman Fort at Gellygaer*, 1903; *The Roman Frontier in Wales*, 1969, pp. 88–91, for a modern summary.

[3] *A.A.*, 2nd ser., 25 (1904), pp. 193–300.

[4] P. K. Baillie Reynolds, *Arch. Camb.*, 82–91 (1927–36); these reports were subsequently bound together, *Excavations on the Site of the Roman Fort of Kanovium*, Cardiff, 1938; *The Roman Frontier in Wales*, 1969, pp.56–59, for a modern summary.

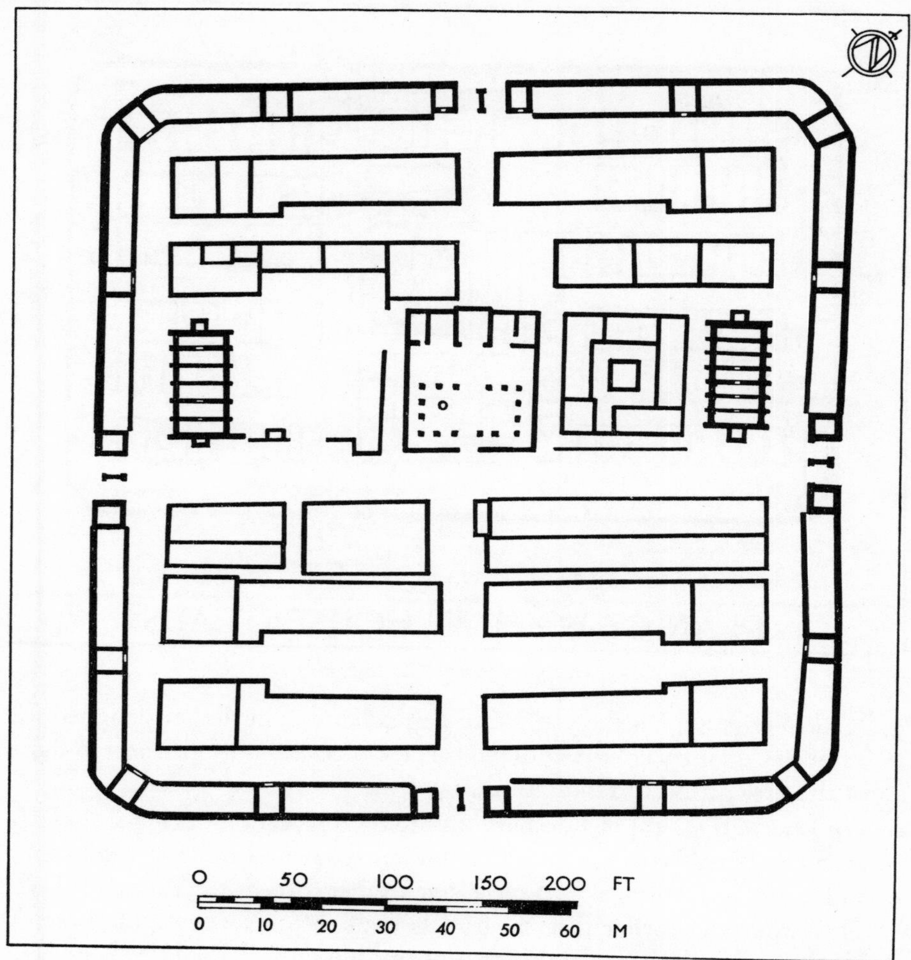

FIG. 43 *Gellygaer*

corner which is occupied by the church. It is 4.86 acres and there are at least five barrack blocks and four buildings in the north-east corner which cannot be precisely identified. This does not give enough room for an *ala*, but rather too much for a *cohors equitata*.

4. *Birrens* (Fig. 46) in Dumfriesshire,[1] in its Antonine form, appears to be for a *cohors milliaria equitata*, but the identification of buildings is not precise.

[1] *P. Soc. Ant. Scot.*, 30 (1896), pp. 81–199; for 72 1937–8, pp. 275–347.

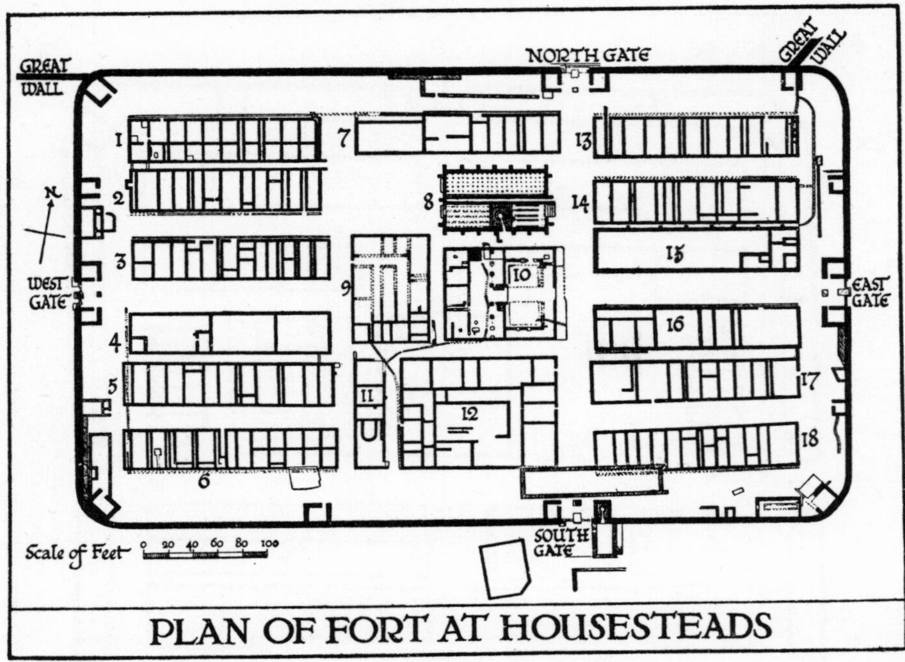

FIG. 44 *Plan of Housesteads*

5. *Fendoch* (Fig. 47),[1] is the plan of an Agricolan fort in timber designed for a milliary infantry cohort with ten barrack blocks almost identical to those of the first period of Housesteads,[2] and there is also a hospital.
6. *Pen Llystyn* (Fig. 48) in Caernarvonshire.[3] Observation during soil clearance combined with some excavation has recovered most of the history and some of the plan of two successive timber forts of the Flavian period. The larger and earlier fort has no less than eleven recognizable barrack blocks, and in the complete absence of anything which could be associated with horses, the unit was most probably infantry, and the likely explanation would appear to be that it housed two cohorts, 500 strong.

There are other examples of forts of which there are almost complete plans like Chesters,[4] but most of them provide merely tantalizing glimpses. This is

[1] *P. Soc. Ant. Scot.*, 73 for 1938–9, pp. 110–54.

[2] *A.A.*, 4th ser., 38 (1960), p. 65.

[3] *Arch. J.*, 125 for 1968 (1969), pp. 102–192.

[4] *Handbook to the Roman Wall*, 12th ed., 1966, p. 85; this fort was occupied by the Second *Ala* of Asturians.

Plate XXII

(a) The cavalry parade helmet from Ribchester (p. 154). (Photograph, British Museum)

(b) Parade armour for a horse from Straubing, Bavaria (p. 154).

Plate XXIII
Cast of part of a cavalry tombstone at Mainz.

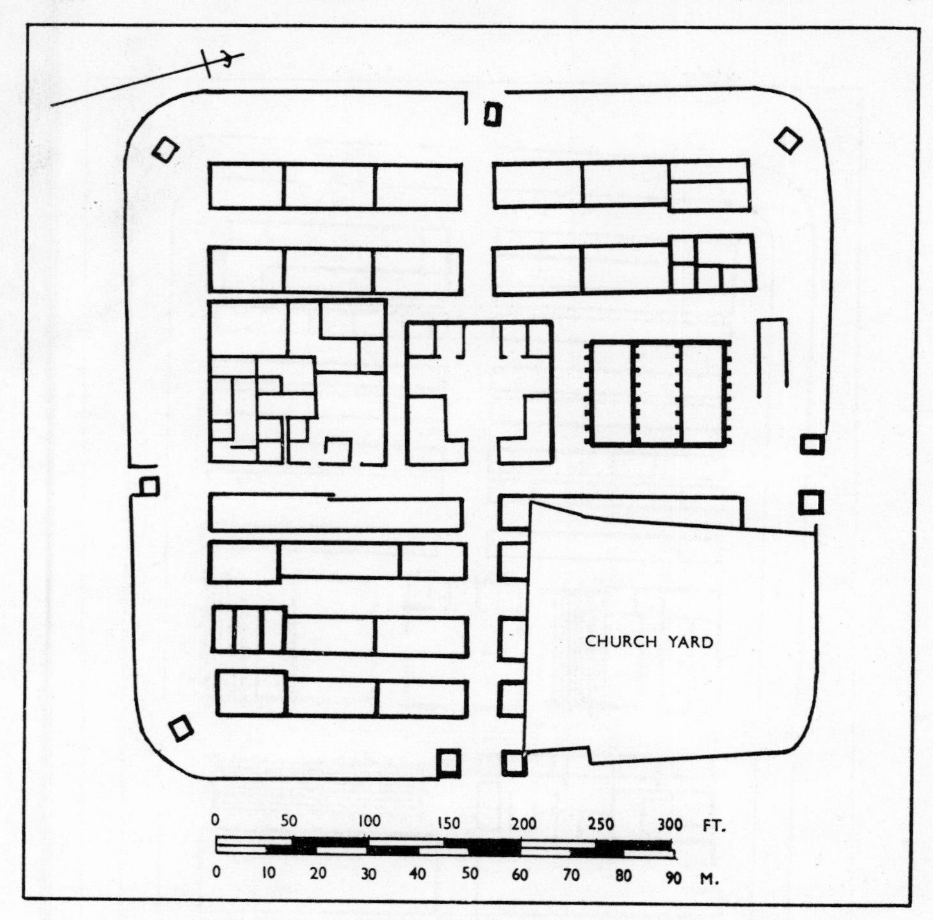

FIG. 45 *Caerhun*

true also of the forts on the Rhine and Danube,[1] although more recent work on these frontiers is making good this deficiency. An exception is Valkenburg

[1] Most of the work in Germany and Austria was done in the last century, before methods of identifying timber buildings were understood. Even the detailed work at Saalburg with its scheme of reconstruction under the patronage of Kaiser Wilhelm revealed only the plans of the *principia* and fragments of a few other buildings (L. Jacobi, *Das Römerkastell Saalburg*, 1897). Among the more revealing are Kastell Wiesbaden (*O.R.L.*, Nr. 31) and Niederbieber (*B.J.*, 120, Taf. xvi); occasionally reconstruction is suggested as at Urspring (*Germania Romana*, 1922, Taf. 6, Nr. 1).

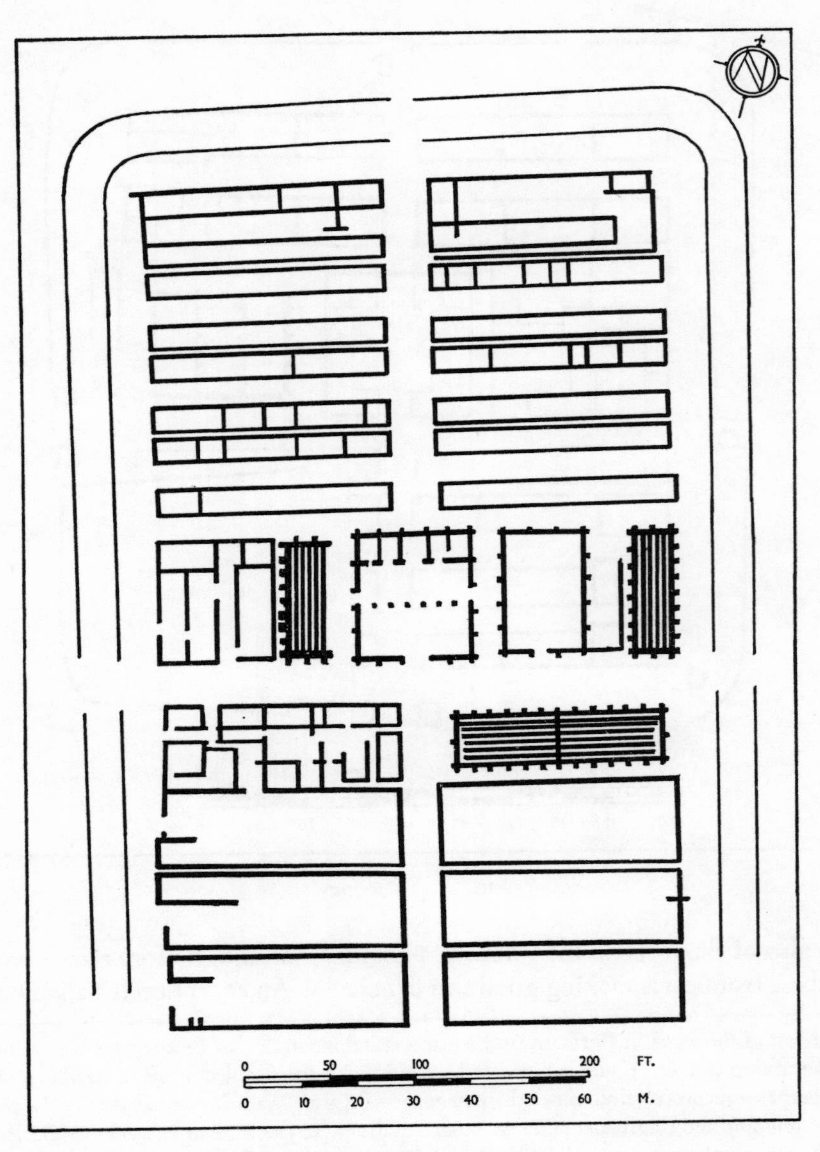

FIG. 46 *Plan of Birrens*

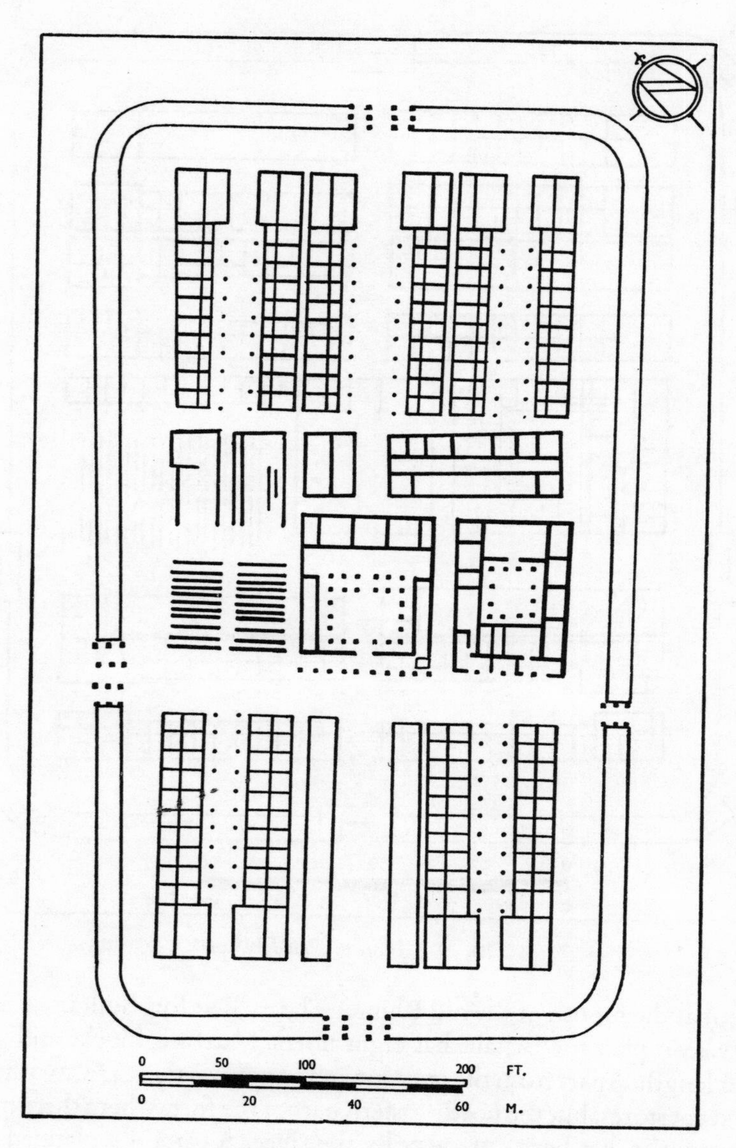

FIG. 47 *Plan of Fendoch*

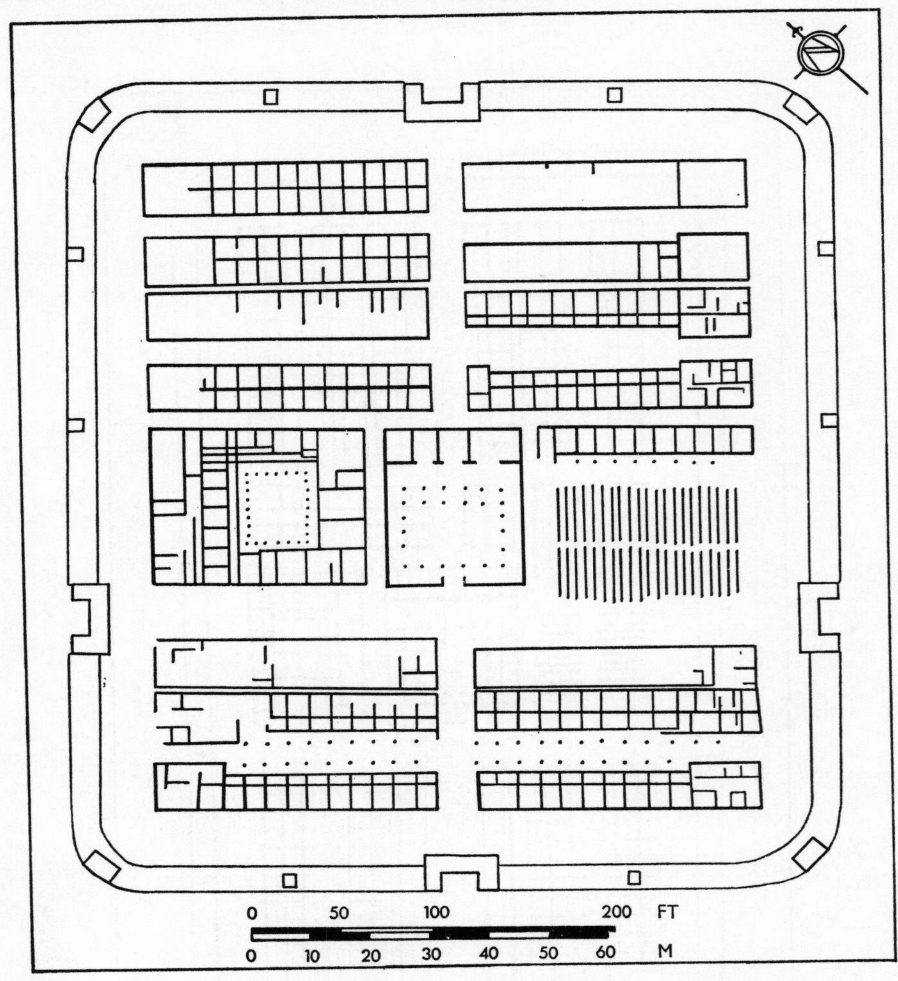

FIG. 48 *Plan of Pen Llystyn*

(Fig. 49) at the mouth of the old Rhine.[1] The earliest fort built in 42 as part of the invasion plan for Britain has eight normal barrack blocks and a pair of greater length. Apart from the *principia* and *praetorium* there are no other buildings except stores, but the north-eastern part of the fort, where there may have been granaries, has been cut away by the river. A carefully planned series of

[1] W. Glasbergen, 'Het eerste jaartal in de geschiedenis van West-Nederland', *Jaarboek der Koninklijke Nederlandse Akademie van Wetenschappen* (1965–6), pp. 1 ff.

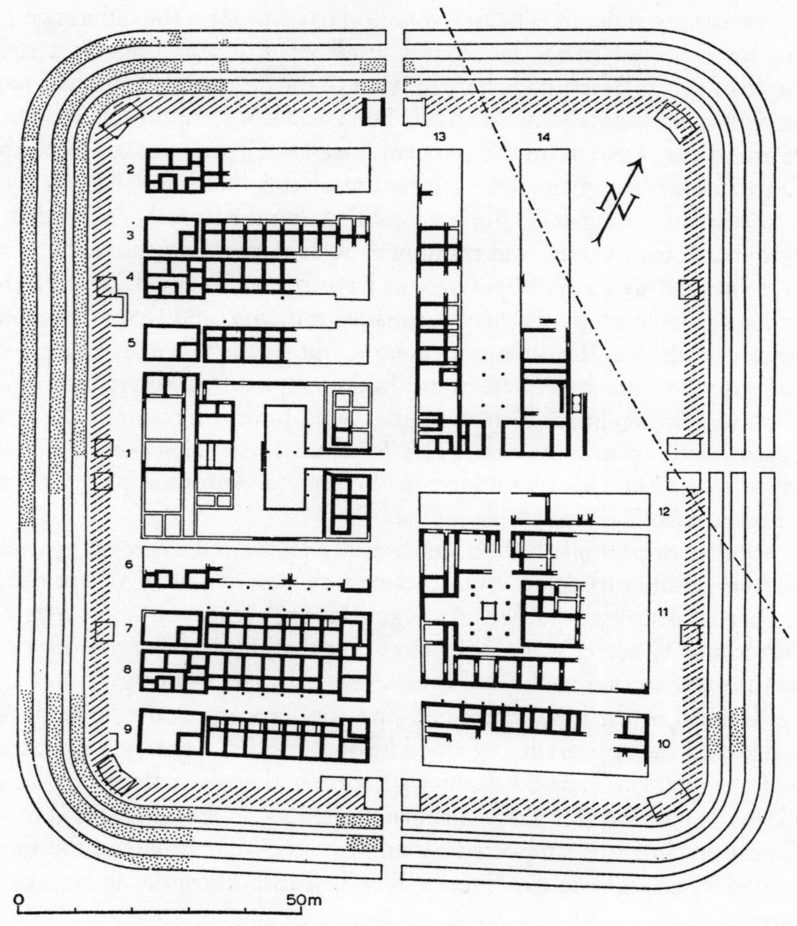

FIG. 49 *Plan of Valkenburg I*

excavations is badly needed to give precise information about the internal planning of forts of different kinds, but the total excavation which this would involve appears to be beyond the present resources of Britain.

The Buildings in the Auxiliary Forts

Principia

While embodying all the main features of the great legionary headquarters, the *principia* in the auxiliary forts is scaled down considerably and there are many variations in detail. The first question to consider is the difference which might be expected from the several kinds of unit and there is a striking example of this at Brecon, for here in front of the *principia* is a forehall 147 feet by 40 feet straddling the main street. Such additions are rare in Britain, but there are twenty known on the German *Limes*. An inscription from Netherby in Cumberland has given the Latin name of this building—*basilica equestris exercitatoria*[1] or riding-hall. Such a building would have been needed by a mounted unit for exercise and training as well as assembly, since the normal-sized courtyard in the *principia* would have been too small. The Netherby building was erected by *I Aelia Hispanorum milliaria,* and that at Brecon was associated with *Ala Hispanorum Vettonum,* but in the German examples it is not always possible to equate these buildings with mounted units. Other examples in Britain are at Halton on the Wall, built as a result of the Severan reconstruction[2] to accommodate *Ala I Pannoniorum Sabiniana*, and at Newstead, where a forehall was added in the second Antonine phase when the fort was reorganized for an *ala milliaria.*[3]

A number of *principia* have been carefully excavated in recent years and it should be possible to detect examples of a change in policy where the plans have been drastically modified. At Segontium (Caernarvon) in North Wales, Sir Mortimer Wheeler was able in 1921–3 to distinguish four periods ranging from the early second to the late fourth century. The most significant change is the division of the great cross-hall into three rooms in the Severan reconstruction and the appearance of an additional room heated by a hypocaust. It is as if the spacious martial dignity of former days is replaced by an office block made comfortable for chairborne bureaucrats. Perhaps this is too bold a suggestion, but it is supported by an inscribed altar found in the building dedicated by an *actuarius* to Minerva, who had administrators and clerks under her special care.

The situation on the Wall does not appear to be quite the same. When it has been possible to study the many changes reflected by the vicissitudes of this frontier, the introduction of amenities like heated rooms is evident, and at Chesterholm one was actually converted into living quarters at the end of the fourth century.[4] But the massive dignity of the cross-hall and its attendant

---

[1] *C.I.L.*, vii, 965., *R.I.B.*, 978.

[2] *A.A.*, 4th ser., 14 (1937), p. 168.

[3] *Newstead*, Fig. 2; *P. Soc. Ant. Scot.,* 84 (1949–50), p. 24.

[4] *A.A.*, 4th ser., 13 (1936), p. 225 and Fig. 2.

shrine with its stone screens are maintained as they appear to be at the other Wall forts and those beyond.[1] Perhaps there is something significant in the situation in Wales where military command may have been more closely linked with the civil administration of a district, but our knowledge is at present too inadequate to form conclusive judgement.

Praetorium

When the commander's house is reduced to its basic essentials, these emerge with a greater clarity than is possible in the larger legionary buildings. One of the best examples for study is the interpretation of the timber structure at Fendoch (Fig. 50). The house was arranged round a central open court with

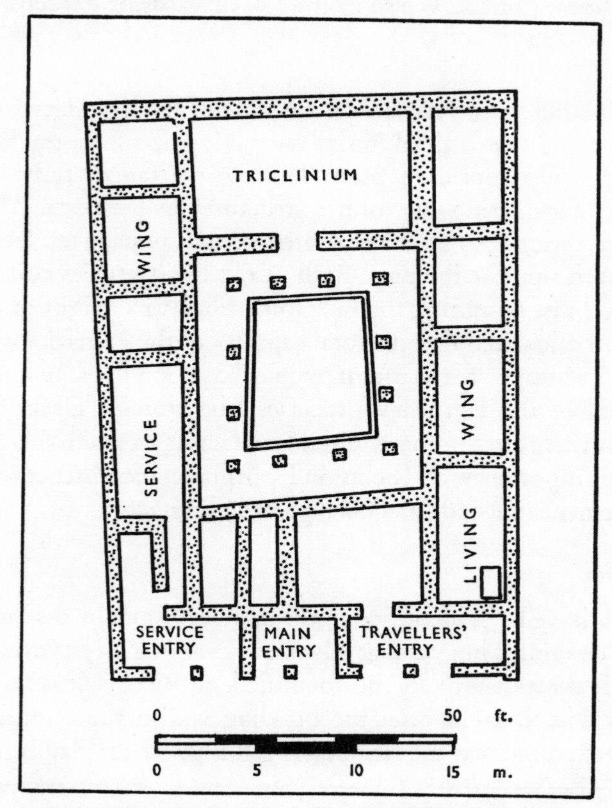

FIG. 50 *Plan of the* praetorium *at Fendoch*

[1] Cf. Bremenium, *Hist. of Northumberland*, 15 (1940), Fig. 18.

a veranda on all sides. The dining-room (*triclinium*), being the principal room, was placed at the rear. On one side of the court was the private suite which includes a bathroom, and on the other side the servants' wing with the kitchen at the end near the dining-room. The front was occupied by the three entrances. One was for the servants; a central one was for distinguished visitors and led by a passage with its attendant cloakroom into the court and so across to the dining-room, and the third one was for minor officials and had an external waiting-room with a reception hall beyond. The whole was planned with neat economy, yet the court gives a certain spaciousness which added to the dignity of the commander's residence. A study of the plans of other *praetoria* shows that these basic features appear in most of them, but there are additions, in the Welsh examples, of yards or gardens.[1]

Horrea

Auxiliary granaries were very similar to those of the legionary fortresses, but only slightly more than a third in size (averaging about ninety feet by thirty feet), and normally appear in pairs. The most detailed study has been by Richmond in considering his timber structures at Fendoch. The width of thirty feet was thought to be divided into a central passage ten feet wide, with ten feet on each side for the bins. Each of the ten centuries could have been allocated two bins. Assuming the bins to be filled to a height of five feet, Sir Ian Richmond calculated that the total capacity of the granaries was 370 cubic yards. A year's supply for the unit would have required at least 400 cubic yards, so, unlike the legionary granaries, the storage space seems barely adequate, but there are too many unknown factors to make this a valid comparison. The importance of Richmond's brilliant reconstruction lies in the structural elements needed for this type of building.

Valetudinarium

Few hospitals have been identified in the auxiliary forts in Britain, only three seem to be certain. These are at Housesteads,[2] Pen Llystyn and Fendoch, and a fourth may eventually be identified at Birrens in the complex of Buildings ix and x. In all cases the building was in the central range. The Housesteads example was the largest (*c.* 60 ft x 90 ft). Its plan is similar to the

---

[1] Nash-Williams, *The Roman Frontier in Wales*, 1954, Fig. 54.

[2] Building No. 9, Fig. 44; in the original excavation report (*A.A.*, 2nd ser., 25 (1904)), this building is shown, but not identified.

legionary type with an internal corridor circulating on three sides round a central open space. The one at Fendoch was a simple rectangular building (40 ft x 106 ft) with a central corridor. On one side there were eight rooms and on the other one large ward with two small rooms at each end. At Pen Llystyn it was a simple range of ten rooms opening on to a veranda with another room at one end extending over the veranda space (16 ft by 108 ft). There is little to distinguish it from a barrack block except for part of the length there was a screen wall three feet beyond the colonnade. Another fort where a building has been tentatively identified as a hospital is Benwell[1] on Hadrian's Wall.

Barrack Blocks

Auxiliary barrack blocks display the same characteristics as those of the legionaries, but are only about five-eighths of their length. Unfortunately in the older excavation reports it is unusual for auxiliary blocks to show internal partitions. This is because the traces of timberwork were not recognized. Knowledge of the number of *contubernia* in each block might be helpful in understanding how some of the units were organized. Where the internal partitions had stone sills, the situation has often been confused by long occupation involving many alterations. Such is the case at Housesteads, but fortunately recent work by Dr John Wilkes has clarified the arrangements in one of the blocks.[2] In the early phases there were ten compartments (11 ft x 25 ft) for the eighty men of the century, very similar in arrangement and size to the blocks at Fendoch. The centurion's quarters of two modest rooms is the most striking difference from that provided for the legionary equivalent.

The Bath-house

It is normal in auxiliary forts to find the bath-house in the annex, where this extension is provided. To facilitate drainage the most usual place was on a downward slope away from the fort towards a river or stream. There are exceptions to this, as at Mumrills[3] and Balmuildy,[4] at each of which a small

---

[1] *A.A.*, 4th ser., 19 (1941), pp. 22–23.

[2] No. xiv, *A.A.*, 4th ser., 38 (1960), pp. 61–71; 39 (1961), pp. 279–99. The *contubernia* could have been subdivided with timber partitions to separate the sleeping place from that for storage of arms and equipment.

[3] *P. Soc. Ant. Scot.*, 63 for 1928–9 (1929), pp. 449–62.

[4] S. N. Miller, *The Roman Fort at Balmuildy*, 1922, p. 43; where larger bath-houses appear inside the fort, as at Brecon, it probably indicates a reduction of the fort, so that in effect the part occupied by the bath-house may be regarded as an annex.

bath-house was found behind the rampart. Since these buildings were constructed of tile and stone, their remains have tended to survive and many plans have been recovered by investigations. Basically all auxiliary bath-houses have some provision of rooms of different heat and humidity, although obviously they are smaller and less ornate than those found in legionary fortresses and large towns. It is possible to recognize certain features as indications of chronology, as in the case of the circular *laconicum*, a pattern which appears to have gone out of use by the end of the first century. A splendid example of this is in the Flavian bath-house at Red House, Corbridge (Fig. 51), associated with the *Ala Petriana*.[1] The compact bath-houses on Hadrian's frontier form a recognizable group.[2]

The Annex

The annex, a feature of many auxiliary forts, is a fortified enclosure attached to one of the sides of the fort, usually that towards the stream. The defences are never as strong as those of the fort itself and apart from the bath-house the space appears to be vacant. This feature has never been subject to thorough exploration and presents a number of problems. Its function can be explained in a variety of ways and one or all may be found to be correct.

There must be cases where the fortified area thought to have been an annex represents, in fact, the remains of an earlier and larger fort projecting beyond the limits of the reduced area. It is more likely that the annex proper was a camp to house the unit which constructed the permanent fort. Once such an enclosure was in existence there would have been several uses for it. A unit on the move, for example, could camp in it, as it is unlikely that there would have been any room for them inside the fort if the garrison was at full strength. Here, as indicated above, the men could have had use of the bath-house. The annex could also have been used as a wagon park and for stores.

There is a small annex at Caerhun which appears to be associated with metal-working. The Gellygaer annex is two acres and is surrounded by a stone wall; it has a large and elaborate bath-house[3] and other buildings so spaced as to leave little room for further features.

---

[1] By C. M. Daniels, *A.A.*, 4th ser., 37 (1959), pp. 85–176.
[2] As at Chesters, Benwell, Carrawburgh, Netherby and Bewcastle.
[3] John Ward, *Cardiff Naturalists' Soc. T.*, 42 (1910), pp. 1–45.

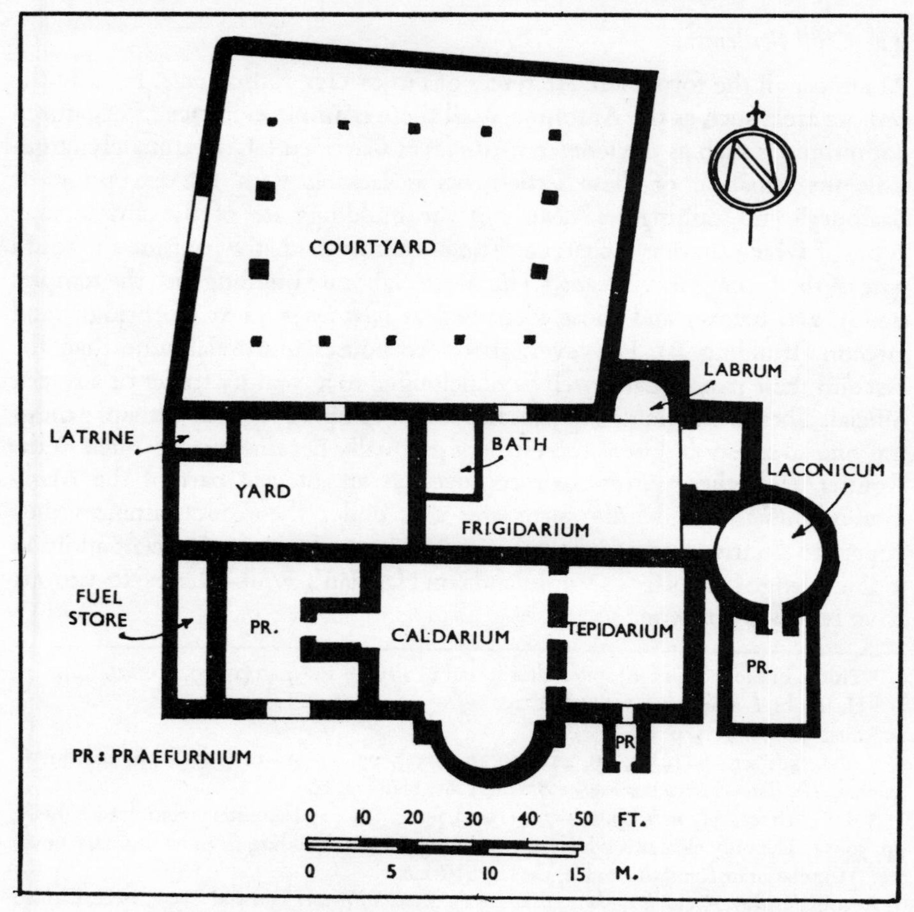

FIG. 51 *Plan of the bath-house at Red House, Corbridge*

Beyond the fort and its annex one might expect to find additional features. Every unit would have had a parade ground, the only trace of which would be a levelled surface and, all too rarely, the altars carefully buried to avoid desecration (see p. 268). There would also have been a small cemetery which at least produced tombstones giving information about the men and their unit. At Tomen-y-Mur on a high mountain pass in Wales, the remoteness of which has helped in preservation, there seems to be a row of actual tombs. At this site there is also a small amphitheatre which might perhaps be considered as a cock-pit.

The Civil Settlement

At almost all the forts there are traces of *vici* or civil settlements. Even in the remote areas such as the Antonine Wall there is ample evidence of organized communities such as the *vicani consistentes* at Carriden.[1] Unfortunately large-scale investigation of these settlements is lacking with the exception of Saalburg[2] and Zugmantel.[3] Most of the buildings are of the simple strip type, of which there is a characteristic cluster immediately outside the south gate of the fort at Housesteads.[4] The more elaborate buildings are the temples (see p. 270 below) and those identified as *mansiones*, as at Corbridge[5] and Brecon (Building B). However, there are houses more elaborate than the rest and these may equally well have belonged to a wealthy trader or a minor official. These settlements began as modest developments with perhaps grudging official sanction, but as recruitment gradually became concentrated in the frontier areas these *vici* were recognized as an integral part of the Army establishment. It is hardly surprising that under these circumstances they appear to flourish from the early second century onwards. A recent addition to knowledge comes from Vindolanda on Hadrian's Wall, where excavations have revealed a *mansio*.[6]

---

[1] This is in the form of a Jupiter altar found in 1956 (*J.R.S.*, 47 (1957), pp. 229–30).

[2] H. Jacobi, *Die Saalburg*, 12th ed., 1929.

[3] *Saal.-J.*, 10 (1951), p. 50.

[4] *A.A.*, 4th ser., 9 (1932), pp. 225–32; 10 (1933), pp. 85–91; 12 (1935), pp. 204–59; P. Salway, *The Frontier People of Roman Britain*, 1965, Plan Fig. 10.

[5] *A.A.*, 3rd ser., 3 (1907), pp. 174–7; 4 (1908), pp. 215–40, and a reinterpretation by Salway, pp. 50–54. The only elements which might distinguish this building from an ordinary house are (1) the elaborate fountain, and (2) the large latrine.

[6] Robin Birley, *A Guide to the Remains of the Roman Frontier Fort and Town*, 1972, Cameo Books. *A.A.*, 4th ser., 48 (1970), pp. 116–123; *Britannia*, 3 (1972), 306.

Chapter 5

THE ARMY IN THE FIELD

TACTICS

Information about tactics can be gleaned from accounts of battles, but the military manuals known to have existed and to have been used extensively by commanders, have not survived. Perhaps our greatest loss is the book of Sextus Julius Frontinus[1] but parts of his work were incorporated in the compilation of Vegetius, who readily admitted that his study was merely a collation of ancient authorities.[2] The difficulty with Vegetius is that from his viewpoint at the end of the fourth century 'the ancients' covered a wide field of history from the Greek historians to those of early Imperial times, and he rarely makes any effort to identify his source. The sections dealing with battle tactics (iii, 11–22) seem to be coherent enough to have been taken entirely from Frontinus and it is difficult to imagine any other source.

The main points brought out by Vegetius/Frontinus are as follows:

Before any engagement with the enemy it is necessary for the troops to be rested and adequately fed so that they are ready for the rigours of the day. The men must also be of good heart and instilled with a rage against the foe by exhortation. The importance of the choice of ground is underlined. There is an advantage of height over the enemy and if you are pitting infantry against cavalry, the rougher the ground the better. The sun must be behind you and dazzle the enemy and if there is strong wind and dust it should blow away from you and give advantage to your missiles and blind the enemy with dust. Then the battle line is considered and the space occupied by the men, each of whom requires a lateral space of three feet, while the distance between ranks is given as six feet. Thus 10,000 men can be placed in a rectangle about

---

[1] Governor of Britain A.D. 74–78; it is ironic that of his great work *The Art of War,* only the *Strategemata,* an appendix, has survived. In his introduction to Book I he makes the claim that his was the only systematic study.

[2] He lists Cato the Elder, Cornelius Celsus, Paternus, Frontinus, the ordinances of Augustus, Trajan and Hadrian, but curiously omits any reference to Caesar (ii, 3).

1,500 yards by twelve yards, and it is advisable not to extend the line much beyond this. The disposition of particular units depends on the circumstances and above all on the enemy's formation. The normal arrangement was to place the infantry in the centre and the cavalry on the wings. The function of the latter was to prevent the centre from being outflanked and once the battle turned and the enemy started to retreat the cavalry moved forward and cut them down. Thus the horsemen were always a secondary force in ancient warfare, the main fighting being done by the infantry. It is recommended that if your cavalry is weak it should be stiffened with lightly armed foot soldiers.[1] A notable exception to this use of cavalry was made by Pompey at Pharsalus, where his superiority in this arm caused him to use his horsemen as shock troops against Caesar's right wing, but with unforeseen results.

In the next section of Vegetius the need for adequate reserves is stressed. Then the positions to be taken up by the commander, normally on the right wing,[2] and his subordinates in the battle line are discussed. Then the tactics known as the wedge and the saw are described. The wedge was commonly used by attacking legionaries; this enabled small groups to be thrust well into the enemy and, when these formations expanded, the enemy troops were pushed into restricted positions, making hand-to-hand fighting difficult. This is where the short legionary *gladius* was useful when held low and used as a thrusting weapon, while the long Celtic sword became impossible to wield. The opposing tactic was to form indentations in the line to accommodate the wedges and prevent penetration. The saw appears to have been a detached

[1] This tactic seems to emanate from Germany and is described by Caesar (*Bell. G.* i, 48) in his engagement against Ariovistus. Six thousand horsemen were accompanied by as many infantry, one for each horseman for his personal protection and working together in the encounters. The foot-soldiers would protect horsemen who had been unseated, and the horsemen would protect footmen under pressure. The function of the men on foot was presumably to hamstring the enemy horses and dispatch their riders who could not cope so easily on the ground. The footmen were lightly armed and chosen for their speed. They helped themselves along by clutching the manes of the horses. Caesar himself subsequently used similar units and they played a significant part in his victory at Pharsalus. But here Caesar ordered the footmen to thrust their javelins upwards into the faces of the horsemen who, being young and inexperienced, drew away in disorder enabling Caesar's men to outflank Pompey's centre. Subsequently the *cohors equitata* became a regular unit of the *auxilia*.

[2] As the soldiers were trained to fight with weapons in the right hand, the left, or shield side, was regarded as the defensive side, thus the attack was normally mounted by the right wing.

unit immediately behind the front line capable of lateral movement to block any holes which might appear or to develop a thrust where there might be a sign of weakness. Seven specific battle tactics are then studied in detail:

1. On level ground the force is drawn up with a centre, two wings and reserves in the rear. The wings and reserves must be strong enough to prevent any enveloping or outflanking manoeuvre.
2. An oblique battle line with the left wing held back in a defensive position, while the right advances to attempt to turn the opponent's left flank. Opposition to this move is to strengthen the left wing with cavalry and reserves, but if both sides are successful the battle front would tend to move in an anti-clockwise direction, the effect of which would vary with the nature of the ground. With this in mind it is as well to attempt to stabilize the left wing with the protection of rough or impenetrable ground, while the right wing should have unimpeded movement.
3. The same as No. 2 except that the left wing is now made the stronger and attempts a turning movement and is to be tried only when it is known that the enemy's right wing is weak.
4. Here both wings are advanced together, leaving the centre behind. This may take the enemy by surprise and leave his centre exposed and demoralized. If, however, the wings are held, it could be a very hazardous manoeuvre, since your army is now split into three separate formations and a skilful enemy could turn this to advantage.
5. The same tactic as No. 4, but the centre is screened by light infantry or archers who can keep the enemy centre distracted while the wings engage.
6. This is a variation of No. 2 whereby the centre and left wing are kept back while the right wing attempts a turning movement. If it is successful, the left wing, reinforced with reserves, could advance and hope to complete the enveloping movement which should compress the centre.
7. This is the use of suitable ground on either flank to protect it, as suggested in No. 2.

All these tactics have the same purpose, that of breaking the enemy battle line. If a flank can be turned, then the strong centre has to fight on two fronts or is forced to fight in a restricted space. Once an advantage like this has been gained it is very difficult to correct the situation. Even in the highly trained Roman Army it would have been difficult to change the tactics during the course of the battle and the only units which can be successfully deployed are those in the reserves or in that part of the line not yet engaged. Thus the most important decisions a general had to make concerned the disposition of the troops. If a weakness could be detected in the enemy line, it was exploited by using a stronger force to oppose it. Likewise, it was necessary to disguise one's battle line—even troops were disguised to delude the enemy. Often the very size of the army was skilfully hidden, troops packing tightly together to make it appear small or spreading out to appear large. There were also many examples of surprise tactics made by detaching a small unit which suddenly emerged from a hidden place with much dust and noise to make the enemy believe that reinforcements had arrived. Frontinus is full of the oddest stratagems to mislead the enemy or demoralize his troops.

Once the enemy cracked, however, they were not to be surrounded, but an easy escape route left open. The reasons for this were that trapped soldiers would fight to the death and if they could get away, they would, and then were exposed to the cavalry waiting on the flanks. This important section of Vegetius closes with the tactics to be used in the case of a withdrawal in the face of the enemy. This highly difficult operation requires great skill and judgement. Both your own men and those of the enemy need to be deceived. It is suggested that your troops be informed that their retirement is to draw the enemy into a trap and the movement can be screened from the enemy with the use of cavalry moving across the front. Then the units are drawn off in a regular manner, but these tactics can only be employed if the troops have not yet been engaged. During a retreat units are detached and left behind to ambush the enemy if there is a hasty or incautious advance, and in this way the tables can often be turned.

There is another useful section on the march (III, 6–7) which is also probably taken from Frontinus. The point is made that the army is in greater danger on the march than at any other time and that extra care should be taken. Careful reconnaissance should be made and if possible a plan prepared of the country through which the army is passing. If local guides are employed they should be kept under guard and carefully watched, for 'sometimes the common sort of people imagine they know what they really do not, and through ignorance promise more than they can perform'.[1] The route must be kept secret and strong detachments should be sent out in advance to reconnoitre the way and ensure that there are no traps or ambushes.

---

[1] John Clark translation, 1767; this has been republished by the Military Service Publishing Company, Harrisburg, Pennsylvania, in 1944, but in an incomplete form.

(a)

(b)

Plate XXIV
Scenes from Trajan's Column

(a) ships in a harbour of a legionary fortress on the Danube (p. 162);

(b) showing a clavicular entrance and two auxiliary guards with oval shields (p. 171).

(a) Defensive traps (*lilia*) at Rough Castle, on the Antonine Wall (p. 176).

Plate XXV

(b) a scene from Trajan's Column showing the construction of defences (p. 177).

The order of the march is for the cavalry to go first, then the infantry, then the baggage train in the centre, followed by the best units of the force, since the force is more often attacked from the rear, than from any other direction, presumably in the hope of capturing the baggage train. This should also be protected by flank guards. The servants were carefully drilled in keeping their positions in the train and following procedure at the onset of an attack to avoid their giving way to panic and disturbing the troops. The speed of the march must be properly co-ordinated so that the force remains in a compact body and cannot be split up into sections. Distance must be carefully calculated so that, with knowledge of the difficulty of the terrain, there is adequate time available to reach the destination, and so that water is available for men and horses, especially during the summer.

There is an interesting passage on the crossing of rivers. When there is a ford, lines of cavalry are arranged across the river both above and below the ford. Those above break the force of the water and those below recover items of baggage that become detached and men who are swept away. If the river is too deep, the level is lowered by cutting trenches and allowing the water to flow into the plain. Large rivers require temporary plank bridges—these are made by driving piles into the river bed and attaching a floor of planks[1] or, as an emergency, binding together empty casks and covering them with boards. The horsemen are advised to make little rafts of reeds on which they place all their weapons and their cuirasses and then swim across with their horses, pulling the rafts with them. But the best means is by the pontoon bridge made of light boats hollowed out of logs. The army, it is said, always carries a number of these, together with planks and nails, so that a bridge can be put together very quickly. As enemy attacks are likely at river crossings where the troops are divided, it is necessary to build bridge-head defensive positions against such contingencies.

This is all very sound practical advice clearly based on long practical experience, but it is nevertheless difficult to find specific examples in the first two centuries of the Empire. Tacitus, the only reliable source, tended to shape his descriptions of battle scenes into short dramatic episodes and his cursory accounts of campaigns has led Mommsen to make the serious charge that Tacitus was the most unmilitary of historians, lacking in strategical insight and arbitrary in his selection of material.[2] His most detailed battle accounts are

[1] This method is illustrated on Trajan's Column (Taf., xcviii, 350–3).

[2] R. Syme (*Tacitus*, i, p. 157), quoting Mommsen (*Röm. Geschichte* (1885), p. 165 = *The Provinces of the Roman Empire*, i (1886), p. 181). Surprisingly little is known of the career of

those of the Civil War of A.D. 69, when Roman armies were matched against each other and one might expect sophisticated tactics skilfully applied. But this is far from the case. The situation throughout is one of muddle and confusion. The forces of Otho owed their lack of success to the division of command and the failure of their generals to co-ordinate the various units, and one suspects that many would have had divided loyalties. A Roman commander like Suetonius Paulinus would have had cause to hesitate before crushing another Roman force which he could so easily have done at the Battle of Ad Castores.[1] Vitellius had tried to trap the Othonian force into an ambush, but this was betrayed and the planned retirement of the latter had the same effect once the Vitellian Army had swept forward. The heavy right wing with its large cavalry contingent enveloped them, but Paulinus held his infantry centre back and so lost an opportunity of complete victory. Instead the two Vitellian armies were allowed to join forces. Now Otho's best strategy would have been to delay until the Danubian legions arrived and this was the best advice urged upon him by his experienced generals, but he stubbornly insisted on a pitched battle which was a ragged and confused affair in an area thick with vineyards and any use of tactics was sadly lacking on both sides. The Vitellians won because of their superiority in numbers and experience and because the Othonian Army had a spiritless and divided command whose heart was not in the cause. Exactly the same situation repeated itself later when the Flavian advance guard under Antonius Primus defied the orders to wait and consolidate and swept to victory in the second Battle of Cremona. The Vitellian general Caecina Alienus was defeated and the Flavians pressed forward more eager for booty than victory. The main Vitellian force had just arrived at Cremona after a forced march of thirty miles. Prudence should have demanded rest within the walls of the town while their enemy spent the night outside, but lack of leadership and any kind of tactical planning was replaced by a senseless battle lust which pitched the tired army forward to an all-night battle in which, as Tacitus says, 'clear heads and strong arms availed nothing, and even eyes were helpless in the dark. On both sides weapons and uniforms were the same, frequent challenges

---

Tacitus and nothing of his military experience. He must have served as a tribune and probably as a legionary commander, and became proconsul of Asia in 112–13 (*Orientis Graeci inscriptiones selectae*, 1905, 487; R. Syme, *Tacitus*, i, p. 72).

[1] The attitude of Paulinus and other leaders receives special comment from Tacitus (*Hist.*, ii, 37 and 38).

and replies disclosed the watchword and standards were inextricably confused as they were captured by this group or that and carried hither and thither.'[1] The light of the moon helped them to sort things out, but the critical stage was at dawn, when both sides must have been near exhaustion. The Third Legion from Syria turned to hail the sun in their customary way and this created rumours of reinforcements, heartening the Flavian forces and striking dismay into those of Vitellius which, still without proper leadership, 'were grouped or spread out in their battle lines according to individual impulse or panic'. These episodes stress the need for a calculating and controlling hand to plan battle tactics in advance of the clash, and to manipulate sections of the troops engaged, and those in reserve, as the battle develops, so that as enemy weaknesses are revealed they can be exploited. There are some outstanding examples of the clash of experienced troops where both sides were led by able generals. In the Battle of Pharsalus, Pompey displayed a curious lethargy and relied entirely on his larger numbers and his strong cavalry force which he placed on his left wing to carry out an outflanking movement. Caesar's position was very difficult, since his forces were so much smaller, but he knew that his legionaries were all tough and dedicated fighters. Pompey ordered his troops to stand still and wait for Caesar's men to move forward in order to make his flank attack more effective,[2] but Caesar anticipated this and had given special instructions to his infantry reserve. The success of the latter in routing the splendid cavalry decided the battle and Pompey made no attempt to rally his troops, but retired, a defeated man.

Caesar was not an orthodox, text-book commander, and his actions cannot be used to illustrate normal Roman tactics. He relied greatly on surprise and speed and very early in his military career realized the need for highly trained men of unquestioned devotion, so that, when ordered to do the apparently impossible, they had no hesitation or doubt. In his first major engagement, that against the Helvetii,[3] he sent his own and other horses out of sight to demonstrate that there would be no retirement of the commander and his staff. After a purely frontal battle which forced the Helvetii to retreat, the enemy rearguard delivered a strong flank on the Roman's right. Caesar's

---

[1] *Hist.*, iii, 22.

[2] Plutarch (*Pompey*, lxviii–lxxii) states that this was done to steady the troops, who were lacking in experience, and to stop them moving about too much. Caesar considered it as a wrong tactic, adding that the army which moves forward carries greater force and initiative (*Bell. C.*, iii, 89).

[3] *Bell. G.*, i, 24–26.

forces executed a very difficult manoeuvre at this critical juncture: his right flank wheeled to meet the new threat while the main body continued to engage the Helvetii, who were overcome only after a long struggle.

Against Ariovistus and his Germans, Caesar gained a psychological advantage by forcing a battle when it was inauspicious for the Germans,[1] otherwise this battle was quite orthodox. Caesar saw that the German left wing was the weakest, so he strengthened his right and turned the flank. In the next season he fought the large army of the Belgae on the river Aisne.[2] Here the odds were so much against him that he took unusual precautions in choosing a site and strengthening it artificially. His army was drawn up so that the rear was protected by the river and his front by a marsh, but he also dug a ditch on both flanks and constructed artillery posts. This was so effective that the Belgae refused a frontal engagement and crossed the river to try to cut the Romans off and overrun a fort on the opposite bank which was connected by a bridge to the main Roman position. Caesar was able to attack the Belgae with his cavalry and archers while they were on the move and crossing the river, and inflicted so much damage that they decided to retire. Caesar may have hoped for a frontal attack with his forces at an advantage, but the way he dealt with the changed tactics of the enemy clearly shows him as a brilliant opportunist exploiting the enemy's weakness.

The Nervii showed skill and enterprise in attacking Caesar's army at a time when it was most disorganized, after a day's march and while the camp was under construction. It is one of the most dramatic scenes from Caesar's pen and demonstrates how Roman steadiness and discipline won the day when lesser armies might have given way to panic in the face of this ferocious and unexpected attack.

Caesar's accounts may, at times, be biased, but they show a firm grasp of complicated situations and an eye for detail which put them in a class of their own. Tacitus may be a gifted historian and his reporting of events reasonably accurate, but he is always at pains to create dramatic passages suitable for public readings. His battle scenes suffer from this reduction, making tactical assessments difficult. Of all the battles which must have been fought in Britain, only two are at all informative.[3] The first is the critical battle fought

---

[1] *Bell. G.*, i, 50; a point of interest about this battle is that the Germans attacked so quickly that the Roman legionaries did not have time to discharge their *pila*.

[2] *Bell. G.*, ii, 9–10.

[3] i.e. battles in the open; the attack on the position held by Caratacus does not come into this category. The most important battle in Britain was the initial invasion at the crossing of

by Suetonius Paulinus against the hordes of Boudicca.[1] The Roman force consisted of XIV *Gemina,* part of XX *Valeria* and auxiliaries, altogether some 10,000 men, and while the total British figures are not given, 80,000 were slain. These numbers must have been exaggerated by report. With the greater part of two legions with auxiliary support, the Roman Army would have been at least 15,000 and possibly 20,000, and there may have been two or three times this number in British fighting strength. Paulinus had chosen his site with some care so that although greatly outnumbered he could fight from a position of advantage. His fear was that his smaller force might be outflanked and surrounded, so he picked a defile with thick woods at his back and sides and so forced a frontal attack. The Britons also had to do all the running and charging up the slope, while the Roman legions stood silent and rigid until the signal to discharge *pila.* Then they drew their swords and charged into the Britons in their typical wedge-shaped formations, pushing the enemy down with their shields and butchering them with their short swords. Tacitus makes it seem simple and quickly over, but Dio's account,[2] though lacking in clarity, states that the battle lasted all day—a more likely story.

The other Tacitean account is Agricola's battle against the Caledonian tribes under Calgacus.[3] This is a much more satisfactory account and must have come from the lips of Agricola himself. There were more than 30,000 Britons, while the Roman force consisted of 8,000 auxiliary infantry in the centre and 3,000 cavalry on the flanks and 2,000 in reserve in addition to the legionaries. The British had the advantage of higher ground and Agricola, fearing that he might be outflanked, stretched his line until it was quite thin.[4] After an exchange of missiles Agricola ordered his infantry to close in and crowd the Britons, just as Suetonius's troops had done in 60. The auxiliaries advanced up the slope and the cavalry, after routing the charioteers, followed. But the main British force still held the higher ground and they

the Medway, but this has only survived in a garbled form in Dio (lx, 20), although Mr A. R. Burn has made some sense of it (*History,* 39 (1953), pp. 105–15; see also *The Roman Conquest of Britain,* 1965, pp. 67–70).

[1] *Annals,* xiv, 34–37.

[2] lxii, 12.

[3] *Agricola,* 29–37.

[4] Agricola himself, emulating Caesar (see p. 227 *supra*), sent his horse away and took up his position on foot with the colours, thus denying himself the privilege of height from which to survey the battle, a brave but pointless gesture.

moved down one side to carry out an outflanking manoeuvre. This was countered by four *alae* of cavalry held in reserve which broke the assault and then pushed through and fell upon the British rear. The British now broke and fled, only to be cut down by the Roman horsemen. The casualty figures seem incredible for such a battle, 10,000 Britons, 360 Romans! The most extraordinary feature of this battle is the exclusive use of *auxilia*; was this intended or did victory come more quickly and easily than the commander had anticipated? What is surprising is that the Caledonians submitted to a pitched battle in which they were so clearly outclassed. Had they kept to their glens and forests and pursued guerrilla warfare as did the Silures earlier, the situation might have been different. The real achievement of Agricola seems not so much that he won a victory but that he forced a battle. It could be argued that the Silures at least still had their farms and homesteads, whereas the Roman Army's occupation of the eastern plain of Scotland drove the Caledonians into the mountains and a strategy of desperation.

Siege Tactics

In conducting sieges the Romans showed their practical genius combined with a patient obstinacy and thoroughness. Frontinus[1] complacently stated that the invention of siege weapons had long since reached its limit and he could see no prospect of improvements. If a place could not be overcome by the initial assaults or the inhabitants persuaded to surrender, it was the practice of the Roman Army to surround the whole area with a defensive wall and ditch and spread their units round the circumvallation, one of the best examples of which is still to be seen at Masàda. This had the dual function of preventing the besieged from receiving supplies or reinforcements, and also from making sorties or breaking out. Some places were so strong that a long siege had to be planned and supplies of food and water became important. There are several examples of efforts being made to cut off the water supply. Caesar was able to take Uxellodunum by concentrating on this target.[2] First of all he stationed archers, who maintained a constant fire on the water carriers who went to draw from the river which ran round the foot of the hill on which the citadel stood. The besieged then had to rely entirely on a spring at the foot of their wall. But Caesar's engineers were able, under the cover of mantlets, to undermine the spring and draw the water off at a lower level, thus reducing the town without a serious fight.

---

[1] In his preface to *Strat.*, iii. [2] *Bell. G.*, viii, 40–43.

Siege Engines

Siege weapons were varied and ingenious inventions, their main object being to effect an entrance through the gates or walls. Gateways were usually the most heavily defended positions, so that it was often better to select a point along the walls. First, however, the ditches had to be filled with hard-packed material to allow the heavy machinery to approach the foot of the wall. But the soldiers manning the wall would try to prevent this by firing their missiles at the working party. To counteract this the attackers were provided with protective mantlets (*musculi*) which were screens sheeted with iron plates or hides, but this was hardly enough. Constant fire had to be directed against the men on the wall to harass them. This was managed by bringing up stout timber towers higher than the wall so that men on their tops could pick off the defenders. The ram was a heavy iron head in the shape of a ram's head fixed to a massive beam which was constantly slung against the wall until it was breached. According to Vegetius[1] there was also a beam, to which was fastened an iron hook which was inserted into the hole made by the ram and the stones dragged out. There was also a smaller iron point (*terebus*) used for dislodging individual stones. The beam and frame from which it was slung were enclosed in a very strong shed covered with hides or iron plates and often mounted on wheels. This was called a tortoise (*testudo arietaria*), since it resembled this creature with its heavy shell and head which moved in and out. Similar sheds which merely enabled soldiers to work at the base of the walls were called the *cattus, pluteus* or *vinea*. Vitruvius describes a number of these movable sheds of different sizes, some quite large. The most important part of the construction was the roof, since the defenders would drop heavy stones over the wall or toss down burning tar and brushwood. One type of material recommended for the roof was clay kneaded with hair to such a thickness as to be able to prevent fire from reaching the timbers or the men below. The roof was sloped from the ridge on each side to help throw off the materials thrown down from above. Providing the shed was strong enough and protected also by towers, or with the ram itself fixed at the base of a tower, a constant fire could be maintained against the wall top. These towers could also be fitted with fixed bridges or drawbridges. The former thrust out from the tower (*exostra*) and the latter was let down by pulleys (*sambuca*, so called because of its resemblance to the harp). There was also a

[1] His descriptions of these machines are given in Book iv, 13 and 14. Vitruvius gives the history of these machines in x, 13–15.

kind of crane (*tolleno*) which raised or swung small parties of attackers on to the wall. Titus in the Siege of Jerusalem had three iron-clad towers erected, each about eighty feet in height.[1] Under the protection of the towers or sheds, a gang of men could work at the foot of the wall, making a hole through it or digging down to get underneath. Excavating saps or galleries under defences was common practice. The purpose of these was to weaken walls or towers at the foundations so that they collapsed, or to enable a force to penetrate the defences and emerge inside the town. This was, of course, much more difficult to do without the enemy becoming aware of it. The besiegers themselves could more easily and secretly mine outwards below their own walls, below the wooden towers, so that they sank to one side and became unusable. Remains of saps may have been found and excavated at Dura-Europos below the south-west bastion.[2] There is also the strange incident in the siege of Marseilles recorded by Vitruvius[3] which is not in Caesar's account. The defenders, concerned over the saps which were being driven towards the city, dug a large basin inside the walls which they filled with water. When the mines approached the basin the water flowed out, flooding them and causing them to collapse.

The only defence against these massive siege engines was to destroy them either by fire-missiles, or by sallies made by a small, desperate body of men who would try to set fire to them or turn them over. The attackers, however, were covered by archers or catapults from a distance.

Catapults

The Roman Army used several types of powerful siege weapons for discharging missiles, the largest of which was the *onager* (named after the wild ass because of its kick). According to Ammianus, this name had been applied 'only in recent times' (i.e. in the fourth century) to a machine which had previously been called the scorpion (*scorpio*).[4] The latter word certainly appears in earlier writers like Caesar[5] and Vitruvius,[6] while the term *onager* is confined to Ammianus and Vegetius. But the *scorpiones* of the late Republic and

---

[1] One of them unaccountably collapsed one night (Josephus, v, 7, 1).

[2] How much of this was a natural cave and how much the work of the Persians is not easy to determine (*Report on the Fifth Season*, 1934, pp. 13–15).

[3] x, 16, 11.

[4] xxiii, 4, 7, *cui etiam onagri vocabulum indidit aetas novella ea re.*

[5] *Bell. G.*, vii, 25.

[6] x, 10.

early Empire were much smaller machines.[1] *Onagri* were of various sizes from earliest times and some very large indeed. Vegetius merely comments that they threw stones of such enormous weight that they not only crushed men and horses, but broke enemy machines. It was a weapon used in siege-works to make holes in the walls, and by the defenders to smash towers and wooden machines and could be very effective against massed ranks of attackers. One used by the Massiliotes discharged beams twelve feet long with iron points at the end against Caesar's attackers in the siege of 49 B.C.[2] At the second Battle of Cremona in A.D. 69 *Legio* XV had a very large machine (*magnitudine eximia ballista ingentibus saxis*)[3] which hurled massive stones, while Josephus states that at the Siege of Jerusalem the machines of X *Fretensis* hurled stones which weighed a talent (about 55 lb) a distance of two furlongs (440 yards) or further.[4] Actual stones found on British forts at High Rochester and Risingham weigh about a hundredweight each.[5]

The *onager* is described by Ammianus as follows:[6]

'Two posts of oak or holm-oak are hewn out and slightly bent, so that they seem to stand forth like humps. These are fastened together like a sawing-machine, and bored through on both sides with fairly large holes. Between them, through the holes, strong ropes are bound, holding the machine together, so that it may not fly apart. From the middle of these ropes a wooden arm rises obliquely, pointed upward like the pole of a chariot,[7] and is twined around with cords in such a way that it can be raised higher or depressed. To the top of this arm, iron hooks are fastened, from which hangs a sling of hemp or iron. In front of the arm is placed a great cushion of hair-cloth stuffed with fine chaff, bound on with strong cords, and placed on a heap of turf or a pile of sun-dried bricks; for a heavy machine of this kind,

---

[1] Vegetius says the *manuballistae* (presumably hand-operated guns), because they kill with small and slender darts, were formerly called *scorpiones* (iv, 22), but Ammianus thought it got its name from the upright sting, referring to the hurling arm of the *onager*. Clearly the source of Vegetius is much earlier—probably Frontinus.

[2] *Bell. C.*, ii, 2.

[3] *Hist.*, iii, 23.

[4] *Bell. J.*, v, 6, 3.

[5] 'The Romans in Redesdale', *Hist. of Northumberland*, 15 (1940), p. 98, and *A.A.*, 4 ser, 13 (1936), Pl. xv, Fig. 2.

[6] xxiii, 4, 4–7 (Loeb translation). Vitruvius has a more detailed description, giving the rules for determining the size of the machine with a range of missiles from 2 lb to 360 lb, see E. W. Marsden, *Greek and Roman Artillery, Technical Treatises*, 1971, pp. 185–205.

[7] i.e. when the horses are not harnessed to it.

if placed upon a stone wall, shatters everything beneath it by its violent concussion, rather than by its weight. Then, when there is a battle, a round stone is placed in the sling and four young men on each side turn back the bar with which the ropes are connected and bend the pole almost flat. Then finally the gunner, standing above, strikes out the pole-bolt, which holds the fastenings of the whole work, with a strong hammer. Thereupon the pole is set free, and flying forward with a swift stroke, and meeting the soft hair-cloth, hurls the stone, which will crush whatever it hits. And the machine is called *tormentum* because all the released tension is caused by twisting (*torquetur*); and scorpion, because it has an upraised sting; modern times have given it the new name *onager,* because when wild asses are pursued by hunters, by kicking they hurl back stones to a distance, either crushing the breasts of their pursuers, or breaking the bones of their skulls and shattering them.'

According to Vegetius each legion had ten *onagri,* one for each cohort, drawn on carriages by oxen.[1] These large machines were also used in the defence of forts, especially in the Late Empire. Richmond has described the massive stone platforms packed in stiff clay, twenty-five feet wide and thirty-two feet long, behind the third-century wall at High Rochester (Bremenium),[2] their function attested by dated inscriptions.[3] Elsewhere Richmond demonstrated diagrammatically the fire cover from these platforms.[4]

Several attempts have been made at reconstructing this and other types of machines. In France, General de Reffye made models at Mendon under the instructions of Napoleon III,[5] and these are still to be seen in the National Museum of Antiquities, St Germain-en-Laye; while in Germany von Schramm carried out similar work, but in a more thorough manner and some of his reconstructions[6] are in the Saalburg Museum. There is a model of an *onager* also in the Museum of Roman Civilisation at Rome (Pl. XXVIIIa).

The smaller machines, to which Vitruvius gives the general term of catapult (*catapulta*), were in various sizes. The smaller ones appear to have

---

[1] ii, 25.

[2] *A.A.,* 4, 13 (1936), pp. 180–1.

[3] *R.I.B.,* Nos. 1280 and 1281; each platform had its own dedication, since each refers to a single *ballistarium,* but one was built under Elagabalus (A.D. 220), and the other under Severus Alexander (A.D. 225–35). Presumably, however, they are part of the same scheme.

[4] *Hist. of Northumberland,* 15 (1940), Fig. 21; also in *P. Brit. Acad.,* 41 (1955), Fig. 4.

[5] *Études sur le passé et l'avenir de l'Artillerie,* 1846–71, *Catalogue Illustré du Musée des Antiquités Nationales,* i, 1917, Figs. 52 and 53.

[6] *Griechische—römische Geschütze*, 1910.

been called *scorpiones* and the larger *ballistae*, but the names are used by various ancient writers without much precision. They operated in much the same way as the later cross-bow. The pair of vertical coil chambers was at the front, the bow being drawn back by windlass to the required limit of tension and held by a rack and pinion. The bow was released by a trigger and the bolt shot along a trough through an aperture in the front.[1] The tension-frame of a *ballista* was found at Ampurias, the ancient Emporion, in Spain in 1912.[2] It was not, however, identified until 1914 by Barthel. The size of the frame compares very favourably with the measurements given by Vitruvius, and the range has been calculated as being about 300 metres. According to Vegetius[3] there was a *carro-ballista*[4] to each century of the legion, mounted on a cart drawn by mules and operated by ten men, which seems an excessive number. There is ample confirmation of Vegetius from Trajan's Column, where these guns are seen in action[5] (Pl. XXIXa). Two scenes show the *ballistae* in their carts, while in a third the guns have been taken out and mounted on platforms of logs, presumably during a siege.[6] The obvious use of the *carro-ballistae* would have been on the flanks of the advancing legionaries. They laid down a barrage, shooting over the heads of the troops. The bolts were about nine inches to a foot long and there were two types of iron head. One was of pyramidal form like the head of a *pilum* and the other a simple flat, long, triangular point with a socket formed by making cuts at the shoulder and turning over the lower ends (Fig. 52). They varied in size from three to four and a half inches and there were also variations in weight, but this may not necessarily indicate machines of different calibre. Many bolts of both types are found on Roman sites,[7] but the most dramatic discovery was an

---

[1] Vitruvius (Chap. x) gives a detailed description indicating that the machines were of varying sizes.

[2] E. von Schramm, *Die Antiken Geschütze der Saalburg*, 1918, pp. 40–46; 75–77. There is a model at Saalburg (Pl. XXIXb).

[3] ii, 25.

[4] *Carrus* was a small cart, and as Richmond indicated, it was specially made for this particular purpose (*Pap. Brit. School of Rome*, 13 (1935), p. 13).

[5] Taf. 31, 104, shows the cart and the two mules, but only two men; there is a good view of another example in Taf. 46, 164, which is a front view showing the coil chambers and the bow arched upwards, but the artist, not understanding its function, has only one man operating it.

[6] Taf. 47, 165–7.

[7] Sixty-seven are recorded from Hod Hill (*Durden Coll.*, p. 6, and Pl. vi, Nos. 70, 117–83).

example still wedged in the backbone of a British warrior at Maiden Castle.[1]

There are several detailed eyewitness accounts of Roman siege tactics and these convey the scenes most vividly. It may be fitting to conclude this chapter with some extracts, first from Caesar:

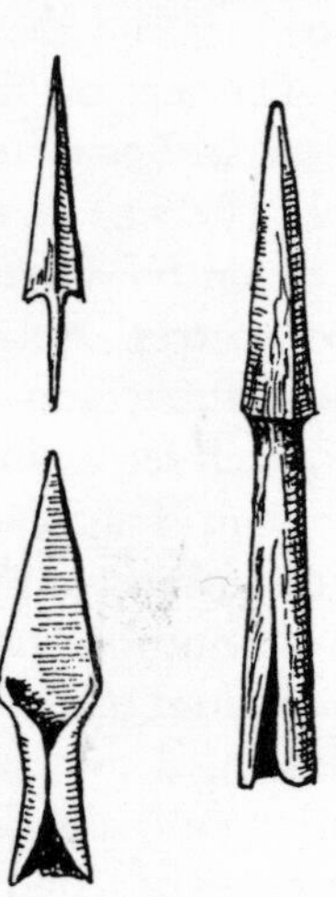

FIG. 52 *Heads of ballista bolts (half-size)*

The Siege of Alesia[2]

'The actual stronghold of Alesia was set atop of a hill, in a very lofty situation, apparently impregnable save by blockade. The bases of the hill were washed on two separate sides by rivers. Before the town a plain extended for a length of about three miles; on all the other sides there were hills surrounding the town at a short distance, and equal to it in height. Under the wall, on the side which looked eastward, the forces of the Gauls had entirely occupied all this intervening space, and had made in front a ditch and a rough wall six feet high. The perimeter of the siege-works which the Romans were beginning had a length of eleven miles. Camps had been pitched at convenient spots, and three-and-twenty forts had been constructed on the line. In these picquets would be posted by day to prevent any sudden sortie; by night the same stations were held by sentries and strong garrisons.

'When the siege-work had been started, a cavalry encounter took place in

[1] Now to be seen in the Dorchester Museum, Dorset (*Maiden Castle*, 1943, Pl. lviii A and Fig. 93).

[2] *Bell. G.*, vii, 69–74, Loeb Translation; see also J. Harmand, *Une Campagne Césarienne—Alesia*, Paris, 1967.

the plain which we have described above as set between hills and extending to a length of three miles. Both sides strove with the utmost vigour. When our men were distressed Caesar sent up the Germans, and posted the legions in front of the camp to prevent any sudden inrush on the part of the enemy's footmen. With the reinforcement of the legions behind them our men's spirit was increased; the enemy were put to flight, and, hampering one another by sheer numbers, as the gates were left too narrow, were crowded together in a press. The Germans pursued most vigorously right up to the fortifications. A great slaughter ensued; some of the enemy abandoned their horses, and tried to cross the ditch and scale the wall. Caesar ordered the legions posted in front of the rampart to advance a short distance. The Gauls inside the fortifications were in just as great a confusion as the rest; believing that the enemy were coming on them at once, they shouted the call to arms, and some in panic burst into the town. Vercingetorix ordered the gates to be shut, lest the camp should be deserted. After much slaughter and the capture of many horses the Germans retired.

'Vercingetorix now made up his mind to send away all his horsemen by night, before the Romans could complete their entrenchments. His parting instructions were that each of them should proceed to his own state and impress for the campaign all men whose age allowed them to bear arms. He set forth his own claims upon them, and adjured them to have regard for his personal safety, and not to surrender to the torture of the enemy one who had done sterling service for the general liberty. He showed them that if they proved indifferent eighty thousand chosen men were doomed to perish with him. He had calculated that he had corn in short rations for thirty days, but that by economy he could hold out just a little longer. After giving these instructions he sent the horsemen silently away in the second watch, at a point where a gap was left in our works. He ordered all the corn to be brought in to his headquarters; he appointed death as the penalty for any disobedience of the order; the cattle, of which great store had been driven together by the Mandubii, he distributed man by man; he arranged that the corn should be measured out sparingly and gradually; he withdrew into the town all the force which he had posted in front of it. By such measures did he prepare for the conduct of the campaign, in anticipation of the succours from Gaul.

'Caesar had report of this from deserters and prisoners, and determined on the following types of entrenchments. He dug a trench twenty feet[1] wide

---

[1] All measurements are in Roman feet.

with perpendicular sides, in such fashion that the bottom thereof was just as broad as the distance from edge to edge at the surface. He set back the rest of the siege-works four hundred paces from the trench; for as he had of necessity included so large an area, and the whole of the works could not easily be manned by a ring-fence of troops, his intention was to provide against any sudden rush of the enemy's host by night upon the entrenchments, or any chance of directing their missiles by day upon our troops engaged on the works. Behind this interval he dug all round two trenches, fifteen feet broad and of equal depth; and the inner one, where the ground was level with the plain or sank below it, he filled with water diverted from the river. Behind the trenches he constructed a rampart and palisade[1] twelve feet high; to this he added a breastwork and battlements,[2] with large platforms[3] projecting at the junctions of palisade and rampart, to check the upward advance of the enemy; and all round the works he set turrets at intervals of eighty feet.

'As it was necessary that at one and the same time timber and corn should be procured, and lines of such extent constructed, our forces, having to proceed to a considerable distance from camp, were reduced in number; and sometimes the Gauls would try to make an attempt upon our works by a sortie in force from several gates of the town. Caesar, therefore thought proper to make a further addition to these works, in order that the lines might be defensible by a smaller number of troops. Accordingly, trunks or very stout branches of trees were cut, and the tops thereof barked and sharpened, and continuous trenches five feet deep were dug. Into these the stumps were sunk and fastened at the bottom so that they could not be torn up, while the bough-ends were left projecting. They were in rows of five fastened and entangled together, and anyone who pushed into them must impale himself on the sharpest of stakes. These they called 'markers'.[4] In front of these, in diagonal rows arranged in a quincunx pattern,[5] pits three feet deep were dug, sloping inwards slightly to the bottom. In these, tapering

---

[1] *Aggerem ac vallum,* probably a bank surmounted by a palisade.

[2] *Loricam pinnasque.*

[3] *Pluteorum,* the same word is used of siege equipment, a movable mantlet, but Caesar uses it here to mean a forward projection of the palisade, a rudimentary kind of tower. It is, however, also reminiscent of the hoardings built in advance of medieval battlements to allow defenders to drop objects on to attackers at the foot of the wall.

[4] The word *cippus* means a wooden stake, a boundary marker and a tombstone, and its double meaning obviously appealed to the rough humour of the Roman soldier.

[5] i.e. like the figure five on a dice.

stakes as thick as a man's thigh, sharpened at the top and fire-hardened, were sunk so as to project no more than four fingers' breadth from the ground; at the same time, to make all strong and firm, the earth was trodden down hard for one foot from the bottom, and the remainder of the pit was covered over with twigs and brushwood to conceal the trap. Eight rows of this kind were dug, three feet apart. From its resemblance to the flower the device was called 'lily'.[1] In front of all these, logs a foot long, with iron hooks firmly attached, were buried altogether in the ground and scattered at brief intervals all over the field, and these they called 'goads'.[2]

'When all these arrangements had been completed Caesar constructed parallel entrenchments of the same kind facing the other way, against the enemy outside, following the most favourable ground that the locality afforded, with a circuit of fourteen miles. This he did to secure the garrisons of the entrenchments from being surrounded by a host, however large it might chance to be. And in order that he might not be constrained to dangerous excursions from camp, he ordered all his men to have thirty days' corn and forage collected.'

The Gallic chiefs took counsel and levied a large force to relieve the fortress and when this large host approached[3] . . .' Caesar disposed the whole Army on both faces of the entrenchments in such fashion that if occasion should arise, each man could know and keep his proper station; then he ordered the cavalry to be brought out of camp and to engage. There was a view down from all the camps which occupied the top of the surrounding ridge, and all the troops were intently awaiting the issue of the fight. The Gauls had placed archers and light-armed skirmishers here and there among the horsemen to give immediate support to their comrades if driven back and to resist the charge of our cavalry. A number of men, wounded unexpectedly by these troops, began to withdraw from the fight. When the Gauls were confident that their own men were getting the better of the battle, and saw ours hard pressed by numbers, with shouts and yells on every side—those who were confined by the entrenchments as well as the others who had come up to their assistance—they sought to inspirit their countrymen. As the action was proceeding in sight of all, and no deed of honour or dishonour could

---

[1] As at Rough Castle on the Antonine Wall (Pl. XXVa).

[2] There is a wealth of double meaning here, since *stimulus* was not only a cattle goad but used against slaves as well; it also indicates incitement in the amatory sense and doubtless had even broader significance in vulgar parlance.

[3] Continuing *Bell. G.*, vii, 80–88.

escape notice, both sides were stirred to courage by desire of praise and fear of disgrace. The fight lasted, and the victory was doubtful, from noon almost to sunset; then the Germans in our part of the field massed their troops of horse, charged the enemy and routed them, and when they had been put to flight the archers were surrounded and slain. Likewise from the other parts of the field our troops pursued the retreating enemy right up to their camp, giving them no chance of rallying. But the Gauls who had come from Alesia, almost despairing of victory, sadly withdrew again into the town.

'After one day's interval, in the course of which they made a great number of hurdles, ladders, and grappling-hooks, the Gauls left camp silently at midnight and approached the entrenchments in the plain. Raising a sudden shout, to signify their coming to the besieged inside the town, they began to fling down the hurdles,[1] to dislodge our men from the rampart with slings, arrows, and stones, and to carry out everything else proper to an assault. At the same moment, hearing the shout, Vercingetorix gave his troops the signal by trumpet, and led them out of the town. Our troops, as on previous days, moved each to his appointed station in the entrenchments; with slings for one-pound stones,[2] stakes set ready inside the works, and lead sling bullets,[3] they beat off the Gauls. As the darkness made it impossible to see far, many wounds were received on both sides. A number of missiles were discharged by the artillery. Then Marcus Antonius and Gaius Trebonius, the lieutenant-generals to whom the defence of these sections had been allotted, withdrew troops from forts farther away, and sent them up to bring assistance wherever they remarked that our men were hard pressed.

'While the Gauls were some distance from the entrenchment they had more advantage from the quantity of their missiles; then, when they came up closer, they were soon caught unawares on the "goads", or they sank into the pits (*lilia*) and were impaled, or they were killed by stakes thrown[4] from the rampart and the turrets, and so perished on every side. Many a man

---

[1] *Crates,* a wickerwork hurdle used here as fascines to fill in and bridge the ditch.

[2] *Fundis librilibus* is a phrase peculiar to Caesar, and probably means slings capable of throwing stones of a pound weight.

[3] *Glans,* an acorn, is used here of the similar-shaped lead bullets used by slingers (*Bell. Jug.*, 57; Livy, xxxviii, 20, 21, 29); a secondary meaning was part of the male organ, and doubtless relished by the troops.

[4] *pilis muralibus,* these were wooden stakes pointed at each end and used for the palisade but were also useful in defence (see p. 171).

Plate XXVI A timber building showing after excavation at Valkenburg, Holland.

Plate XXVII
A section through the defences of the legionary fortress at Chester showing the turf rampart and log corduroy (p. 178).

was wounded, but the entrenchment was nowhere penetrated; and when daybreak drew nigh, fearing that they might be surrounded on their exposed flank by a sortie from the camps above them,[1] they retired to their comrades. Meanwhile the inner force brought out the hurdles which had been prepared by Vercingetorix for a sortie, and filled in the nearer trenches; but they lingered too long in the execution of the business, and, or ever they could get near the entrenchments, they learnt that their countrymen had withdrawn. So without success they returned to the town.

'Twice beaten back with great loss, the Gauls took counsel what to do. They called in men who knew the locality well, and from them they learnt the positions and the defences of the upper camps. On the north side there was a hill, which by reason of its huge circumference our troops had been unable to include within the works; they had been obliged to lay out the camp on ground gently sloping, which put them almost at a disadvantage. This camp was held by Gaius Antistius Reginus and Gaius Caninius Rebilus, *legati,* with two legions. Having reconnoitred the locality by means of scouts, the commanders of the enemy chose out of the whole host sixty thousand men belonging to the states which had the greatest reputation for courage: they determined secretly together what should be done and in what fashion, and decided that the advance should take place at the moment when it was seen to be midday. In charge of this force they put Vercassivellaunus the Arvernian, one of the four commanders, a kinsman of Vercingetorix. He left camp in the first watch, and having almost completed his march just before dawn, he concealed himself behind the height and ordered his soldiers to rest after their night's work. When at last it was seen to be near midday he moved with speed on the camp above mentioned and at the same moment the horsemen began to advance towards the entrenchments in the plain, and the rest of the force to make a demonstration before the camp.

'When from the citadel of Alesia Vercingetorix observed his countrymen, he moved out of the town, taking with him the hurdles, poles, mantlets, grappling-hooks, and all the other appliances prepared for the sally. The fight went on simultaneously in all places, and all expedients were attempted, with a rapid concentration on that section which was seen to be least strong. With lines so extensive the Roman Army was strung out, and at several points defence proved difficult. The shouting which arose in the rear of the fighting

---

[1] One of the main camps of the Roman legions, here referred to, lay on a height to the southward of Alesia: the other on a height to the northward.

line did much to scare our troops, as they saw that the risk to themselves depended on the success of others,[1] for, as a rule, what is out of sight disturbs men's minds more seriously than what they see.

'Caesar found a suitable spot from which he could see what was proceeding in each quarter. To parties distressed he sent up supports. Both sides felt that this was the hour of all others in which it was proper to make their greatest effort. The Gauls utterly despaired of safety unless they could break through the lines; the Romans anticipated an end of all toils if they could hold their own. The hardest struggle occurred by the entrenchments on the hill, whither, as we have mentioned, Vercassivellaunus had been sent. The unfavourable downward slope of the ground had great effect. Some of the enemy discharged missiles, others moved up in close formation[2] under their shields (Pl. XXVIIIb); fresh men quickly replaced the exhausted. Earth cast by the whole body together over the entrenchments gave the Gauls a means of ascent and at the same time covered over the appliances which the Romans had concealed in the ground; and our troops had now neither arms nor strength enough.

'When Caesar learnt this, he sent Labienus with six cohorts to support them in their distress. He commanded him, if he could not hold his ground, to draw in the cohorts and fight his way out, but not to do so unless of necessity. He himself went up to the rest of the troops, and urged them not to give in to the strain, telling them that the fruit of all previous engagements depended upon that day and hour. The enemy on the inner side, despairing of success on the level ground, because of the size of the entrenchments, made an attempt to scale the precipitous parts, conveying thither the appliances they had prepared. They dislodged the defenders of the turrets by a swarm of missiles, filled in the trenches with earth and hurdles, tore down rampart and breastwork with grappling-hooks.

'Caesar first sent young Brutus with some cohorts, and then Gaius Fabius, a *legatus,* with others; last of all, as the fight raged more fiercely, he himself brought up fresh troops to reinforce. The battle restored, and the enemy repulsed, he hastened to the quarter whither he had sent Labienus. He withdrew four cohorts from the nearest fort, and ordered part of the cavalry to follow him, part to go round the outer entrenchments and attack the enemy in the rear. Labienus, finding that neither ramps nor trenches could resist the rush

---

[1] i.e. that, if the line were broken elsewhere, they themselves would be in peril. *Salus* = 'safety' secured by success.

[2] *alii testudine facta subeunt,* i.e. in 'tortoise' formation.

of the enemy, collected together eleven cohorts, which had been withdrawn from the nearest post and by chance presented themselves, and sent messengers to inform Caesar what he thought it proper to do. Caesar hurried on to take part in the action.

'His coming was known by the colour of his cloak[1] which it was his habit to wear in action as a distinguishing mark; and the troops of cavalry and the cohorts which he had ordered to follow him were noticed, because from the upper levels these downward slopes and depressions were visible. There-upon the enemy joined battle: a shout was raised on both sides, and taken up by an answering shout from the rampart and the whole of the entrenchments. Our troops discarded their javelins and got to work with their swords. Suddenly the cavalry was noticed in the rear; other cohorts drew near. The enemy turned to flee; the cavalry met them in flight, and a great slaughter ensued. Sedulius, commander and chief of the Lemovices, was killed; Vercassivellaunus, the Arvernian, was captured alive in the rout; seventy-four war-standards were brought in to Caesar; of the vast host few returned safe to camp. The others beheld from the town the slaughter and rout of their countrymen, and, in despair of safety, recalled their force from the entrenchments. Directly they heard what had happened the Gauls fled from their camp. And if the troops had not been worn out by frequent reinforcing and the whole day's effort, the entire force of the enemy could have been destroyed. The cavalry were sent off just after midnight and caught up the rearguard: a great number were taken and slain, the rest fled away into the different states.'

The Siege of Jotapata[2]

Jotapata was built on a precipice with steep valleys on every side but the north, where a wall had been built to prevent easy access, and it was on this side that Vespasian pitched his camp. After several days in which the Jews made a number of sorties against the Romans, Vespasian decided to prosecute the siege with vigour and he caused a bank of earth and wood to be built up against the wall. Though the soldiers working on the bank were protected

---

[1] He wore the scarlet cloak (*paludamentum*) of a commander-in-chief.

[2] Josephus, iii, 7, 3–36. Précis of the Whiston translation. Jotapata, known today as Mizpe Yodefat, is a hill guarding the route from the Bay of Haifa to the Sea of Galilee. There are no visible remains of the Roman works, as at Masàda, since in a continuous occupied area any structural elements of stone would have been removed for building.

by hurdles, they were greatly impeded by the huge stones and darts aimed at them by the Jews. Vespasian then set 160 engines to work to dislodge the enemy from the walls by aiming stones and lances at them, but when the Jews were driven from the battlements they made sallies out of the city and demolished the protecting hurdles. However, the work on the bank continued, and Josephus, the leader of the Jews, decided to build the wall higher at this point, and he was able to do this by having a screen made of the raw hides of newly killed oxen strung along the top of the wall to protect the workmen. The hides broke the impact of the Roman stones and darts, and being moist, they quenched the fire missiles.

The besieged had plenty of corn within the city, but there was little water and so Josephus, foreseeing the awful consequences of such a shortage, caused water to be rationed at an early stage. The Romans heard about this and took heart, thinking the siege almost over. But the Jews confounded them by wringing out their clothes in their precious water and hung them on the battlements until the walls were running with water, so that the Romans thought they must have some secret source of supply.

At this stage Josephus decided to leave Jotapata, thinking that this would draw the Romans off, but the people pleaded with him not to leave them, and so he stayed and organized many sorties against the Romans, causing them much distress. The Romans counteracted with their great offensive engines, and the Arabian archers and the Syrian slingers were brought into action, but all this made the Jews fight all the harder.

Vespasian now brought in the battering ram, and at the very first stroke the wall was shaken and a terrible clamour was raised by the people within the city. Josephus tried to defeat the ram by ordering sacks to be filled with chaff and lowering them down over the wall so that they would receive the strokes of the battering ram; as fast as the Romans moved the great engine, so the Jews moved the sacks of chaff to that new place. In the end the Romans contrived to cut the sacks from off the ropes and so continued their battering of the wall. Three parties of the Jews then rushed out of the gates and set fire to the protective hurdles and skins on the ram, and the materials, being dry with the bitumen and pitch, were soon well ablaze. While this was taking place a Jew, renowned for his strength, cast a huge stone down from the wall and on to the ram and broke off the head of the engine. Before the Romans could recover, the Jews set fire to many of the other engines and their coverings, but all this did not prevent the besiegers from erecting the ram again and continuing their battering of the wall.

About this time Vespasian was wounded in the foot and the Romans were so incensed that they renewed their attack on the city, and all the night through continued their bombardment with great stones and other missiles which brought havoc upon the besieged Jews. By morning the wall had yielded to the battering of the ram. The trumpets of the several Roman legions sounded, the army gave a terrible shout and their darts flew so fast that they intercepted the light of day. However, Josephus's men stopped their ears at the sounds and covered their bodies against the darts. Then, while the Roman archers paused to reload, they charged out through the breach, and heavy fighting ensued. Then the Romans attempted to scale the unbreached part of the wall, and the soldiers were joined side to side with their shields so that they could not easily be broken. But Josephus caused scalding oil to be poured down upon them so that it burnt the Romans, and as the men were cooped up in their helmets and breast-plates, they could not get free of the burning oil and so they were beaten back. The Jews followed this up by pouring boiling fenugreek upon the boards which the Romans were using to scale the walls, and this made them so slippery that the soldiers could not get footholds.

Vespasian then decided to raise the banks they had built against the wall, and to erect three towers each fifty feet high upon them. The towers were to protect the engines and the men operating them, and from their battlements the Roman soldiers were able to discharge their missiles down on to the heads of the now unprotected enemy in the city.

By the forty-seventh day of the siege a deserter went to Vespasian and told him how few were left in the city and how weak they were. He also told how the Jews, worn out by their constant fighting and vigilance, usually slept during the last watch of the night and that this was the hour to attack the city. And so at the appointed hour the Roman Army marched silently to the wall, cut the throats of the watch and entered the city without waking the sleeping Jews. So quickly and quietly did the Romans enter the city that the Jews were taken by surprise, and they were further confounded by a thick mist which arose so that they could not be sure what was happening. The Romans, remembering all that they had suffered at the hands of the Jews, drove many of them down the precipice and slew all the multitude that appeared openly, but the woman and infants, who numbered about 1,200, they took into slavery. In all about 40,000 were slain at the siege of Jotapata. Vespasian gave the order that the city should be entirely demolished and all the fortifications burnt down, but there were few left to care.

TRANSMISSION OF INFORMATION BY SIGNALS

The development of radio systems in the modern army for immediate transmission of commands and information has made it difficult to appreciate a situation where methods were elementary or non-existent. Much of the muddle and confusion of medieval and later warfare was due to lack of information and inability to redeploy forces easily. This is such a basic need in any well-organized army that one must expect Roman staff officers to have given it their serious attention, yet there is very little mention of any systems in the literature. On the other hand, the whole arrangement of the frontiers was based on information concerning enemy forces being passed to rearward areas to enable major troop concentrations at threatened points. These could only have been effected by a system of signals passing along main lines of communication.

The earliest possibility of an advanced form of signals comes from Polybius.[1] This method, invented by Cleoxenus or Democlitus, and perfected by Polybius himself, was based on two sets of torches. Having divided the alphabet into groups of five letters, the first set of torches indicated the groups, i.e. first, second, third, fourth or fifth, and the second set, appearing at a short distance from the first, indicated the position of the letter in the group, i.e. one to five. Thus a message could be spelt out letter by letter, presumably using torches by night and smoke columns by day. Simple code books could be introduced to prevent the enemy from understanding the messages. Vegetius describes a method of communication by raising and lowering a beam of wood (*trabes*) from a tower which would appear to indicate a semaphore for spelling out messages.[2] The only additional information comes from Trajan's Column and the actual remains of signal stations.

On the Column three watch-towers are shown along the banks of the Danube.[3] From windows of each tower signal torches project and these may belong to the semaphore system similar to that described by Vegetius. It is possible, however, that a single torch if moved in predetermined movements could give a wide variety of different signs and cover most of the alphabet. At the side of one of the towers is a pile of logs carefully built into a platform, each layer of logs being laid at right-angles to its neighbour. It was the view

---

[1] x, 43–47. [2] Vegetius, iii, 5. [3] Cichorius, Scenes 4–6.

of von Domaszewski,[1] followed by two eminent scholars,[2] that this was a beacon carefully set for kindling in the event of any serious alarm. Nevertheless it would seem an odd way of arranging timbers to burn quickly; far better to have them in a more vertical column to enable the air to move upwards. The log pile of the Column is exactly like those on which *ballistae* are mounted and has more the nature of a platform. Near by are two hay-cocks, and it would seem that a more likely explanation is that the function of the pile was to raise the fire or smoke column well above the ground clear of vegetation and ground mist, a common feature of river valleys. If the torch in the watch-tower was used for semaphore signals, the platform and straw must have been used for basic signals in an emergency. A similar explanation is possible for the three pairs of semicircular platforms projecting from the inner side of the Antonine Wall.[3] Pairs of platforms have also been noted in Syria along Roman roads,[4] but on Dere Street, seven miles beyond Hadrian's Wall, there exists no less than seven in a row carefully sited in the lee of a hill to the north, but in direct line with Limestone Bank.[5] If these were all in use at the same time, considerable variations could have been played. These circular platforms, ten feet in diameter, may have been the foundations for the log piles discussed above. Signal stations were quite small and could have vanished almost entirely from the landscape, especially if they were only timber-built in the first place. The best description of one is that at Martinhoe on the north coast of Devon by Lady Fox.[6] This station, like its neighbour at Old Burrow, was associated with the early conquest of South Wales begun by Ostorius Scapula. Both had a single surrounding ditch with a bank and a single entrance, and an overall diameter of 275 feet. There was a square inner enclosure with two ditches round a space about seventy feet square in which were found two small barrack blocks, sufficing perhaps for a century with its officers. There is no suggestion here of a tower for a semaphore, but in the

---

[1] *Die Marcus-Säule auf der Piazza Colonna in Roma*, p. 109, fn. Very similar examples are portrayed on the Marcus Column.

[2] Sir George Macdonald in *The Roman Wall in Scotland*, 1934, p. 356, and Sir Ian Richmond in *Pap. Brit. School of Rome*, 13 (1935), p. 36.

[3] K. Steer, 'The nature and purpose of the expansions on the Antonine Wall', *P. Soc. Ant. Scot.*, 90 (1959), pp. 161–9.

[4] A. Poidebard, *La trace de Rome*, Pl. xxxix, lix, xcv and civ.

[5] Richmond, 'The Romans in Redesdale', *Hist. of Northumberland*, 15 (1940), p. 101.

[6] A. Fox and W. Ravenhill, 'Early Roman outposts on the North Devon coast, Old Burrow and Martinhoe', *P. Devon Arch. Expl. Soc.*, 24 (1967), pp. 3–39.

outer enclosure there was an area of twenty-four square feet which had been much reddened by burning as if from bonfires.

The signal sites, singly or in pairs, can be assumed to have given a limited amount of basic information alerting forces to the rear. The semaphore system, on the other hand, passed detailed messages along main routes. It is clear from the Column that this was the function of the watch-tower and a number of examples have been noted (Pl. XXXa). On Hadrian's Wall this was one of the functions of the milecastles, but there is an independent structure, Pike Hill tower, which may have started as a feature on Stanegate, but continued in use in the Wall system. There are other towers like Robin Hood's Butt[1] and Mains Rigg[2] which must have been parts of a considerable system, and traces of further towers may yet be discovered.[3]

It is rare that signal posts can be linked together into a coherent system; apart from the two Walls there are but two in Britain, one along the Stainmore Pass[4] and the other on the Gask Ridge in Scotland.[5] In the former case the timber towers at Bowes Moor and Roper Castle occupied a space thirty feet by thirty-five feet protected by a ditch and bank. The Gask stations are ten in number, spaced at approximately a Roman mile apart. They consist of circular platforms thirty-five to fifty feet in diameter, protected by a ditch and outer bank; excavations inside have revealed the presence of four post-holes of a timber tower ten feet square, the whole being very close to the Trajanic model. This suggests that they may have belonged to the main line of communication established by Agricola.

MEDICAL SERVICES

The attention paid by the Army to its medical services deserves a small section to itself, since so many of the ideas and practices have a modern ring. On the whole the attitude of the upper classes in the classical world to illness and

[1] *T. Cumb. and West.*, 33 (1933), pp. 241–5; 38 (1938), p. 198.

[2] *T. Cumb. and West.*, 29 (1929), p. 314.

[3] There is another isolated station at Brownhart Law on Dere Street which must have connected with a series, *P. Soc. Ant. Scot.*, 83 for 1948–9 (1951), pp. 167–74.

[4] I. A. Richmond, 'A Roman arterial signalling system', *Aspects of Archaeology in Britain and Beyond*, 1951, pp. 293–302.

[5] O. G. S. Crawford, *Topography of Roman Scotland*, 1949, pp. 51–55, and A. L. F. Rivet in *Arch. J.*, 121 for 1964 (1965), pp. 196–8.

injury was full of common sense and sound practice. Most of the best doctors were Greek and they looked back to their great practitioner Hippocrates (*c.* 460 B.C.), although ideas were much entangled with superstition.[1] At first Rome was resistant to Greek medicine as to other ideas[2] and the mass of the population no doubt continued to use the time-honoured therapeutic methods assisted by incantations and charms. The Greek deity Asclepius, who was associated with a snake, was established in Rome at the time of a plague in 293 B.C., and a temple was built on the island in the Tiber. Asclepiades[3] (*c.* 100 B.C.) had a great reputation in Rome, basing his treatment on diet, exercise, massage and a severe cold-water treatment. The Roman attitude to medicine in the early Empire was best summarized by Cornelius Celsus, an educated layman who published a work of general knowledge, probably a compilation of which only the section *de Medicina*[4] survives. His advice follows in the Asclepian tradition of living a healthy varied life with reasonable exercise, proper sleep and rest and a well-regulated diet. No one would quarrel with this even today. His section on diseases places great emphasis on proper diagnosis, treatment relies on diet, hygiene and drugs, the latter being natural agents which had a long history of efficaciousness. It is his section on surgery which has received most attention, but some of his operation techniques are couched in such obscure language that it has been thought that his accounts are second-hand and not fully understood by the writer. Some simple operations such as amputation are, however, clear enough. 'A scalpel (*scapellus*) is used to cut the flesh to the bone, but not at a joint. When the bone is reached the flesh is pulled back from it and undercut from around it so that part of the bone is made bare. The bone is now cut with a little saw (*serrula*)[5] as close as possible to the flesh. Then the end of the bone is smoothed and the skin drawn

---

[1] The chief Roman god of health was Salus, but for stronger protection against pestilence they turned to Mars. Each obvious symptom like fever, inflammation and skin irruptions had its own deity.

[2] Pliny, who gives the best general account of early Roman medicine appears to jibe at the large number of Greeks practising in Rome in his own day (*Nat. Hist.*, xxix, 11). Cato, the stern old Roman, typically stated that Rome was *sine medicis sed non medicina* and advised cabbage for all internal ailments, but for dislocations could only suggest the recitation of a meaningless jingle.

[3] A native of Bithynia and like other Greeks became named after the God of healing.

[4] Published in the Loeb series in three volumes in 1935.

[5] Illustrated in the *B. M. Guide to Greek and Roman Life*, 1929, Fig. 200; see also J. S. Milne *Surgical Instruments in Greek and Roman Times*, 1907.

over it. The skin must be loosened enough in this operation to cover the bone as completely as possible' (vii, 33).[1]

A Greek doctor Scribonius Largus, probably a freedman, at the court of Tiberius and Claudius came with the latter to Britain at the time of the invasion and a short treatise of medical recipes by him has survived.[2] But by far the greatest of all the medical men of the Roman times was Galen who was practising in Rome *c.* A.D. 164. He was a prolific writer and may have published as many as five hundred works, but only about a hundred which can be considered genuine have survived.[3] One of the most famous of these books is *de usu partium corporis humani* in which he sought to demonstrate the perfection of every part of the human anatomy but his knowledge of bone structure and muscles was mainly derived from a study of animals. He also wrote on hygiene (*de sanitate tuenda*), and general aims and principles (*methodus medendi*) which remained for centuries the standard work, becoming in the process fossilized into an established system which denied the possibility of progress.

It is against this general background of common-sense and sound practice that one must view the army medical service.[4] It was not a lucrative profession and hardly likely to attract an eminent practitioner who could build up a practice in Rome or in one of the great cities in the Empire. A man of wealth or connections might have taken his own doctor as a servant while on a tour of duty but the *medicus* did not rank very high in the military hierarchy.[5] A

---

1 Although anaesthetics were unknown, except for the liberal use of alcohol, Celsus lists a number of antiseptics which include pitch, turpentine, salt, disulphide of arsenic, silphium and a variety of oils. Most of these judiciously used would have been helpful, but mixed with them are other remedies which are purely magical.

2 *De compositione medicamentorum.*

3 The most complete edition is that in twenty volumes edited by C. G. Kühn, Leipzig, 1821–33; some smaller works appear in *Galeni scripta minora*, Leipzig, 1884–93.

4 The best introduction is that of I. A. Richmond, 'The Roman Army Medical Service', in *The University of Durham Medical Gazette*, June 1952; see also F. H. Ganison, *Notes on the History of Military Medicine*, Washington, 1922. For doctors in the Republican Army see O. Jacob, in *L'Antiquité classique*, 2 (1933), pp. 313–29.

5 Since this rank does not appear in a career sequence it is difficult to equate it to those of the other officers (*Rang.* pp. 15 and 45). The fact that they dedicated altars on behalf of the units suggests that they could not have been classed below a centurion. One *medicus*, M. Ulpius Sporus, owned slaves who as freedmen erected his tombstone (*C.I.L.*, xi, 3007); other *medici* appear on xiii, 6621; iii, 10854 and *R.I.B.*, 1028, *medicus* of the *Ala Vettonum* stationed at Binchester, Co. Durham. M. Rubrius Zosimus actually states that he was *medicus cohortis* (I.L.S. 2602).

few are known through their writings.[1] This officer must, however, be distinguished from the *medicus ordinarius*, who had the rank of a centurion,[2] the *medici* were *immunes*.[3] Below this rank came the *capsarii*,[4] or dressers, so named for the round box of bandages (*capsa*) they carried with them in the field. In the Fleet the orderlies were classed as *medicii duplicarii* (i.e. with double pay).[5] The medical services in the legions were under the command of the *praefectus castrorum*, but presumably the *medicus legionis*[6] would have been a more important officer than those attached to auxiliary units.

The basis of good health is a high standard of personal hygiene and effective sanitation and in both of these the Roman Army ranked highly. Every permanent fort had its bath-house which with its pattern of rooms of different temperatures and humidities, was far better for the promotion of health and cleanliness than our modern practice of wallowing in dirty water. In sanitation the Romans exercised their engineering abilities, drains and latrines were constantly flushed with the overflow from the water supply, making it an integrated system. Trouble was taken to seek out a copious supply of water at a spring[7] and carefully graded ducts led it to the fort, which was sometimes several miles away.[8]

Drains discharged into rivers at points well below all watering points for animals and when this was not possible there were large soak-aways which functioned much as the modern septic tank. Where the site of the fort was on a hill and such arrangements impossible, water tanks were constructed below ground; these were presumably lined with lead or clay and had detachable

---

[1] Dioscorides who served in the army under Nero, wrote a pharmacopoeia which was held in great respect by Galen. Lucian mentions the *medicus* of a cavalry regiment who was a historian (*de hist. conscrib.*, 24).

[2] Roy W. Davies, 'The Roman Medical Service', *Saal. J.*, 27 (1970), pp. 84–104.

[3] *Digest*, 1, 6, 7.

[4] This was also the name given to the slaves who acted as cloak-room attendants at public baths.

[5] *C.I.L.*, vi, 3910; xi, 29; and *R.I.B.*, 2315. Galen mentions an oculist attached to the British Fleet, but he may have been a *medicus* who specialized in treatment of the eyes.

[6] An altar in Greek from Chester is dedicated to the 'mighty saviour gods' by Hermogenes who was probably a *medicus* to XX Legion (*R.I.B.*, 461), or he may have been the personal physician of a legionary commander or senior officer.

[7] Vitruvius, viii, 4.

[8] For a study of settling tanks at Benwell see *A.A.*, 4th ser., 19 (1941), pp. 12–17.

wooden covers,[1] and were kept fresh and clean by fatigue parties. Latrines were deep slots cut four or five feet into the ground and fitted with wooden covers; flies do not normally penetrate into such dark depths[2] and the buckets would have been emptied regularly.

Thus the men were carefully selected in the first place for their good physique; they were kept in excellent condition by tough training and had a sensible plain diet, while personal hygiene and effective sanitation lessened the spread of disease.

There was always the hazard of battle and the medical service was organized to deal mainly with casualties. There is a scene on Trajan's Column showing the corps in the field.[3] A wounded legionary is sitting on a rock being attended by an orderly who appears to be carrying out an examination. Next to him is an auxiliary having a thigh wound bandaged by a *capsarius* the soldier grits his teeth in agony and the veins of his arm are swollen as he clutches the rock on which he rests (Pl. XXXb). The main task of the field orderlies was to patch up the wounds and get the men into a hospital (*valetudinarium*) as soon as possible. The main tasks of the *medici* would have been the cleaning and stitching of gashes and sword cuts and extraction of missiles, on which Paul of Aegina gives useful advice.[4] In his chapter on extraction of missiles (vi, 87), he mentions a special kind of forceps; 'If the head of the weapon has fixed in the flesh, it is to be drawn out with the hands or by seizing the shaft, if it has not fallen off and which is commonly made of wood. When it has fallen off we make the extraction by means of a tooth forceps or a stump forceps, or a tool for extraction of weapons (literally a weapon extractor). . . .' If the head of the missile had penetrated too far into the body to make extraction feasible, Paul suggests pushing it forward to emerge from the body at a point opposite the entry—'we make an opening by means of the weapon itself, pushing it by the shaft or, if that has come away, by an impellent, taking care not to divide a nerve, artery or vein or any important

---

[1] As in the Claudian fort on Waddon Hill, Dorset (*P. Dorset Nat. Hist. and Arch. Soc.*, 86 (1965), pp. 138–9); similar tanks have been noted at Hofheim, Abb. 14 and Fendoch, *P. Soc. Ant. Scot.*, 73 (1938–9), p. 129.

[2] As at Waddon Hill. This practice of deep trenches is apparently still considered sound military practice.

[3] Cichorius, scenes 102 and 103.

[4] *de re medica;* an English translation was published by Adams of Banchory for the Sydenham Society in 1846; the sixth book dealing with surgery has also been published in French in 1855 by Briau.

part, for it would be malpractice if in extracting the weapon we should do more mischief than the weapon itself had done. If the weapon has a tang which is ascertained by examination with the probe, having introduced the female part of the impellent and engaged it we push the weapon forward, or if it has a socket, the male part.' It is clear from this that the impellent is a metal rod with a hollow at one end for engaging in a tang, and a point at the other for thrusting into a socket. If the injuries were found to be too serious, the weapon being lodged in a vital place, 'and fatal symptoms have already shown themselves, as the extraction would occasion much laceration, we must decline the attempt, lest while we do no good we expose ourselves to the reprobation of ignorant people. But if the result be dubious, we must make the attempt, having first given warning of the danger.'

Celsus, who also has some pertinent advice on the extraction of weapons (vii, 5), describes in some detail an implement for extracting arrows; '. . . if a broad weapon is buried it should not be extracted by means of another opening lest to one large wound we add another, therefore it is to be extracted with a special instrument known to the Greeks as the Scoop of Diocles. Its blade is iron or bronze and it has two hooks at one end, one at each side turned backwards. The other end is folded over at the sides and slightly curved up towards the part that is bent and in it there is a perforation. It is introduced crosswise near the weapon and when it comes near the point it is twisted a little to receive the point in the hole, two fingers are then placed under the hooks at the other end and both instrument and weapon extracted.' This seems a simple enough operation, but it is difficult to reconstruct in practice and would seem to have little advantage over forceps. In the same passage Celsus describes the use of split reeds, similar to those used for writing, in shielding the barbs of a weapon to facilitate extraction. Paul devotes further sections (89–122) to bone and skull fractures and dislocations which would have been very relevant to the work of the military surgeon. They show great common sense and a considerable knowledge of the human anatomy.

Military surgeons obviously needed special tools or normal ones specially adapted for the work, like the scoop of an ear probe which Paul recommended for dealing with sling stones (vi, 88). If the missile was buried deeply into a bone Paul advocated the use of a drill to widen the aperture in the bone.

Actual implements, usually in bronze, have been found,[1] although there has been a practice of identifying long scoops as surgical, whereas they were

---

[1] *B.M. Guide to Greek and Roman Life,* 1929, Fig. 200.

more likely for extracting ointment out of long-necked bottles. One of the best discoveries has been the instruments at Baden, in Switzerland, found in 1896.[1]

Each legionary fortress was provided with a hospital, as were some of the larger auxiliary forts. The legionary hospital was a large building with an internal courtyard and continuous circulating corridor. The sixty wards arranged in pairs were probably associated with the sixty centuries.[2] The example which has produced the most interesting details is that at Inchtuthil (see p. 195 and Fig. 41).[3] This timber building has double internal walls for insulation against noise and extremes of temperature. The building at Xanten on the Rhine has a large basilica hall at the front which may have been for reception and what appears to be its own small bath suite.[4] In auxiliary forts the building is basically on the same plan, but smaller, and normally it is found only in those of milliary units,[1] as at Housesteads, Fendoch, Birrens and Pen Llystyn (see p. 216 above).

Diet

An important aspect of army health which can hardly be overlooked is the diet. The literary evidence suggests that the principal food was derived from cereals. Caesar invariably uses the word *frumentum* when referring to food supplies.[5] Vegetius lists as necessities, 'corn, wine, vinegar and even salt in plenty at all times'.[6] By corn is meant wheat and not barley, the attitude towards which is indicated by its issue as a punishment to units which had disgraced themselves.[7] On campaigns the corn-meal was baked in the form

---

[1] J. S. Milne, *Surgical Instruments in Greek and Roman Times*, 1907, p. 22. These are now in the Municipal Museum at Baden (Aquae Helveticae) which was an important spa in Roman times. When discovered it was thought that the building in which the instruments were found was a military hospital, but it is now considered more likely to have been a doctor's house.

[2] The officer in command was the *optio valetudinarii* (*C.I.L.*, viii, 2553 and 2563).

[3] *J.R.S.*, 47 (1957), pp. 198–9.

[4] *B.J.*, 137 (1932), pp. 273 ff. Other legionary hospitals occur at Haltern (*Germania*, 12 (1928), p. 74); Novaesium (*B.J.*, 111–12 (1904), p. 180; Carnuntum (*R.L.Ö.*, 7 (1906), p. 47; Caerleon (*J.R.S.*, 55 (1965), p. 199 and Fig. 10).

[5] *Bell. G.*, v, 8, etc.

[6] iii, 3.

[7] Suetonius, *Aug.*, 24; Polybius, vi, 38, 3. According to Columella, *de re rustica*, barley was mainly grown for feeding geese and oxen (viii, 14, 8, and xi, 2, 99), although when mixed with wheat it made excellent food for the domestic staff (ii, 9, 16).

of hard biscuits known as *buccellata*.[1] There is much evidence, summarised by Mr R. W. Davies[2] that the Army diet was well mixed corn, meat, including poultry, fish, cheese and vegetables. On special festivals there would have been extra quantities and special delicacies of which pork was highly favoured. Meat was certainly eaten on special occasions, festivals and feasts, but was not regarded as a daily need. There is an interesting reference by Tacitus to Corbulo's army in Armenia reduced to short rations and keeping starvation away with a meat diet.[3] Caesar's men were in similar straits at the siege of Avaricum; when the corn failed to arrive they could only be saved from starvation by bringing cattle in from distant villages.[4] Army units are known to have possessed *prata* (meadows),[5] which must have been grazing land for cattle or milk animals, since there were soldiers designated as *pecuarii*,[6] who presumably attended to the animals. There is little doubt also that hunting wild boar and deer, apart from the sporting aspects, provided a useful addition of fresh meat.[7] There was probably an increase in the consumption of meat due to the influx of Teutonic elements[8] into the Army, especially in the third and fourth centuries. Excavators of military sites are accustomed to recovering large quantities of animal bones and if more detailed studies could be done more light could be thrown on the soldiers' diet.[9]

---

[1] Ammianus, xvii, 8, 2. Equivalent, no doubt, to the modern 'iron rations'. The Latin word was derived from *buccella* (a small mouthful).

[2] R. W. Davies, 'The Roman Military Diet', *Britannia*, 2 (1971), pp. 122–142.

[3] *Annals*, xiv, 24, *ita per inopiam et labores fatiscebant, carne pecudum propulsare famem adacti.*

[4] *Bell. G.*, vii, 17.

[5] A. Mócsy, 'Zu den prata legionis', *Studien zu den Militärgrenzen Roms*, 1967, pp. 211–14.

[6] *C.I.L.*, xiii, 8287, a soldier of XX from Cologne.

[7] There is an altar from Birdoswald (*R.I.B.*, 1905) set up by *venatores Banniesses* (the huntsmen of Banna, which may have been Bewcastle); see *Rang.*, p. 46, for legionary *venatores* and *C.I.L.*, iii, 7449.

[8] The evidence for the Germans preferring meat is not conclusive and habits varied from tribe to tribe. But they were all keen hunters and fresh meat was much enjoyed according to Tacitus (*Germania*, 23, of the tribes nearest the Rhine, whose food was plain wild fruit, fresh meat and curdled milk) and Caesar (*Bell. G.*, iv, 1, of the Suebi); there are other references assembled by E. A. Thompson, *The Early Germans*, 1965, pp. 3–6, which give a general picture of a pastoral hunting people with grain as a minor part of their diet.

[9] A modest attempt has been made by the author in the case of a Claudian fort in Dorset, but the number of bones is too small to give any reliable statistics. Cattle and sheep were present, but very little pig. Bird bones included doves and pigeons and a fair amount of fowl. In the latter case, however, there were six cocks' spurs indicating gaming birds and five neatly

SOLDIERS' PAY

One of the most difficult aspects of army service to understand is that of the soldiers' pay. It is complicated by the lack of knowledge of what the men received as equipment and rations and precisely how much was compulsorily deducted for various purposes. The situation changed from time to time and with gradual inflation the pay progressively increased.

Basic facts about pay are few and far between. Caesar doubled the daily pay of legionaries according to Suetonius[1] and it can be shown[2] that this increase was from five to ten *asses* a day, 225 *denarii* a year. When Augustus left in his will 300 *sestertii* (75 *denarii*) to all legionaries[3] this was a third of the annual amount and most probably indicates that the troops were paid three times a year and Augustus merely added an extra pay-day. Confirmation comes from Tacitus in a speech by Percennius in the Pannonian Revolt following the death of Augustus. 'Ten *asses* a day,' cried the mutineer, 'was the value of body and soul and out of this we have to buy clothes, weapons and tents, bribe the centurion and pay for release from fatigues. . . .'[4] But all in vain—the basic rate remained unchanged until Domitian, who increased it from nine to twelve gold pieces a year[5] (i.e. to 300 *denarii*). This has been assumed[6] to indicate that there were now four instalments instead of three, but this is not so according to Dio.[7] In spite of the steady inflationary tenden-

on account of its large number of bones. There were also the inevitable remains of shellfish (*P. Dorset Nat. Hist. and Arch. Soc.*, 86 (1965), pp. 142–4).
severed skulls suggesting sacrificial victims. What was interesting and unusual was the quantity of fish bones, probably of the Giant Wrasse, which is today considered to be virtually inedible

[1] *Caesar*, 26.

[2] The most up-to-date study is by G. R. Watson (*Historia*, 5 (1956), pp. 332–40; 7 (1958), pp. 113–20; 8 (1959), pp. 372–8). *The Roman Soldier*, 1969, pp. 89–114.

[3] *Annals*, i, 8. During his reign he had standardized pay and allowances (Suetonius, *Aug.*, 49), but this does not suggest any increase.

[4] *denis in diem assibus animam et corpus aestimari; hinc vestem, arma, tentoria; hinc saevitiam centurionum et vacationes munerum redimi* (*Annals*, i, 17); all this was echoed by the mutineers in Germany; *pretia vacationum, angustias stipendii* (*Annals*, 1, 35).

[5] Suetonius, *Dom.*, 7.

[6] By Domaszewski ('Der Truppensold der Kaiserzeit', *Neue Heidelberger Jahrbücher*, 10 (1900), pp. 218–41), and which Parker followed (p. 216); this has been disputed by P. A. Brunt (*Pap. Brit. School at Rome*, 18 (1950), p. 54).

[7] 67, 3, 5; 'He (Domitian) ordered that 400 *sestertii* should be given to each man in the place of 300 he had been receiving.' This would suggest that it was merely the instalment which was increased and not an additional pay-day.

(a)

Plate XXVIII

(a) Model of a large ballista (*onager*) in the Museum of Roman Civilization, Rome (p. 234).

(b) Legionaries forming a *testudo*, from Trajan's Column (p. 242).

(b)

(a) A mobile field-gun (*carro-ballista*) on Trajan's Column (p. 235).

Plate XXIX

(b) Model of a field-gun in the Saalburg Museum, Germany.

cies during the second century, there is no further increase until the time of Severus.[1]

Occasionally there were bounties and donatives. Caligula after the abortive invasion of Britain gave all legionaries four gold pieces (100 *denarii*)[2] and even Tiberius rewarded the troops of Syria.[3] Claudius started an unfortunate precedent in giving a donative to the praetorians on his accession, and it can be assumed that equivalent amounts would have been given to the legionaries.[4] Later emperors felt obliged to follow this example to secure the loyalty of the troops.[5] The inevitable result was that it was expected,[6] until Vespasian, who had been able to satisfy part at least of his victorious army with booty, quietly dropped the idea, and it does not appear to have been revived until Marcus Aurelius did so.

Apart from the bounties and donatives, the legionaries could look forward to substantial grants on their discharge either in cash or land (*praemia*). Augustus fixed the amount in A.D. 5 at 3,000 *denarii*[7] and by the time of Caracalla it had risen to 5,000 *denarii*.[8] The normal land allotment in the early Empire was 20 by 20 *actus* (200 *iugera*),[9] but this amount would only reflect

---

[1] Herodian, iii, 8, 4; this indicates that Severus was the first to raise the soldiers' pay, from which Domaszewski argued that in all previous cases it has been the number of *stipendia* that had been increased and not the actual amount of each *stipendium*.

[2] Suetonius, *Gaius.*, 46.

[3] Suetonius, *Tib.*, 48; it was a small sum as a reward for their refusal to allow the statues of Sejanus to be placed with their standards.

[4] Suetonius, *Claud.*, 10; the sum of 150 gold pieces (3,750 *denarii*), about seventeen years' pay for a legionary, seems somewhat excessive and may reflect the Emperor's gratitude for being found behind the curtain and spared. He may have considered he had a precedent in Tiberius, who doubled the Augustan bequest (Suetonius, *Tib.*, 48), but only after the mutiny in Germany (*Annals*, i, 36). According to Vegetius, half the donatives were put to the soldiers' savings account (ii, 20). The troops' nickname for the donative was *clavarium* (nail money). The danger of having too much money on deposit in the fortress was recognized by Domitian who restricted the individual amount to ten gold pieces (250 *denarii*) (Suetonius, *Dom.*, 7). This would have created difficulties and, it must be presumed, could not have lasted.

[5] The donative of Nero came when he assumed the *toga virilis* a year ahead of normal citizens (*Annals*, xli, 41).

[6] The troops on the Rhine clamoured for the donative which Vitellius was reputed to have sent and Hordeonius Flaccus gave it them, but in the name of Vespasian; morale had, however, sunk so low that they immediately launched into debauchery (*Hist.*, iv, 36).

[7] Dio, lv, 23.

[8] Dio, lxxviii, 36.

[9] John Bradford, *Ancient Landscapes*, 1957, p. 151.

the cash grant in the areas where land was either appropriated through rights of conquest or purchased cheaply.[1]

The real difficulty in assessing the soldiers' pay is that of the stoppages and deductions. This practice dates back to the origins of the Army. Polybius records that the soldiers had to purchase their corn and clothes and some of their arms, presumably replacements, at a set price which the *quaestor* deducted from their pay.[2] Although attempts were made to alleviate this burden, it remained a source of grievance in the early Empire, as the complaints of the mutineers testify. The details are shown in the much-quoted document from Egypt which appears to show the annual account of Q. Iulius Proculus from Damascus, a soldier in A.D. 81.[3]

		Stip. II dr.	*Stip. III dr.*
Accepit stip. I an. III Do. (Received the first salary payment for our lord's third year)	*dr. 248*	*248*	*248*
Ex eis faenaria (bedding?)	10	10	10
in victum (rations)	80	80	80
caligas fascias (boots)	12	12	12
Saturnalicium K. (annual feast)	20	—	—
ad signa (burial club)	—	4	—
in vestimentis (clothes)	60	—	146
Expensas (expenses)	182	106	248
Reliquas deposuit (remainder deposited to his account)	66	142	—
Et habuit ex priore (balance carried forward)	136	202	344
Fit summa (total)	202	344	344

In Egypt the pay is reckoned in *drachmae*, 248 of which were tariffed at 75 *denarii*. The stoppages include fodder for the horse or mule and rations

[1] Tenny Frank calculated that the *praemia* of 12,000 *sestertii* would have purchased only 8–10 *iugera* (*Economic Survey*, v, 1920, p. 170).

[2] vi, 39.

[3] *Pap. Lat.* I, Geneva; there are actually two accounts, identical but for one item.

and boots for himself,[1] all regular amounts. The small amount put to the standards went towards the burial club and was in the charge of the chief *signifer*.[2] This went to the feast and funeral expenses. The item for clothes is irregular and presumably this was related to replacements. The extraordinary feature about this account is that the soldiers do not seem to have spent any money; their balances are carried forward. Watson argues very plausibly that this document is a record kept by the clerks in charge of the *deposita* and merely shows the sums standing to the men's credits. This is confirmed by other papyri from Fayûm, but of later date,[3] which include a note of withdrawal and debtors. It also contains an item not present in the Geneva papyrus, a deduction for arms for the large amount of 103 *denarii*. There is also evidence that soldiers could obtain advances ahead of pay-day.[4]

In the higher ranks of the legion pay was assessed at one and a half and two times that of the ranker.[5] According to an inscription of the Severan period in a *schola principalium*[6] a *cornicen* who would rank merely as an *immunis* received 500 *denarii*, no more than an ordinary ranker, but the *librarii* and *exacti* received 800 *denarii*, one and a half times as much, the *cornicularius praefecti* double and the *optio ad spem ordinis* three times basic pay. There is no evidence on the pay of centurions, but it seems likely that it was at least five times the soldiers' rate and may even have been more. One of the main privileges of the centurion's position was the practice of levying fees for exemption from fatigues. Tacitus has a passage about this abuse among praetorians and tells us that Otho tried to correct it by making a grant from the treasury of an equivalent amount which would have had the effect of raising centurions' pay. Later he says it became an established rule under good emperors.[7] This suggests that this abuse may well have continued and was only suppressed when emperors like Hadrian imposed stricter discipline. The

---

[1] Nero had allowed praetorians *frumentum sine pretio* (*Annals*, xv, 72), but this evidently did not extend to other troops.

[2] Vegetius, ii, 20.

[3] *Roman Military Records on Papyrus*, No. 68.

[4] An ostracon from a quarry in Egypt where a unit was stationed records a receipt for an advance at the time of Hadrian (*Chronique d'Égypte*, 26 (1951), pp. 354–63; also in Lewis and Reinhold, No. 151).

[5] Vegetius, ii, 7.

[6] *I.L.S.*, 2354, 2438, 2445, 9097, 9099, 9100; it seems highly probable that the *annularium* mentioned here was equivalent to a year's pay.

[7] *Hist.*, i, 46, *et a bonis postea principibus perpetuitate disciplinae firmatam.*

primi ordines probably received twice as much as ordinary centurions,[1] and the *primus pilus* twice as much again. Bounties, donatives and discharge grants would all be in proportion.[2] A *primus pilus* on discharge would receive enough to acquire equestrian status (a property qualification of 400,000 *sestertii*).

The pay of the *auxilia* poses difficult questions through absence of reliable evidence. It seems clear from Hadrian's *allocutio* at Lambaesis that there were basic differentials between units. The cavalry of the *alae* were better paid than the men in the *cohortes* and in the *cohortes equitatae* mounted men got more than the foot soldiers. One can start with the proposition that a legionary considered it a promotion to become a *duplicarius* of an *ala*. This suggests that it was the *sesquiplicarius* who may have received the legionary equivalent in which case an *eques* would have got a *stipendium* of 200 *denarii* under Domitian.[3] By pursuing this ingenious line of argument G. R. Watson scales down the humble foot soldier to 100 *denarii*.[4] This scheme at least has the virtue of simplicity which is lacking in those of others.

---

[1] Domaszewski, *Rang.*, p. 111, suggested that donatives are a reflection of salary scales and this doubling can be shown to follow.

[2] Caligula reduced the discharge grants for senior centurions to 600,000 *sesterces* (Suetonius, *Gaius*, 44).

[3] During the Revolt of Civilis the Batavians claimed the double pay (*duplex stipendium*) promised by Vitellius; this would have increased it to beyond the legionary rate, but may be merely an aspect of the disturbed conditions of A.D. 69.

[4] G. R. Watson, 'The Auxiliary Forces' in his series on pay (*Historia*, 8 (1959), pp. 372–8).

Chapter 6

THE ARMY IN PEACEFUL ACTIVITIES

INTRODUCTION

It is characteristic of historians and commentators of all ages to dwell on the exciting and unusual, and while delighting in wars and battles to ignore the normal pursuits of peaceful citizens as unworthy of note. One may gain a superficial view of the Roman Army engaged continuously in conquests, civil wars and frontier battles, but a closer examination discloses that there were always long periods of peace, and a Roman soldier might well spend his whole lifetime without seeing a hostile face. But does peace mean inactivity for the Army? Here our sources are very inadequate and one has to depend on scraps of papyri of army records, some precious fragments of which have survived in the dry sands of Egypt and the Near East.[1] Soldiers behave much the same the world over and throughout the ages, and the Roman soldiers were often in conflict with the civil authorities. A petition was presented to Gordian in A.D. 238 by the inhabitants of Scaptopara, a small place in Northern Thrace.[2] Their village was near some hot springs, but midway between two forts, and during an annual festival the soldiers were wont to descend on the village and demand hospitality and food for which they were not prepared to pay, in spite of the governor's orders, and the villagers implored the divine Emperor to stop this before they were forced to flee from their homes. Soldiers when on the move and far from their base must have continually helped themselves to the basic needs of life, and the terrified civilians gave in, hoping at least to escape more serious injury. There are examples of attempts to stop this practice. Germanicus on his visit to Egypt in A.D. 19 had occasion to deal with these abuses as a papyrus attests: 'I have deemed it

---

[1] There is much useful source material in Ramsay MacMullen's *Soldier and Civilian in the Later Roman Empire*, 1963, in which the relationship between the Army and the civil population is explored, but the argument that there was greater integration in the third and fourth centuries remains unconvincing.

[2] *C.I.L.*, iii, 12336.

necessary to make known that I wish neither boat nor beast of burden to be seized nor quarters to be occupied by anyone except on the order of my friend and secretary Baebius. . . .'[1] These words were echoed in A.D. 42 by Lucius Aemilius Rectus, prefect of Egypt: 'No one shall be permitted to requisition transportation facilities from the people in the country districts nor demand *viatica* (money for travel) or anything else gratis without a permit from me and those possessing such permits may take sufficient supplies on payment of the price therefor. But if any of the soldiers or police or anyone at all among the aides in the public services is reported to have acted in violation of my edict or to have used force against anyone of the country people or to have exacted money, I shall visit the utmost penalty upon him.'[2] A stronger line was taken by Avidius Cassius when governor of Syria, for he crucified soldiers who had robbed provincials.[3] It is quite clear that had one more evidence from other parts of the Empire it would show the same pattern—soldiers taking what they wanted by threats and also using their position for extortion. It became the accepted order of life and protests were made only when the demands or violence became excessive. A second-century trader's account from Egypt lists payments and bribes as normal running costs.[4] It is reflected even in the Gospels: 'The soldiers also asked him "and what shall we do?" He said to them "Never extort money or even lay a false charge but be content with your pay." '[5]

The Army was responsible for law and order in the provinces as well as their defence, and in effect acted as a police force. Such a force, in the modern sense, was unknown in the ancient world and the protection of citizens and their property was a constant problem. Most of the town authorities owned slaves whose duties included street cleaning, garbage collection and the stoking of bath-houses. Among their other duties was the maintenance of the

---

[1] Hunt and Edgar, *Select Papyri*, ii, 1934, No. 211.

[2] Abbott and Johnson, *Municipal Administration in the Roman Empire*, 1926, No. 162; another document dated A.D. 49 lays down the same requirements (*Supplementum Epigraphicum Graecum*, viii, No. 794); prefects were still issuing orders couched in the same language a century later (see *Select Papyri*, ii, No. 221, dated A.D. 133–7).

[3] This savagery was prompted by the need for support from the wealthy Syrians when Cassius was making a bid for the purple (*S.H.A., Cassius*, iv, 2).

[4] *Revue de Philologie*, 17 (1943), pp. 111–19, list given also in Lewis and Reinhold, ii, pp. 402–3, there are two items 'To the soldier on his demand' for 500 and 400 drachmas, as well as smaller amounts to guards and police spies.

[5] *Luke*, 3, 12–14 (Moffatt translation). See also C. B. Welles, 'The *Immunitas* of the Roman legionaries in Egypt', *J.R.S.*, 28 (1938), pp. 41–49.

local prison, but it seemed unusual even to Pliny that they should also act as guards and warders and he asked Trajan if it would not be better to use troops.[1] But the Emperor was reluctant to detach soldiers from their normal stations and in using them for these kind of services he saw the danger of oppression or extortion by unscrupulous agents as Tiberius had done before him.[2] Trajan was even unwilling to allow the creation of a fire brigade for fear that a group of aspiring young citizens might use such an association for political ends.[3]

There were nevertheless troops legitimately serving on detached duties, but Pliny's letters show how watchful Trajan was in controlling the numbers.[4] A provincial governor needed a large staff for all the administration, guards and special services. His *officium*,[5] normally drawn from the legions, was under a centurion (*princeps praetorii*) and consisted of three secretaries (*cornicularii*), three judicial officers (*commentarienses*),[6] ten messengers (*speculatores*)[7] for every legion in the province, and a considerable number of guards or orderlies (*beneficiarii*)[8] and specialists such as interpreters (*interpretes*),[9] torturers (*quaestionarii*),[10] clerks of different kinds and grooms. The governor's

---

[1] *Ep.*, 10, 19–20. Most prisoners were awaiting trial and sentences did not normally involve a stay in prison, but execution, mutilation, fines or forced labour.

[2] *Annals*, iv, 15; an official in Asia felt the heavy hand of the Emperor for using soldiers to enforce his will.

[3] *Ep.*, 10, 33–34.

[4] *Ep.*, 10, 22.

[5] As reconstructed by A. H. M. Jones from Domaszewski (*J.R.S.*, 39 (1949), pp. 44–45), see also Ruggiero, *Dizionario Epigrafico*, under Legio, iv, pp. 603–5.

[6] They were also in charge of the prisons. These two chief officers would have had deputies (*adiutores*).

[7] Originally, as the name implies, they were scouts, but they became used as dispatch riders and they also served as executioners. In relating the episode of the beheading of John the Baptist at the court of Herod, the word used to identify the executioner is *speculator* (*Mark* vi, 27; also Sherwin White, *Roman Society and Roman Law in the New Testament*, 1963, p. 137). Ulpian in considering the personal belongings of those executed ruled that they should not be taken by the *speculatores, optiones* or *commentariensis*, but should go into a special fund (*Digest*, xlviii, xx, 6). There is an early third-century tombstone of a *speculator* from London who was presumably on the governor's staff (*R.I.B.*, 19).

[8] Under the late republic and early Empire this title covered most of the grades of orderlies, but later became limited to those of the lower ranks. Pliny uses the term for soldiers detached for escort duties (*Ep.*, 10, 27).

[9] *C.I.L.*, iii, 10505 and 14349.

[10] *C.I.L.*, viii, 2568.

bodyguard was a separate unit of *equites* and *pedites singulares*, usually taken from the *auxilia*. Lesser officials had correspondingly smaller staffs.[1]

It was clearly a great advantage for a soldier to secure such a staff posting, since it brought him in personal contact with influential people and it is hardly surprising to find examples of rapid promotion, often into the centurionate.

Hadrian, with his passion for rationalization, reorganized the higher ranks of the civil service in such a way that the effect was to produce specialists. The normal equestrian and senatorial careers of the first century provided experience in both fields, but from Hadrian's day it was possible to specialize in one or the other. But precisely how deeply this went and the effects of this change are difficult to determine. It has been argued that in the absence of a strong central authority it was disastrous, since the Army had no knowledge of, or sympathy with civil administration;[2] this surely is an extreme view. There appears to be little difference in the provincial *officia* except that there is an increase in the number of officials such as the *beneficiarius*, an officer who by the late second century is found apparently associated with places where customs could be collected.[3] A British example can be cited from Dorchester on Thames, where a *beneficiarius consularis*[4] may have been a customs officer at a port on the Thames where goods may have been transferred from river to road or *vice versa*.[5] Although this post could have been purely civilian in function, the officer was still attached to his legion, as inscriptions elsewhere attest.[6] Similarly there are the *stationarii*, who were in charge of the small posts along the main Imperial routes and presumably responsible for toll-collection and police work.[7] In the law digests there are repeated enactments which

---

[1] Pliny records a procurator being allocated ten *beneficiarii* under a centurion. It is interesting to note that when the British procurator Catus Decianus sent a force to help the veterans at Colchester in A.D. 60 it was a force of two hundred men without proper arms (*sine iustis armis*), and it is likely that he retained some for his own safety as he made his escape to Gaul (*Annals*, xiv, 32).

[2] *C.A.H.*, xi, p. 432.

[3] Patsch, *Mitteilungen des deutschen archäeologischen Instituts*, 8 (1893), p. 193 and Domaszewski, *Westd. Z.* 21 (1902), pp. 158 ff.

[4] *R.I.B.*, 235; the stone has been lost, but a drawing by Horsley (Pl. 76) shows a style of ligature which would fit a late second–early third century dating. Most of these officials belong to the post-Severan period (Ramsay MacMullen, p. 68).

[5] Richmond, *Roman Britain*, 1955, p. 96.

[6] *C.I.L.*, iii, 2023; 6376 = 8656; 14703, etc.

[7] For main sources, mostly from the Near East, see Ramsay MacMullen, p. 56, fn. 20. A

state that these officials cannot exact taxes[1] or act as judges or put people in prison. Clearly the latter acts are for higher authorities, but the prohibition of tax-collecting may reflect the changed conditions in the very late period of the Empire when there may have been a further division of officials distinguishing between those handling money and those engaged in law enforcement. Certainly by this time there were *stationarii* whose duty was to report crime in the towns.[2] Like the *beneficiarii*, they had quasi-military functions under the governor or procurator. An official whose functions underwent a more definite change was the *frumentarius*. Originally this officer was responsible for the collection of corn in the provinces. His travels among the provincials gave him opportunities of exchanging and collecting gossip which he no doubt reported to his commander on reaching base. This useful role was developed and Hadrian turned them into agents,[3] but the name remained unchanged. To pursue this interesting development of the provincial civil service takes us beyond the Army of the first two centuries A.D., since the best picture of the peacetime activities of the Army comes from the Fayûm in Egypt, where a set of documents has survived from the archives of a cavalry *ala* in the middle of the fourth century.[4] The procurator of the Imperial estates writes asking for a detachment to aid his official to collect taxes; a priest requests the loan of some nets to catch gazelles which are destroying the crops; an official requests the commander to impound all the natron;[5] a request is made for a military escort for an official; a tax-collector (*exactor*)[6] requests help in an obscure situation in which he is in danger; the same official asks for justice to be done in the case of thirty villains who

---

suggestion has recently been made by R. W. Davies that the Army may have helped in the running of the *cursus publicus* (*Latomus*, 26 (1967), pp. 67–72).

[1] *Cod. Just.*, 4, 61, 5; see also De Laet, *Portorium*, pp. 209 and 268.

[2] A. H. M. Jones, *The Later Roman Empire* A.D. *284–602*, 1964, pp. 521 and 600.

[3] *S.H.A.*, Spartianus, *Hadrian*, 11; the use of *frumentarii* as spies may have become established earlier, but Hadrian extended the practice to the surveillance of the senatorial families.

[4] These documents have been published in a single volume, *The Abinnaeus Archive*, 1962.

[5] A mineral occurring in Egypt, chemically sodium sesquicarbonate, used as fertilizer and an important and basic constituent of soap and some cosmetics. It was an imperial monopoly, hence the need to protect the supplies. Today this substance is mainly used as a water softener.

[6] These letters show clearly the relationship between the Army and the civil official in the gathering of taxes. The accountancy and collection was the responsibility of the *exactor*, but he expected armed support from the unit and the abrupt tone of his letters indicates his sense of equality or even superiority to the commander.

attacked his grandson; a local dignitary vigorously protests at the soldiers conscripting villagers, looting and committing other outrages; a farmer complains that a soldier with others has sheared eleven of his sheep in the night . . . and so on.

Although these documents belong to the fourth century and apply to the province of Egypt, the unique character of which makes it difficult to draw comparisons, most of the Army's activities and the complaints made against it would have applied to all parts of the Empire at all times. Escorts for officials, escorts for tax-collectors and action against smugglers are commonplace, and always the soldiers are alleged to have seized goods and beaten up citizens. Perhaps the last word can come from Juvenal,[1] 'Let us first consider the benefits common to all soldiers of which not the least is that no civilian would dare to thrash you, if thrashed himself he must hold his tongue and never venture to show his wounds to the Praetor . . .'

RELIGION

Philosophic ideas and beliefs belonged only to the highly educated minority. As distinct from this, the religions of the Near East with their strong ties of brotherhood, mystical experiences and hope of salvation spread rapidly throughout the Empire in the wake of the Army. At a lower level, there was a steady persistence of the many forms of animism. To the untutored soldier the world was full of spirits who in varying degrees menaced all his actions. It was necessary to come to terms with these influences, especially if he was far from home. In new surroundings there were unknown beings and all over the empire there is silent witness in the numerous altars dedicated to the *genius loci,* the spirit whose very name is not known but who presides in the hill, river, swamp or forest where the uneasy soldier pitches his tent. As one could come to terms with the spirits, so logically one could also bargain with them. Imbued in the Romans was a strong streak of common sense stiffened by a sound legal bent. This is manifest in the extraordinary numbers of altars recording the fulfilment of vows. The soldier solemnly swore before witnesses that should he come through the battle unscathed he would erect an altar to Mars or his appropriate local affinity. The officer out hunting swore that should he kill the boar he was then chasing, an altar would be dedicated

[1] *Satires,* xvi, on the advantages of a military life; this refers to praetorians at Rome.

to Silvanus. The precise nature of the oath is rarely stated; merely that the donor freely and willingly fulfilled his vow.[1] The twin themes placation and contract were well developed in Republican times in the official Army ceremonies. The soldiers naturally fought better knowing that the gods were on their side and elaborate arrangements were made to ensure this. Before all campaigns and battles, sacrifices were offered and vows undertaken. Guidance was also sought by means of omens,[2] and this was naturally subject to manipulation. While the troops may have been eager for battle, the commander could well have seen that the moment was inopportune. The omens were then found to be unfavourable and the men satisfied with the delay. Other omens such as the flight of birds, appearance of eagles, winds, storms, cloud formations, could easily affect the soldiers with good or disastrous results and Roman historical accounts are full of such incidents, many, of course, related with the advantage of hindsight.

While units were stationed at their permanent headquarters there was a regular calendar of festivals to be observed with parades and sacrifices. By remarkable fortune an official document setting these out in detail has been recovered from the fort at Dura-Europos on the Upper Euphrates.[3]

The Feriale Duranum

This is a roll of papyrus probably discarded when obsolete in the third decade of the third century. It came from the archives of the *Cohors XX Palmyrenorum* and is a list of festivals to be observed by the unit with an indication of the necessary sacrifices; it is an official document such as would have been issued to every unit in the Army. We learn from the list that the 7th of January was celebrated as one of the three pay-days and also for the discharge of the veterans from the unit. The festivals of leading deities, Jupiter, Apollo, Mars, Neptune and Hercules, were celebrated with the sacrifice of a bull, the female deities were given cows and past Emperors male oxen. The *dies imperii* of the reigning Emperor Severus Alexander was the 13th of March[4] with

[1] The formula *Votum solvit libens merito* almost invariably occurs.

[2] This involved a study of the internal organs of the sacrificial animals, a lore learnt from the Etruscans. There is a fine example of an Army sacrificial parade or *suovetaurilia* (a corruption of *sus ovis taurus*) on Trajan's Column (Cichorius, Pl. 38).

[3] *Feriale Duranum* was found by the Yale University Expedition in the 1931–2 season and has been published as Volume 7 of the *Yale Classical Studies*, 1940, by R. O. Fink, A. S. Hoey and W. F. Snyder.

[4] This is but one useful piece of new information this document gives, the precise date not being previously known.

appropriate sacrifices to Jupiter, Minerva, Mars and other deities whose names are missing on the document. The wording, however, seems to make it clear that on the same day both the Emperor's elevation and recognition by the Senate was celebrated with the salutation as *imperator* by the troops. The 18th to 22nd of April saw the five-day popular festival, the *Quinquatria*, associated with the opening of the campaign season; at this time in Rome the arms, horses and trumpets of the Army would be purified with ceremony. Of special military significance were the May *Rosaliae Signorum* festivals, when the standards were paraded decorated with wreaths of roses.[1] This appears to be a civil festival, the carnival element having a stronger survival than its original religious purpose. It is interesting to note that the auxiliary *vexilla* and all other standards would be included under the general official heading of *signa*.

It is surprising, for example, to find that the 23rd of June was still celebrated, as late as the third century as the birthday of Germanicus Caesar, nephew of Tiberius. He was always apparently regarded as a great military hero.[2] The 27th of July was the day on which the Emperor had received his *toga virilis*, an important step in a citizen's life and in this case to be celebrated with the sacrifice of a bull, and later in the month there was the day on which he became consul for the first time. The birthdays of all deified Emperors and their deified consorts continued to be festivals from Julius Caesar onwards and there were also the other public holidays which account for twenty-one of the forty-one entries. There were eighteen animal sacrifices and twelve *supplicationes* which would involve a kind of thanksgiving ceremony led by the commanding officer.

The *Feriale* shows very clearly how the official religious ceremonies and festivals could be used to indoctrinate barbarians into Roman ways. There are no concessions here to Eastern ideas and practices, the document is all thoroughly Roman and the cumulative effect of years of ritual on barbarian and Eastern recruits must have been considerable.

Archaeological evidence of the official cults is found at the edge of the unit's parade ground in the form of altars. At convenient intervals the altars were buried to prevent desecration. By chance a collection of stones has been

---

[1] *Harvard Theological Review*, 30 (1937), pp. 15–35.

[2] Modern historians have not assessed his capabilities very highly (F. B. March, *Reign of Tiberius*, pp. 69–104; Sir Ronald Syme sums him up as 'a versatile and amiable mediocrity', *C.A.H.*, 10 (1934), p. 622), but as avenger of the Varian disaster he must have remained a great symbol.

found in Britain, at Maryport in Cumberland.[1] Chief among any such collection would be the altars to Iuppiter Optimus Maximus and it is significant that at Maryport there are twenty-two, and possibly twenty-three, examples.[2] Another interesting collection is the five altars from Auchendavy, on which there are no less than twelve dedications.[3] This group of deities is only otherwise found among the troopers of the *equites singulares*, and as Professor Birley has indicated, the centurion was probably a member of the corps earlier in his career. Viewed as an insurance policy, it was wise to cover all contingencies, and it is worthy of note that the goddess of victory appears twice, a double indemnity. Inside the forts evidence of religion is rare apart from the chapel of the standards in the *principia* (p. 189) and an altar in the courtyard of the same building. There is no space for further shrines or temples, which tend in consequence to appear in the *canabae* and civil settlements.[4] However, in the Severan legionary works depot at Corbridge was found evidence of these practices and this has been the subject of a penetrating analysis by Sir Ian Richmond.[5] Remains of six temples or shrines were identified. No dedication could be directly associated with any of these

---

[1] *R.I.B.*, Nos. 815–17, 819, 822, 824–8, 830–1, 838–42 and 843; it is possible that some of the other altars found earlier may have originally belonged to the same group (*T. Cumb. and West.* 39, (1939), pp. 19–30.

[2] It is thought that it was customary for the commander to dedicate a new altar to Jupiter every year. It has also been argued that as each new annual dedication took place when the soldiers renewed their oath of loyalty to the Emperor so the old altar was carefully buried, but the evidence is far from conclusive and fails to explain the failure to discover many more altars at other forts (*T. Cumb. and West.* 65 (1965), pp. 115–17). The worship of Jupiter, closely associated with the head of the State, would be primarily based in any fort on the *principia* (*C.I.L.*, viii, 1839; iii, 13443), but the parade ground can be seen as a natural extension of the courtyard of this building and where the whole unit could be drilled and paraded. It is interesting to note that there are at least twenty-two Jupiter altars from Birdoswald (*R.I.B.*, 1874–96, except 1884, the dedication of which does not survive).

[3] *R.I.B.*, 2174–8. No. 2178 is only a fragment; the deities are Jupiter, Diana, Apollo, Victory, Mars, goddesses of the parade ground, Hercules, Epona and Silvanus and the spirit of the land of Britain and also the welfare of the Emperor and his family. All the altars were erected by Marcus Cocceius Firmus, whom Professor Birley very persuasively identified with a centurion mentioned in Justinian's Digest and a quaestor of a town on the Black Sea (*P. Soc. Ant. Scot.*, 70 (1936), pp. 363–77, reprinted in *Roman Britain and the Roman Army*, 1953, pp. 87–103).

[4] P. Salway, *The Frontier People of Roman Britain*, 1965, pp. 18–21; see also M. J. T. Lewis, *Temples in Roman Britain*, 1966.

[5] *A.A.*, 4th ser., 21 (1943), pp. 127–224, 'Roman legionaries at Corbridge, their supply-base, temples and religious cults'.

buildings, but Corbridge has produced a wealth of inscriptions and sculptured detail amply reflecting both the official military choice and that of the soldiers. The Capitoline triad Jupiter, Juno and Minerva would have been recognized within the compound and probably Victory also. Other deities which must surely have been official are Discipulina and Concordia, and as Richmond reminded us, the different legionary detachments working in the compound probably needed the occasional divine intervention.

There were also those religions which, as indicated above, touched the deeper emotional levels of the soldiers and the chief among these was Mithraism.[1] It is by no means an accident that this Persian cult was so well adapted to the Army. Only men could be initiated and in small groups forming a close-knit brotherhood with grades of membership, akin to Army ranks. There were training periods and initiation rites and meetings took place in dark underground caverns built to simulate the original cave in which Mithras was born. In this deep gloom the thoughtful application of lighting, scent-laden smoke from burning pine-cones or herbs and liturgical intonation worked on the minds and senses of the brotherhood, inducing deep emotional fervour and mystical revelations. Remains of these small temples have been found along all the frontiers and one which has been most carefully studied is that near the fort of Carrawburgh on Hadrian's Wall.[2]

The first building of the early third century is only eighteen feet by twenty-six feet externally and could not have held more than ten or twelve men. One of its most interesting features was a small bunker for pine-cones which had been carbonized by careful roasting, and when burned in this state would have given off a strong pine aroma, clearly an emotional stimulant. The temple

---

[1] There is a large body of literature on this subject, but the only general account in English is *The Mysteries of Mithras* by Franz Cumont (republished 1956); for inscriptions and buildings see M. J. Vermaseren, *Corpus Inscriptionum et Monumentorum Religionis Mithriacae*, i–ii, (1956–60).

[2] I. A. Richmond and J. P. Gillam, 'The Temple of Mithras at Carrawburgh', *A.A.*, 4th ser. 29 (1951), pp. 1–92; for a short general account there is a booklet published by the Museum of Antiquities of the University of Newcastle upon Tyne by Charles Daniels, *Mithras and his Temples on the Wall*, 1962; for other Mithraic temples in Britain see *Arch. Camb.*, 109 (1960), pp. 136–72 (Segontium); *A.A.*, 4th ser. 40 (1962), pp. 105–15 (Housesteads); *J.R.S.*, 45 (1955), pp. 137–8 and Pl. xlii–xlviii, and *Recent Archaeological Excavations in Britain*, 1956, pp. 136–43; *A.A.*, 4th ser. 32 (1954), pp. 176–218 (Rudchester). It has been shown that the temple at Burham in Kent was a Roman cellar (*Arch. Cant.*, 70 (1956), pp. 168–71); there is also some doubt about a building at Colchester which has been so identified (*Roman Colchester*, 1958, pp. 107–13).

was soon enlarged by fourteen feet and to this was added a square apse projecting six feet which held the reredos with its Mithraic relief. Benches supported by timber wattling stood on each side of the nave and on these the worshippers reclined. The next stage was a thorough remodelling of the interior. The introduction of posts suggests a change in the roof structure. The narthex was now fitted with an ordeal pit lined with stone and sunk into the floor, where initiates could be subjected to heat and cold while lying entombed. In the fourth stage the ordeal pit was replaced by a bench and wood floooring put down in the nave as if the trials were no longer as rigorous as before. The building was destroyed by fire and deliberately wrecked in the troubles at the end of the third century. A new temple was erected on the old foundations differing in plan only at the apse end, where the reredos was now housed, and which at this stage became merely a niche. The side benches were shorter, the room for reclining worshippers reverting, in fact, almost to that of the very first building, although there was space for lower grades to stand at the back of the nave. It was this temple which was finally abandoned, but not before it suffered desecration. This took the form of breaking some of the statuary, but leaving the altars untouched. This selective destruction is dated by Richmond to 324, when the struggle between Mithraism and Christianity had political aspects. Once the building ceased being used the drain became blocked and the water table rose. It was this event which preserved the remains and enabled its excavators to make such an excellent attempt at reconstruction. The three altars found undisturbed had been dedicated by three prefects of the unit. On one of these altars was carved a relief of the great Mithras with a radiate crown, the rays of which were cut through the stone so that a lamp placed in a niche at the back would have lit the crown like a halo.

Christianity was another mystic religion which came from the East, but appears to have had little success in the military areas, although it was adopted as the official Imperial religion in the fourth century.[1]

Burial Customs

It was the common practice in the ancient world to honour the dead, since it was firmly believed that their spirits continued life on another plane and un-

[1] The best general account is by J. M. C. Toynbee, 'Christianity in Roman Britain', *J. Brit. Arch. Ass.*, 3rd ser. 16 (1953), pp. 1–24. The evidence for the northern area is listed by John Wall in *A.A.*, 4th ser. 43 (1965), pp. 201–25. There is a tradition of two martyrs, Aaron and Julius, suffering their fate at a military station in Britain. See also G. R. Watson, 'Christianity in the Roman Army in Britain', *Christianity in Britain, 300–700*, 1968.

less the burial was conducted in a proper manner the spirit could return and cause unpleasantness to the neglectful. The Romans viewed this as a solemn contract with the dead. All soldiers were assured of proper rites, since they were all members of a burial club financed by compulsory deductions from pay. This would have covered the funeral feast to speed the departing spirit on its way, and the burial itself. The erection of a monument was the pious duty of the heirs, although occasionally this was an instruction laid down in the will of the deceased.[1]

These tombstones and monuments are very important to students of the Roman Army, because they give us much information—the name of the soldier, his birthplace, his rank and unit, age and years of service, and sometimes the name or names of his heir or heirs. In the case of an officer his previous career is often listed. Thus there is evidence of the presence of units at particular forts. It is also possible to obtain statistics relating to recruitment[2] and the information on steps in the ladder of promotion provides us with the framework for the organization of the Army.[3] Some monuments also bear a relief of the effigy of the deceased as if on parade. Cavalrymen are depicted riding over the prostrate foe, usually a naked, cowering, hairy barbarian. As Richmond has pointed out, this is intended to symbolize victory over death rather than any earthly glory.[4] Although these reliefs were originally covered with gesso and painted in natural colours, the carvings supply us with much accurate detail of uniform and equipment, especially important where the name of the unit is given on the inscription. In well-established forts the troops, many of whom could reckon on being there with their families most of their lives, provided elaborate monuments equivalent to family vaults.[5] On the anniversary of the death there would have been a simple ceremony at the grave or tomb and libations made.[6]

---

[1] A common formula is H.F.C. (*Heres* or *Heredes faciendum curavit* or *curaverunt*, i.e. his heir or heirs caused this to be erected) and a Lincoln stone (*R.I.B.*, 257) reads T(ESTAMENTO) P(ONI) I(USSIT), i.e. He gave instruction in his will for this to be set up.

[2] For example, G. Forni, *Il reclutamento delle legioni da Augusto a Diocleziano*, 1953.

[3] It is the basis of the great work of von Domaszewski, *Die Rangordnung des römischen Heeres*, 2nd ed. by B. Dobson, 1967.

[4] I. A. Richmond, *Archaeology and the After-Life in Pagan and Christian Imagery*, University of Durham, Riddell Memorial Lecture, 1950.

[5] This is clear from the remains of such structures at Chester, R. P. Wright and I. A. Richmond, *Catalogue of the Roman Inscribed and Sculptured Stones in the Grosvenor Museum, Chester*, 1955.

[6] An example of this practice is seen in the pipe burial found at Caerleon. This consisted

(a)

(b)

Plate XXX
Scenes from Trajan's Column

(a) a watch tower and signal station (p. 248).
(b) a field dressing station (p. 252).

Plate XXXI
A forage party from Trajan's Column.

Disposal of the dead underwent a change during the second century. During the early years of the Empire cremation was the normal practice and was gradually replaced by inhumation until by the fourth century this in turn was almost universal in the Roman world. Cremation certainly had advantages for the Army. Soldiers killed on the battlefield far from base could be cremated on the spot and their ashes, often a token amount, placed in a vessel of pottery or glass and then dispatched to the appointed burial place.[1] The remains of high-ranking officers were delivered to the family in Italy or elsewhere and placed in the family tomb. The practice also led to the use of *columbaria*, tombs with many niches where the vessels could be placed, so-called because of their similarity to dovecots. An example has been noted beyond the Newport Arch at Lincoln.[2] Cemeteries and cremation pyres (*ustrina*) were always placed outside the forts and are normally to be found alongside the roads. Many of the tombs suffered despoliation at the end of the fourth century or earlier in Britain and Germany, when the massive stones were removed to strengthen or rebuild fort defences.[3] Those that survived were removed by later stone-robbers or cultivators, especially when the steam plough was introduced in the nineteenth century and stones just below the soil were encountered and dug out.[4]

CIVILIZING INFLUENCES

It is doubtful if the Roman administrators ever saw themselves as civilizing influences any more than the colonial officers in modern times in the more

---

of a cremation in a lead-cist, out of the top of which projected a vertical pipe into which libations could have been poured at ground-level (*Ant. J.*, 9 (1929), pp. 1–7).

[1] There were notable exceptions when the remains could not be recovered as those in the disaster of Varus. Tacitus gives a dramatic picture of the men under Germanicus, six years later, gathering up the remains and the general himself laying down the first sod of a funeral-mound, only to be rebuked by Tiberius, since as an *augur* Germanicus should not have touched any object associated with the dead (*Annals*, i, 61 and 62).

[2] *Assoc. Archit. Soc. Rep.*, 22, 57; *Arch. J.*, 103 for 1946 (1947), p. 52. There is also recorded the small tombstone (*R.I.B.*, 261) of a soldier which might have come from such a structure.

[3] For example, the north wall of the legionary fortress at Chester has yielded a fine collection and it may be supposed that many more fragments are yet to be found in the stretch to the east of the north gate which remains untouched.

[4] It is very rare that even in the highland zone any traces survive of upstanding tomb monuments in the north-western provinces; a notable exception may be at the mountain site of Tomen-y-Mur (*R.C.H.M.*, Merioneth, p. 150; *Arch. Camb.*, 93 (1938), pp. 192–211; *The Roman Frontier in Wales*, 1954, p. 37).

backward areas of Africa or Asia. Nevertheless the upper levels at least of Roman society possessed a deep sense of the destiny of Rome and a feeling that their nation had been singled out from the earliest times as the natural ruler of the world. This is implicit in the great history of Livy and the old stories and legends were adapted and shaped to this end. But it was not until the rise of Augustus that Rome's destiny seemed to have been fulfilled. Now could Vergil put into the mouth of Anchises the true role of the new Empire:

> Let it be your task, Roman, to control the nations with your power (these shall be your arts) and to impose the way of peace; to spare the vanquished and subdue the proud.[1]

Words which are echoed through the centuries from Aelius Aristides[2] to the late fourth-century poet Claudian.[3]

Roman governors and high-ranking officials, military and civil, spread the Roman way of life, not through any sense of duty, but because this was the only acceptable way, and they would fail to understand people who had the means yet rejected the concept.

There was also a strong practical aspect which helped to emphasize this Roman feeling of rightness, since warlike barbarians, humbled by defeat, donned the toga and became respectable citizens. It was this combination of realities and sense of inevitability, so typically Roman, that led to the successful assimilation of alien peoples within the folds of the Empire. But there were Roman cynics like Tacitus who saw this process merely as enslavement, the free men being seduced by the soft pleasure of civilization. The most revealing passage concerns the Britons and comes in the *Agricola*:[4]

'The following winter (i.e. A.D. 78–79) was spent in introducing new schemes. A scattered and savage people, addicted to war, were to be drawn from their habits into the pleasures of quietude and peace. (Agricola) gave private encouragement and official help towards the building of temples, *fora*

---

[1] *Aeneid*, vi, 851–3.

> Tu regere imperio populos, Romane, memento
> (Haec tibi erunt artes) pacisque imponere morem,
> Parcere subiectis et debellare superbos.

[2] This panegyric *To Rome* was delivered in Rome *c.* A.D. 150, see J. H. Oliver, *T. American Phil. Soc.*, 43 (1953), Pt. 4.

[3] *de cons. Stilichonis*, iii, 150–9.

[4] *Agricola*, 21. Perhaps Tacitus was reacting to the death of Domitian two years before this work was published.

and private houses, praising those who quickly responded and castigating the slothful, the desire to gain honour became more effective than compulsion. He also arranged for the sons of the chiefs to be educated in the liberal arts and expressed a preference for the natural ability of the Britons to the studiousness of the Gauls. A distaste for the Latin tongue gave place to an aspiration to speak it with eloquence, thereafter our national dress became acceptable and the toga was commonly seen. Gradually they were drawn towards the blandishments of vice, colonnades, baths and elegant banquets, but what were regarded as refinements of civilization were in effect their chains of enslavement.'

To Rome the only possible way of life was that of the town and the most remarkable aspect of the Empire is the spread of these urban centres, each consciously modelled, however modestly, on Rome itself. The main features were the gridded plan of the streets, the enormous public buildings such as the forum with its market-place and law-courts, the temples and the bath-houses, many of which survive in part or ruin simply because they were of such durable and solid construction. The wonder is that these were successfully imposed on people totally alien to these conceptions—to the Briton it was a revolution indeed. It has even been considered that it was an entirely artificial scheme which disintegrated when faced with serious economic stresses;[1] such a view is no longer valid. In spite of all the difficulties, the British way of life was totally transformed within a few generations, but how deeply this change penetrated into the lower strata of society is another matter. The mass of peasants tied to the land may have viewed these splendours from a distance and been touched not by the gentle refinements of the new civilization but by the lash of a rapacious new landowner.

The wealth of the Empire was based on agriculture and trade and they are aspects which leave little record in the pages of contemporary history or in inscriptions. It has been left to scholars in this century like Rostovtzeff,[2] Tenney Frank[3] and Charlesworth[4] to show the extensive, closely knit trade relationships throughout the Empire and beyond.[5] The life-blood of the

---

[1] This was the view of R. G. Collingwood, *Roman Britain and the English Settlements*, 1937, pp. 202–4.

[2] In his brilliant survey of vast scope, *The Social and Economic History of the Roman Empire*, 2nd ed., 1947.

[3] *An Economic History of Rome*, 1920.

[4] A work on a more modest scale, but a useful summary, *Trade Routes and Commerce of the Roman Empire*, 1924.

[5] Sir Mortimer Wheeler, *Rome beyond the Imperial Frontiers*, 1954.

Empire was its commerce developed by hard-headed entrepreneurs in Rome and fed by a remarkable network of communications by road, sea and river. But this would have been quite impossible but for the *Pax Romana* maintained by the Army. As long as there were no civil wars and the barbarians were kept firmly beyond the frontiers, the great variety of peoples within the Empire could prosper under the rule of Rome in new and unaccustomed ways of life.

But the responsibility of the Army went far beyond the mere guardianship of the frontiers. It was the army staff officers who were responsible for much of the administration and justice in the imperial provinces. Engineers, architects and surveyors would doubtless have been made available for advice and technical assistance where large public buildings were constructed and water and drainage projects undertaken.[1] This would apply especially to those provinces, such as Britain, where these skills were non-existent. There is, however, little direct evidence of army participation in the early period, except on road and bridge building,[2] which are of strategic importance, and such projects as the canal of Drusus which linked the Rhine with the Zuyder See.[3] When large-scale emergency schemes were carried out, the Army must have taken part at least in planning and supervision; an example could have been the Car Dyke in Britain which helped to bring the Fens into cultivation and also to supply a link by water between this area and first Lincoln via the Witham, the Fossdyke and possibly later York via the Trent and Ouse.[4]

---

1 When Augustus made Egypt a Roman province he used troops to clear out the canals and improve the irrigation of the Nile waters and so increase Egypt's grain yields (Suetonius, *Aug.*, 18). But Egypt was a special case and treated almost as the Emperor's personal estate (*C.A.H.*, x, pp. 284 and 285) and evidence of the use of soldiers at this time or later on maintenance and construction projects (*S.H.A.*, *Probus*, 9) cannot be used to illustrate a general principle.

2 Tacitus, *Annals*, i, 20.

3 Apart from the usefulness of such projects there was probably the need to keep the Army occupied and fit during a lull in hostilities, as when Marius used his waiting army to improve the navigation of the Rhône (Plutarch, *Marius*, xv), and later Corbulo in Germany put his soldiers to dig a canal between the Maas and the Rhine, twenty-three miles long, to keep the troops from idleness (*ut tamen miles otium exueret—Annals*, xi, 20).

4 *Ant. J.*, 29 (1949), pp. 145–63; the canal was found to have been thirty feet wide at the base and fifty feet at the top with a depth of water of four to five feet. It was thus designed for barges, but also had the function of a catch-water drain. The date of construction from the pottery found in the rapid silt would appear to be in the first century and it has been suggested that its main purpose was the need to increase the corn supply for the Army, since the British

Unless masons and architects had been specially imported from Gaul, their army equivalents must have been available for the planning of the early *fora* and *thermae* of Roman Britain. The British *fora* form a distinctive pattern unlike those of other provinces. The striking similarity between the plans of most of the British *fora* and the military headquarters building can hardly be a coincidence,[1] and one may be justified in seeing here the hand of the military technician. Only at Verulamium, the only known first-century forum, is there a lack of conformity and closer approach to the Gallic plan.[2] The use of soldiers as miners was by no means exceptional,[3] but guarding and supervision was a common practice, since prison labour was often used and the mines were in wild mountain areas.[4]

The founding of *coloniae* was one means by which it was hoped that Roman ideas would become established in newly conquered areas. This was an eminently sensible and practical way in which the problems of demobilization could be solved. The veteran settlements were designed to convert professional soldiers into respectable citizens instead of their banding together and resorting to brigandage. In some cases the sites chosen were those from

---

Iron Age economy would not have allowed large surpluses. The date of this canal is now thought by Dr Peter Salway to be early second century.

[1] The most detailed exposition of this is by D. Atkinson, *Report on Excavation at Wroxeter, 1923–27*, 1942, Appendix C, pp. 345–60. The suggestion by Collingwood, however, that the Wroxeter forum was based on the plan of a double legionary fortress can no longer be accepted, as it now seems unlikely that both legions (XX and XIV) were stationed there together at the same time, except possibly as a temporary measure in A.D. 69.

[2] Dated by an inscription (*Ant. J.*, 36 (1956), pp. 8–10; 37 (1957), pp. 216–18). For the *forum* see *St Albans Archit. and Arch. Soc. T.*, (1953), pp. 15–25; the forum at Cirencester may also be of first-century date, but too little is at present known of its plan for useful comment.

[3] There is the interesting case of Curtius Rufus, who may have attempted to emulate Corbulo, whom Claudius compensated with triumphal honours after forbidding a campaign in Germany (*Annals*, xi, 20). Curtius Rufus achieved the same honours by employing his troops to search for silver in the territory of the Mattiaci (near Wiesbaden). But the yield was poor and the legionaries found the underground work exhausting. A secret letter on behalf of the whole Army was sent to the Emperor suggesting that honours might be awarded before a governor took up his command.

[4] A pig of lead, the waste product of the silver-works, stamped by the Second Legion (*C.I.L.*, xiii, 3491) under Nero is clear evidence of military participation in mining in the Mendips in the early conquest period, but this is the only evidence of the Army at this task in Britain (*Flints Hist. Soc.*, 13 for 1952–3, pp. 8–12). Cold-struck stamps on a pig from Alouettes, near Chalon-sur-Saône (*C.I.L.*, xiii, 2612 b) may merely show military ownership of the lead.

which legions had been moved; the veterans might have spent most of their lives there and raised families. This was a technical irregularity which was put right on discharge. The idea of settling there in the old and accustomed haunts would have had a great appeal.[1] When an attempt was made to transplant veterans in Italy at Tarentum and Antium to arrest depopulation, most of them returned to their provinces, 'in which they had served for so long and since they were unable to marry and rear families the homes they had left were childless and without heirs. . . . The settlers now were strangers among strangers; men from totally different maniples; without leaders, with no common cause except that they were all soldiers, put into one place to make up an aggregate rather than a colony.'[2]

The *colonia* had definite advantages; it not only gave a strategic reserve in newly conquered areas at least for a few years,[3] but the sons of the veterans would turn naturally to the colours and so *coloniae* were valuable centres of recruitment. The setting up of a model city was also intended as an example to the natives, but theory, as is often the case, was not always reflected in practice. The colonists settled by Ostorius Scapula at Camulodunum (Colchester) receive a scathing comment from Tacitus: 'They were acting as if they had been given the whole country, driving the natives from their homes, evicting them from their lands, they called them "captives" and "slaves".'[4] These veterans were hardly the model citizens in their model town, exemplars to all aspiring Britons; but the main focus of hatred and anti-Roman feeling was the great temple of Claudius, prematurely deified,[5] which was the *arx aeternae dominationis* (citadel of our eternal tyranny).

One might indeed consider whether the service in the Army could be regarded as a civilizing influence in itself. Many of the recruits to the rank and file were deliberately taken from the barbarous frontier districts, the legionaries being given citizenship on enlistment (see p. 107). They served most of their lives on the frontiers and would rarely have visited any of the large cities. The young recruits were moulded into disciplined soldiers, tied by oath of loyalty to the emperor, forming part of a complex sophisticated

---

[1] This amalgam of native with Roman population is well illustrated at Cologne (*Colonia Claudia Ara Agrippinensis*) by the speech put by Tacitus into the mouths of its envoys in A.D. 70 (*Hist.*, iv, 65).

[2] *Annals*, xiv, 27.

[3] As Cicero envisaged, *Pro Fonteio*, 13.

[4] *Annals*, xiv, 31.

[5] Seneca, *Apocolocyntosis*, 8.

organization, and soon adopting the ideas of personal hygiene reflected in the bath-house. The traditional view of soldiers as unruly and licentious is usually true of times of unrest or when they are far from their base. In the frontier forts and fortresses the young barbarians quickly learnt the basic elements of civilization by becoming members of an organized community and becoming exposed to Roman influences at least in the religious and social spheres. There were opportunities for betterment open to the bright young man with ambition and intelligence. If one accepts the Army as a rough but effective way of introducing fresh blood into the ranks of Roman civilization it is, numerically at least, impressive, for it means on an average about 10,000 new citizens a year, If, however, one adds their wives and families,[1] the total can be at least doubled and probably trebled over a hundred years. The Army could well have been responsible for adding three million citizens to the roll. All this helped to dilute the value of citizenship and it is likely that Caracalla's *constitutio Antoniniana* was not such a sweeping extension of the franchise as might appear at first glance.[2]

While a large number of young men were steadily brought from the fringes of the Empire under the influence of Rome it is clear that it was by no means a one-way effect. The steady flow of barbarians into the Army had an equal consequence of diluting the Roman way of life. The old Roman virtues so evident in the early Empire tended gradually to fade, but the most obvious loss is in the arts; while architecture continued to blossom with technical innovation there was a marked coarsening in sculpture and a more obvious decline in literature.

While these effects were only slow and gradual throughout the second century, the process may have been quickened by Severus, who was not readily accepted by the Senate and sought to bind the Army closer to himself and his family. The Army became the road to advancement even into

---

[1] Up to the time of Pius, the granting of a discharge certificate gave civic rights to the recipient, his wife and his family, both those he already possessed or those he might acquire after discharge. But *c.* 138 there was a change in the formula on the diploma (the omission of the phrase *liberis posterisque eorum*) which implies that children born before discharge no longer enjoyed civic rights. This may not have affected many since, by this time, most of the auxiliaries were citizens on enlistment. Nevertheless it seems to be a peculiar change for this period and may have been due to factors which now elude us (Cheesman, pp. 32–34).

[2] Its main purpose, as Dio underlines, was to bring more people into the range of taxation and civic responsibilities which had hitherto been the privilege of citizenship. Nor is it clear precisely who was now included and who excluded.

the Senate. Severus, according to Rostovtzeff, 'militarized the principate',[1] but at the same time encouraged liberal legislation and protection for the lower classes from which the military caste was drawn. While critics may deplore the dimming of ancient glories, it was this egalitarianism with the continuous introduction of so much new blood which enabled the Empire to hold together for so long and adapt itself to the changing conditions of the third and fourth centuries.

---

[1] *The Social and Economic History of the Roman Empire*, 1957, p. 404.

BIBLIOGRAPHY

MAIN HISTORICAL SOURCES

Ammianus Marcellinus
Arrian, *Tactica*
Caesar, *de Bello Alexandrino*
de Bello Africo
de Bello Civili
de Bello Gallico
de Bello Hispaniensi
Cassius Dio, *Historia Romana*
Celsus
Frontinus, *Strategemata*
Hygini gromatici, *liber de munitionibus castrorum*
Josephus, *Wars of the Jews*
Livy
Paulus of Aegina, *de re medica*
Pliny the Elder, *Naturalis Historia*
Pliny the Younger, *Epistulae*
Plutarch, *Vitae parallelae*
Polybius, *Historiae*
Sallust, *Bellum Jugurthinum*
Scriptores Historiae Augustae
Suetonius, *de vita Caesarum*
Tacitus, *Agricola*
Annals
Germania
Histories
Vegetius, *Epitoma rei militaris*
Vitruvius, *de architectura*

GENERAL

F. E. Adcock, *The Roman Art of War under the Republic,* 1940

A. Alföldi, 'Rhein und Donau in der Römerzeit', G. *pro Vindonissa,* (1949), pp. 3–19

A. Alföldi, 'The moral barrier on Rhine and Danube', *Congress of Roman Frontier Studies, 1949,* 1952, pp. 1–16

H. I. Bell, V. Martin, E. G. Turner and D. van Berchem, *The Abinnaeus Archive,* Oxford, 1962

E. Birley, 'Noricum, Britain and the Roman Army', *Beiträge zur älteren europäischen Kulturgeschichte,* I (1952), pp. 175–88

E. Birley, 'Senators in the Emperors' Service', *P. Brit. Acad.,* 39 (1954), pp. 197–214

E. Birley, *Roman Britain and the Roman Army,* Kendal, 1953

E. Birley, 'The Epigraphy of the Roman Army', *Actes du Deuxième Congrès International d'Épigraphie Grecque et Latine,* 1953, pp. 226–38

E. Birley, 'Hadrianic frontier policy', *Carnuntina, Vorträge beim internationalen Kongress der Altertumsforscher, Carnuntum 1955,* 1956, pp. 25–33

B.M. Guide to Greek and Roman Life, 1929

P. A. Brunt, 'The Army and the land in the Roman Revolution', *J.R.S.,* 52 (1962), pp. 69–86

R. Cagnat, *L'armée romaine d'Afrique et l'occupation militaire de l'Afrique sous les empereurs,* 2nd ed., Paris, 1913

R. G. Collingwood, *The Archaeology of Roman Britain,* London, 1930

R. G. Collingwood and R. P. Wright, *The Roman Inscriptions of Britain,* I, Oxford, 1965

D. R. Dudley and G. Webster, *The Roman Conquest of Britain,* London, 1965

A. M. Duff, *Freedmen in the Early Roman Empire,* revised ed., Cambridge, 1958

J. Filip, *Celtic Civilisation and its Heritage,* Prague, 1960

R. Forrer, *L'Alsace romaine,* 1935

Sheppard Frere, *Britannia,* London, 1967

E. Gren, *Kleinasien und der Ostbalkan in der wirtschaftlichen Entwicklung der römischen Kaiserzeit,* 1941. Deals with economic aspects of the Army in the Balkans and Asia Minor.

S. Gsell, *Les Monuments antiques d'Algérie,* 2 vols, Paris, 1901

J. Guey, 'Inscriptions du second siècle relatives à l'annone militaire', *Mélanges d'archéologie et d'histoire de l'Ecole française de Rome,* 55 (1938), pp. 56 ff.

M. Haberling, *Die römischen Militärärzte,* 1910

B. W. Henderson, *The Life and Principate of the Emperor Hadrian,* 1923

M. G. Jarrett, 'The African contribution to the Imperial Equestrian Service', *Historia,* 12 (1963), pp. 209–26

M. G. Jarrett, '*Legio* II *Augusta* in Britain', *Arch. Camb.,* 113 (1964), pp. 47–63

F. Koepp, *Germania Romana, ein Bilder-Atlas,* 2nd ed., published in six parts, 1924–30

W. Krämer, ed., *Neue Ausgrabungen in Deutschland,* 1958

J. Kromayer and G. Veith, *Heerwesen und Kriegführung der Griechen und Römer,* Munich, 1928

J. Lesquier, *L'Armée romaine d'Égypte d'Auguste à Dioclétien* (Mémoires publiés par les membres de l'Institut français d'archéologie orientale du Caire 41), Cairo, 1918

N. Lewis and M. Reinhold, *Roman Civilisation,* Vol. 1 The Republic, Vol. 2 The Empire, New York, 1955. A useful source-book

R. MacMullen, *Soldier and Civilian in the Later Roman Empire,* Cambridge, Mass., 1963

H. M. D. Parker, *The Roman Legions,* 1928, reprinted 1958

H. M. D. Parker, 'The *antiqua legio* of Vegetius', *Class. Q.,* 26 (1932), pp. 137–49

H. F. Pelham, *Essays on Roman History,* edited by F. Haverfield, 1911

H. von Petrikovits, 'Über die Herkunft der Annäherungshindernisse an den römischen Militärgrenzen', *Studien zu den Militärgrenzen Roms,* 1967, pp. 215–20

H.-G. Pflaum, *Les carrières procuratoriennes équestres sous le Haut-Empire romain,* Paris, 1960–1

G. Ch. Picard and H. Le Bonniec, 'Du nombre et des titres des centurions légionnaires sous le Haut-Empire', *Revue philologique,* 11 (1937), pp. 112–24

I. A. Richmond, 'Recent discoveries in Roman Britain from the air and in the field', *J.R.S.,* 33 (1943), pp. 45–54

I. A. Richmond, 'Gnaeus Iulius Agricola', *J.R.S.,* 34 (1944), pp. 34–45

I. A. Richmond, 'Roman Britain and Roman military antiquities', *P. Brit. Acad.,* 41 (1955), pp. 297–315

I. A. Richmond, 'New evidence upon the achievements of Agricola', *Carnuntina, Vorträge beim internationalen Kongress der Altertumsforscher, Carnuntum 1955,* 1956, pp. 161–7

I. A. Richmond, ed., *Roman and Native in North Britain,* 1958

I. A. Richmond, 'The Roman Frontier Land', *History,* 44 (1959), pp. 1–15.

E. Ritterling, 'Military forces in the senatorial provinces', *J.R.S.*, 17 (1927), pp. 28–32
E. Ritterling, *Fasti des römischen Deutschland unter dem Prinzipat*, 1932
M. I. Rostovtzeff, *The Social and Economic History of the Roman Empire*, 2nd ed., revised by P. M. Fraser, 1967
W. Roy, *The Military Antiquities of the Romans in Britain*, 1793
E. de Ruggiero, *Dizionario Epigraphico di Antichità Romane*
P. Salway, *The Frontier People of Roman Britain*, Cambridge, 1965
F. Staehelin, *Die Schweiz in römischer Zeit*, 3rd ed., 1948
C. G. Starr, *Roman Imperial Navy* 31 B.C.–A.D. 324, 2nd ed., Cambridge, 1960
A. Stein, *Der römische Ritterstand*, Munich, 1927
E. Stein, *Die kaiserlichen Beamten und Truppenkörper im römischen Deutschland unter dem Prinzipat*, Wien 1932
R. Syme, 'Rhine and Danube legions under Domitian', *J.R.S.*, 18 (1928), pp. 41–55
R. Syme, 'Some notes on the legions under Augustus', *J.R.S.*, 23 (1934), pp. 14–33
R. Syme, *Tacitus*, Oxford, 1958
J. H. Thiel, *Studies on the History of Roman Sea-power in Republican Times*, 1946
F. Wagner, *Die Römer in Bayern*, 4th ed., 1928
G. Webster, 'Fort and town in early Roman Britain', *The Civitas Capitals of Roman Britain*, 1966, pp. 31–45
L. Wickert, 'Die Flotte der römischen Kaiserzeit', *Würzburger J.*, 4 (1949–50), pp. 100–25
J. J. Wilkes, 'The military achievement of Augustus in Europe; with special reference to Illyricum', *Uny. Birmingham Hist. J.*, 10. 1 (1965), pp. 1–27
H. Zwicky, *Zur Verwendung des Militärs in der Verwaltung der römischen Kaiserzeit*, Zurich, 1944

FRONTIERS

E. Fabricius, 'Limes', P-W., 13, 1926, cols. 571–671
Congress of Roman Frontier Studies—published proceedings:
E. Birley, ed., *The Congress of Roman Frontier Studies, 1949*, Durham, 1952
E. Swoboda, ed., *Carnuntina, Vorträge beim internationalen Kongress der Altertumsforscher, Carnuntum 1955*, in *Römische Forschungen in Niederösterreich*, Band 3, Graz-Köln, 1956

Limes-Studien, Vorträge des 3. Internationalen Limes-Kongresses in Rheinfelden-Basel 1957, in *Schriften des Institutes für Ur- und Frühgeschichte der Schweiz*, Basel, 1959

Quintus Congressus internationalis limitis Romani studiosorum, Zagreb, 1963

Studien zu den Militärgrenzen Roms, Vorträge des 6. Internationalen Limes-kongresses in Süddeutschland, Beiheft der *B.J.*, Band 16, Köln, 1967

Britain

E. Birley, 'Roman garrisons in the north of Britain', *J.R.S.*, 22 (1932), pp. 53–59

E. Birley, *The Centenary Pilgrimage of Hadrian's Wall*, 1949

E. Birley, 'Roman garrisons in Wales', *Arch. Camb.*, 102 (1952–3), pp. 9–19

E. Birley, *Research on Hadrian's Wall*, 1961. A complete review of all the work on the Wall

J. C. Bruce, *The Roman Wall*, 3rd ed., 1867

J. C. Bruce, *Handbook to the Roman Wall*, 12th ed., by I. A. Richmond, 1965

O. G. S. Crawford, *Topography of Roman Scotland*, Cambridge, 1949

M. G. Jarrett, 'The military occupation of Roman Wales', *Bull. Board Celtic Studies*, 20 (1963), pp. 206–20

M. G. Jarrett, 'The garrison of Maryport and the Roman Army in Britain', *Britain and Rome*, 1966, pp. 27–40

M. G. Jarrett, 'The Roman frontier in Wales', *Studien zu den Militärgrenzen Roms*, 1967, pp. 21–31

M. G. Jarrett, 'Aktuelle Probleme der Hadriansmauer', *Germania*, 45 (1967), pp. 96–105

G. Macdonald, *The Roman Wall in Scotland*, 2nd ed., Oxford, 1934

S. N. Miller, ed., *The Roman Occupation of south-western Scotland*, Glasgow, 1952

V. E. Nash-Williams, *The Roman Frontier in Wales*, Cardiff, 1954

I. A. Richmond, 'The Antonine frontier in Scotland', *J.R.S.*, 26 (1936), pp. 190–4

I. A. Richmond, 'The Romans in Redesdale', *Northumberland County History*, 15 (1940), pp. 63–154

I. A. Richmond, 'Hadrian's Wall, 1939-1949', *J.R.S.*, 40 (1950), pp. 43–56

I. A. Richmond and K. A. Steer, '*Castellum Veluniate* and civilians on a Roman frontier', *P. Soc. Ant. Scot.*, 90 for 1957–8 (1959), pp. 1–6.

Anne S. Robertson, *The Antonine Wall*, 1968. A handbook published by the Glasgow Archaeological Society; includes a bibliography

Grace Simpson, *Britons and the Roman Army*, London, 1964. Concerned almost entirely with the problems of Wales in the second century

K. A. Steer, 'The nature and purpose of the Expansions on the Antonine Wall', *P. Soc. Ant. Scot.*, 90 for 1957–8 (1959), pp. 161–169

K. A. Steer, 'The Antonine Wall, 1934–59', *J.R.S.*, 50 (1960), pp. 84–93

C. E. Stevens, *The Building of Hadrian's Wall*, Cumb. and West., extra series No. 20, 1966

G. Webster, 'The Roman military advance under Ostorius Scapula', *Arch. J.*, 105 for 1958 (1960), pp. 49–98

G. Webster, 'The Claudian frontier in Britain', *Studien zu den Militärgrenzen Roms*, 1967, pp. 42–53

Africa

J. Baradez, *Fossatum Africae*, Paris, 1959

J. Baradez, 'Compléments inédits au *Fossatum Africae*', *Studien zu den Militärgrenzen Roms*, 1967, pp. 200–10

R. Cagnat, *Les Deux Camps de la Légion IIIe Auguste de Lambèse*, 1908

M. Euzennat, 'Le *Limes* de Volubilis', *Studien zu den Militärgrenzen Roms*, 1967, pp. 194–9

R. G. Goodchild and J. B. Ward-Perkins, 'The *Limes Tripolitanus* in the light of recent discoveries', *J.R.S.*, 39 (1949), pp. 81–95

R. G. Goodchild, 'The *Limes Tripolitanus* II', *J.R.S.*, 40 (1950), pp. 30–38

J. Guey, 'Note sur le *limes* romain de Numidie et la Sahara au IVe siècle', *Mélanges d'archéologie et d'histoire de l'Ecole française de Rome*, 56 (1939), pp. 178 ff.

F. de Pachtère, 'Les camps de la troisième légion en Afrique aux premiers siècles de l'empire', *Comptes rendus de l'Académie des Inscriptions et Belles-Lettres*, 1916, p. 273

Rhine

Reports on the German and Austrian frontiers have been published in *O.R.L.* and *R.L.Ö.*

D. Baatz, 'Zu den älteren Bauphasen des Odenwald-Limes', *Studien zu den Militärgrenzen Roms*, 1967, pp. 86–89

J. E. Bogaers, *Militarie en burgerlijke nederzettingen in Romeins Nederland*, 1959

Olwen Brogan, 'The Roman *Limes* in Germany', *Arch. J.*, 92 for 1935 (1936), pp. 1–41

H. Dragendorff, *Westdeutschland zur Römerzeit,* 2nd ed., 1919
U. Kahrstedt, 'The Roman frontier on the Lower Rhine in the Early Roman period', *Congress of Roman Frontier Studies, 1949,* 1952, pp. 41–54
F. Koepp, *Die Römer in Deutschland.* 3rd ed., 1926
H. von Petrikovits, 'Beobachtungen am niedergermanischen Limes seit dem zweiten Weltkrieg', *Saal. J.,* 14 (1955), pp. 7 ff.
H. von Petrikovits, *Das römische Rheinland. Archaologische Forschungen seit 1945,* Beiheft *B.J.,* 8 (1960)
H. von Petrikovits, *Die römischen Streitkräfte am Niederrhein,* Führer d. Rhein. Landesmuseums Bonn, 13, 1967
W. Schleiermacher, *Der römische Limes in Deutschland,* ein archäologischer Wegweiser für Autoreisen und Wanderungen, 3rd ed., 1967
H. Schönberger, 'Neuere Grabungen am obergermanischen und rätischen Limes', *Limesforschungen,* 2 (1962), pp. 69–121
K. Stade, 'Der römische Limes in Baden', *Badische Fundberichte,* 2 (1929)

Danube

G. Alföldy, 'Die Truppenverteilung der Donaulegionen am End des 1. Jahrhunderts', *Acta Arch. Acad. Scient. Hungaricae,* 11 (1959), pp. 113–41
A. Betz, 'Die römischen, Militärinschriften in Österreich', *J. Oest.,* 29 (1934)
E. Condurachi, 'Neue Probleme und Ergebnisse der Limesforschung in Scythia Minor', *Studien zu den Militärgrenzen Roms,* 1967, pp. 162–74
C. Daicoviciu and D. Protase, 'Un nouveau Diplôme Militaire de Dacia Porolissensis', *J.R.S.,* 51 (1961), pp. 63–70
A. Frova, 'The Danubian *limes* in Bulgaria and excavations at Oescus', *Congress of Roman Frontier Studies, 1949,* 1952, pp. 23–30
V. Groh, 'Les Restes du *limes* romain en Tchécoslovakie', *Eos,* 32 (1929), pp. 665 ff.
F. Hertlein, O. Paret and P. Gössler, *Die Römer in Württemberg.* vols 1–3, 1928–32
F. Křížek, 'Das Problem der römischen Grenzen am nordpannonischen Limes', *Limes Romanus Konferenz, Nitra,* 1957
F. Křížek, 'Neue Ergebnisse der römischen Forschung in der Tschechoslowakei', *Limes-Studien, Vorträge des 3. Internationalen Limes-Kongresses in Rheinfelden, 1957,* 1959, pp. 77–84.
E. Paraiteseu, 'Le *limes* dacique', *Bull. Acad. Roum.,* 15 (1929)
G. Pascher, *Römische Siedlungen und Strassen im Limesgebiet zwischen Enns und Leitha, R.L.Ö.,* 19, 1949

R. M. Rohrer, 'Römische Siedlungen und Strassen im Limesgebiet zwischen Inn und Enns (Oberösterreich)', *R.L.Ö.*, 21, 1958

W. Schleiermacher, 'Flavische Okkupationslinien in Raetien', *J.R.G.Z.M.*, 2 (1955), pp. 245–52

C. Schuchhardt, 'Die sogenannten Trajanswalle in der Dobrudscha', *Preuss. Akad. der Wiss. Phil-Inst.*, Kl., 12 (1918)

E. Swoboda, *Forschungen am obermoesischen Limes*, 1939

R. Syme, 'The first garrison of Trajan's Dacia', *Laureae Aquincenses*; (1938), pp. 267–86

R. Syme, 'The Lower Danube under Trajan', *J.R.S.*, 49 (1959), pp. 26–33

J. Szilágyi, 'Roman Garrisons stationed at the Northern Pannonia—Quad Frontier—Sections of the Empire', *Acta Arch. Acad. Scient. Hungaricae*, 2 (1952), pp. 189 ff.

W. Wagner, *Die Dislokation der römischen Auxiliarformationen in den Provinzen Noricum, Pannonien, Moesien und Dakien von Augustus bis Gallienus*, in *Neue Forschungen*, 203, 1938

Near East and Egypt

Y. Aharoni, 'Forerunners of the Limes: Iron Age Fortresses in the Negev', *Israel Exploration J.*, 17 (1967), pp. 1–17

V. Chapot, *La frontière de l'Euphrate*, Paris, 1907

M. Gichon, 'Roman Frontier Cities in the Negev', *Quintus Congressus internationalis limitis Romani studiosorum*, 1963, pp. 195–207

M. Gichon, 'Idumea and the Herodian *Limes*', *Israel Exploration J.*, 17 (1967), pp. 27–42

M. Gichon, 'The Origin of the *Limes Palaestinae* and the major phases of its development', *Studien zu den Militärgrenzen Roms*, 1967, pp. 175–93

R. G. Goodchild, 'The Roman and Byzantine *Limes* in Cyrenaica', *J.R.S.*, 43 (1953), pp. 65–76

R. K. McElderry, 'The Legions of the Euphrates Frontier', *Class. Q.*, 3 (1909), pp. 44 ff.

A. Poidebard, *La Trace de Rome dans le désert de Syrie*, 1934

W. M. and A. M. Ramsay, 'Roman garrisons and soldiers in Asia Minor', *J.R.S.*, 18 (1928), pp. 181–90

A. Stein, 'Note on the Remains of the Roman *Limes* in North-western Iraq', *Geographic J.*, 92 (1938), pp. 62 ff.

R. E. M. Wheeler, 'The Roman frontier in Mesopotamia', *Congress of Roman Frontier Studies, 1949*, 1952, pp. 112–29

LEGIONARY FORTRESSES, THEIR *CANABAE* AND *TERRITORIA*

Lauriacum excavation reports appear in *R.L.Ö.*, Heft. 8–11 and 13–15

Vindonissa excavation reports appear in *Jahresberichte G. pro Vindonissa*, issued by the Brugg Vindonissa-museum

C. Albrecht, 'Das Römerlager in Oberaden', *Veröffentlichungen aus dem Städt. Museum für Vor- und Frühgeschichte Dortmund*, Band 2, Heft 1 (1938)

C. Albrecht (ed.), *Das Römerlager in Oberaden und das Uferkastell in Beckinghausen*, vols 1 and 2, Dortmund 1938 and 1942

D. Baatz, *Mogontiacum*, in *Limesforschungen*, 4, 1962

G. Behrens, 'Neue und ältere Fund aus dem Legionskastell Mainz', *M.Z.*, 12–13 (1917–18)

A. Betz and H. Kenner, 'Ausgrabungen und Funde in Lagerfriedhof von Carnuntum', *R.L.Ö.*, 17 (1937), pp. 23 ff.

J. E. Bogaers, 'Romeins Nijmegen', *Numaga*, 12 (1965), pp. 10 ff.

G. C. Boon, *Isca: The Roman Legionary Fortress at Caerleon, Mon.*, 3rd ed., 1962. Includes a full bibliography. The National Museum of Wales also published a separate plan with comparative material in 1967

H. Brunsting, 'Romeins Nijmegen', *Numaga*, 7 (1960), pp. 6–27

H. Brunsting, 'Romeins Nijmegen—De Nijmeegse Castra', *Numaga*, 8 (1961), pp. 49–67

M. P. M. Daniëls, *Noviomagus—Romeins Nijmegen*, Nijmegen, 1955

A. von Domaszewski, 'Die *Principia* des römischen Lagers', *Neue Heidelberger J.*, 9

A. von Domaszewski, 'Die *Principia* et *Armamentaria* des Lagers von Lambaesis', *Westd. Z.*, 1902

R. Fellmann, 'Das Zentralgebäude der römischen Legionslager und Kastelle', *G. pro Vindonissa*, (1957–8), pp. 75 ff.

R. Forrer, *Das römische Strassburg, Argentorate*, I and II, 1927

W. F. Grimes, 'The Works-Depôt of the Twentieth Legion at Castle Lyons, Denbighshire', *Y Cymmrodor*, 41 (1930)

von Groller, Carnuntum excavation reports, *R.L.Ö.*, Heft. 1–10 and 12

J-J. Hatt, 'Le Passé de Strasbourg Romain d'après les Fouilles de 1947 et 1948', *Revue d'Alsace*, 88 (1948), pp. 81–96

J-J. Hatt. 'Découverte de Vestiges d'une Caserne Romaine dans l'Angle du Castrum d'Argentorate', *Cahiers d'Archéologie et d'Histoire d'Alsace* (1949)

J-J. Hatt. 'Contribution des fouilles de Strasbourg (1947–1957) à l'histoire de la défense romaine sur le Rhin et sur le *Limes*', *Limes-Studien, Vorträge des 3. Internationalen Limes-Kongresses in Rheinfelden, 1957*, 1959, pp. 49–54

W. A. Jenny and H. Vetters, *Forschungen in Lauriacum,* Band 1, 1953

L. Klima and H. Vetters, 'Das Lageramphitheater von Carnuntum', *R.L.Ö.*, 20, 1953

A. Kloiber, *Die Gräberfelder von Lauriacum, Das Ziegelfeld,* 1957

K. Kraft, 'Das Enddatum des Legionslagers Haltern', *B.J.*, 155–6 (1956), pp. 95–111

G. Kropatscheck, 'Ausgrabungem bei Haltern', *Mitteilungen der Altertums-Kommission für Westfalen,* 5 (1909)

V. Kuzsinszky, *Aquincum. Ausgrabungen und Funde,* Budapest, 1934

R. Laur-Belart, *Vindonissa: Lager und Vicus. Römisch-Germanische Forschungen,* 10, Berlin-Leipzig, 1935

R. Laur-Belart, 'Fortschritte in der Erforschung des Legionslagers Vindonissa', *Carnuntina, Vorträge beim internationalen Kongress der Altertumsforscher,* 1956, pp. 91–94

H. Lehner, 'Novaesium', *B.J.*, 111–12 (1904)

H. Lehner, *Das Römerlager Vetera bei Xanten,* 1926. A short guide

H. Lehner, 'Vetera, die Ergebnisse der Ausgrabungen des Bonner Provinzial-museums bis 1929', *Römisch-Germanische Forschungen,* 4 (1930), pp. 40 ff. A summary, for interim reports see *B.J.*

H. Lorenz, *Untersuchungen zum Praetorium,* 1936. (Vetera)

S. N. Miller, 'Roman York, excavations of 1925', *J.R.S.*, 15 (1925), pp. 176–94

S. N. Miller, 'Roman York, excavations of 1926–27', *J.R.S.*, 18 (1928), pp. 61–99

A. Mócsy, 'Das Territorium legionis und die Canabae in Pannonien', *Acta Arch. Acad. Scient. Hungaricae,* 3 (1953), pp. 179 ff.

A. Mócsy, 'Zu den *Prata Legionis*'. *Studien zu den Militärgrenzen Roms,* 1967, pp. 211–14

H. Nissen, 'Geschichte von Novaesium', *B.J.*, 111–12 (1904), pp. 1 ff.

E. Nowotny, 'Das Territorium legionis von Carnuntum', *R.L.Ö.*, 18 (1937), pp. 129 ff.

H. von Petrikovits, *Novaesium, Das römische Neuss,* 1957

H. von Petrikovits, 'Die Legionfestung Vetera II', *B.J.*, 159 (1959), pp. 89–133

H. von Petrikovits, 'Die Ausgrabungen in Neuss', *B.J.*, 161 (1961), pp. 449–85

H. von Petrikovits, 'Mogontiacum—Das römische Mainz', *M.Z.*, 58 (1963), pp. 27 ff.

S. K. Póczy, 'Die Töpferwerkstätten von Aquincum', *Acta Arch. Acad. Scient. Hungaricae*, 7, fac. 1–4 (1958)

H. G. Ramm, 'Roman York: Excavations in 1955', *J.R.S.*, 46 (1956), pp. 76–90

R.C.H.M., *Eburacum: Roman York*, 1962

I. A. Richmond, 'The Agricolan legionary fortress at Inchtuthil', *Limes-Studien, Vorträge des 3. Internationalen Limes-Kongresses in Rheinfelden 1957*, 1959, pp. 152–5

I. A. Richmond, *Hod Hill*, ii, 1968, London

M. I. Rostovtzeff, ed., *The Excavations at Dura-Europos*, 5th season, 1934. Includes the *Principia*

A. Schulten, 'Das Territorium legionis', *Hermes*, 29 (1894), pp. 481 ff.

R. Schultze, 'Das *Praetorium* von Vetera', *B.J.*, 126 (1921), pp. 1 ff.

R. Schultze, 'Die römischen Legionslazarette in Vetera und anderen Legionslagern', *B.J.*, 139 (1934), pp. 54–63

K. Schumacher, 'Das römische Mainz', *M.Z.*, 1 (1906), pp. 19 ff.

G. Stefan, 'Le camp romain de Drajna-de-Sus, Dép. de Prahova', *Dacia* (1948), pp. 115 ff. The fortress of XI Legion from 101 to *c.* 130

P. Steiner, 'Ein römischer Legionsziegelofen bei Xanten', *B.J.*, 110 (1903), pp. 70 ff.

P. Steiner, 'Xanten', *Kataloge West- und Süddeutscher Altertumssammlungen*, 1 (1911)

A. Stroh, 'Untersuchung an der Sudostecke des Lagers der Legio III in Regensburg', *Germania*, 36 (1958), pp. 78–89

P. Stuart, *Gewoon aardewerk uit de romeinse legerplaats en de bijbehorende grafvelden te Nijmegen*, 1963. Contains a bibliography

E. Swoboda, *Carnuntum: Römische Forschungen in Niederösterreich*, Band 1, 1964. The most up-to-date summary

J. Szilágyi, 'Die Bedeutung von Aquincum im Spiegel der neuesten Ausgrabungen', *Carnuntina, Vorträge beim internationalen Kongress der Altertumsforscher*, 1956, pp. 187–94

J. Szilágyi, *Aquincum*, Budapest, 1956

J. Szilágyi, 'Erforschung der Limes-Strecke im Bereich von Aquincum', *Limes-Studien, Vorträge des 3. Internationalen Limes-Kongresses in Rheinfelden 1957*, 1959, pp. 167–74

F. H. Thompson, 'Roman Lincoln, 1953', *J.R.S.*, 46 (1956), pp. 22–36

F. H. Thompson, 'The legionary fortress of "Deva" (Chester), Recent Discoveries', *Latomus,* 58 (1962), pp. 1491–1500

F. H. Thompson, *Deva: Roman Chester,* 1959

G. Ulbert, 'Das römische Regensburg', *Beiheft zum Gymnasium,* 1 (1960), pp. 64–77

G. Ulbert, 'Die römische Keramik aus dem Legionslager Augsburg-Oberhausen. *Materialhefte zur Bayer. Vorgeschichte,* 14 (1960)

G. Webster, 'The Legionary Fortress at Lincoln', *J.R.S.,* 39 (1949), pp. 57–78

R. P. Wright and I. A. Richmond, *The Roman Inscribed and Sculptured Stones in the Grosvenor Museum, Chester,* 1955

AUXILIARY FORTS

D. Baatz, 'Limeskastell Echzell, Kurzbericht über die Grabungen 1963 und 1964'. *Saal.-J.,* 22 (1965), pp. 139–57

E. Birley, 'Excavations at Birrens, 1936–37', *P. Soc. Ant. Scot.,* 72 (1938), pp. 275–347

D. Christison and J. Macdonald, 'The Excavations at Birrens in 1895', *P. Soc. Ant. Scot.,* 30 (1896), pp. 81–199

J. Curle, *A Roman Frontier Post and its People: The Fort of Newstead in the Parish of Melrose,* 1911

C. M. Daniels, 'The Roman bath-house at Red House, Beaufront, near Corbridge', *A.A.,* 4th ser., 37 (1959), pp. 85–176

A. E. van Giffen, 'Opgravingen in de Dorpswierde te Ezinge en de romeinse terpen van Utrecht, Valkenburg Z.H. en Vechten', *Jaarverslag van de Vereeniging voor Terpenonderzoek,* 19–22 (1948). With English summary.

A. E. van Giffen, 'Three Roman frontier forts in Holland at Utrecht, Valkenburg and Vechten', *Congress of Roman Frontier Studies, 1949,* 1952, pp. 31–40

W. Glasbergen, '42 n.C. Het eerste jaartal in de geschiedenis van West-Nederland', *Jaarboek der Koninklijke Nederlandse Akademie van Wetenschappen,* 1965–6

R. Hogg, 'Excavation of the Roman auxiliary tilery, Brampton'. *T. Cumb. and West.,* 65 (1965), pp. 133–68

L. Jacobi, *Das Römerkastell Saalburg bei Homburg vor der Höhe,* Homburg, 1897

G. Mildenberger, 'Untersuchungen im Kastell Risstissen', *Germania,* 39 (1961), pp. 69–87

S. N. Miller, *The Roman Fort at Balmuildy*, Glasgow, 1922

G. Müller, *Untersuchungen am Kastell Butzbach,* in *Limesforschungen,* 2 1962

I. A. Richmond and J. McIntyre, 'The Agricolan Fort at Fendoch', *P. Soc. Ant.Scot.*, 73 (1939), pp. 110–54

I. A. Richmond and F. A. Child, 'Gateways of forts on Hadrian's Wall', *A.A.*, 4th ser., 20 (1942), pp. 134–54

I. A. Richmond, 'Excavations at the Roman fort of Newstead 1947', *P. Soc. Ant. Scot.*, 84 for 1949–50 (1950), pp. 1–38

E. Ritterling, 'Das frührömische Lager bei Hofheim im Taunus', *Annalen des Vereins für Nassauische Altertumskunde und Geschichtsforschung,* 40 (1913)

Anne S. Robertson, *An Antonine Fort: Golden Hill, Duntocher,* Glasgow, 1957

Anne S. Robertson, 'Birrens 1962–63', *T. Dumfriesshire and Galloway Nat. Hist. and Antiq. Soc.*, 41 for 1962–3 (1965), pp. 135–55

H. Schönberger, 'Plan zu den Ausgrabungen am Kastell Zugmantel bis zum Jahre 1950', *Saal.-J.*, 10 (1951), pp. 55–75. Presents the layout of a large vicus of an auxiliary fort

H. Schönberger, 'Das Kastell Altenstadt', *Germania,* 35 (1957), pp. 54 ff., and *Limesforschungen,* 2, 1962, pp. 75 ff. and 92 ff.

H. Schönberger, *Führer durch das Römerkastell Saalburg,* 23rd ed., 1966

H. Schönberger, 'Das Römerkastell Quintana-Künzing'. *Bayerische Vorgeschichtsblätter,* 24 (1959), pp. 109–46

H. Schönberger, 'Römerkastell Künzing, Grabung 1962', *Saal.-J.*, 21 (1963–4), pp. 59–89

H. Schönberger, 'Über einige neu entdeckte römische Lager und Kastelltore aus Holz', *B.J.*, 164 (1964), pp. 39–44

G. Ulbert, 'Zum claudischen Kastell Oberstimm', *Germania,* 35 (1957), pp. 318 ff.

G. Ulbert, *Die römischen Donaukastelle Aislingen und Burghöfe,* in *Limesforschungen,* 1, 1959

G. Ulbert, *Der Lorenzberg bei Epfach: die frührömische Militärstation*, München, 1965

N. Walke, *Das römische Donaukastell Straubing—Sorviodurum,* in *Limesforschungen,* 3, 1966

J. Ward, *The Fort of Gellygaer,* Cardiff, 1903

R. E. M. Wheeler, 'Segontium and the Roman Occupation of Wales', *Y Cymmrodor,* 33 (1923)

R. E. M. Wheeler, 'The Roman Fort near Brecon', *Y Cymmrodor,* 37 (1926)

CAMPS, SIEGE-WORKS, WATCH AND SIGNAL TOWERS

D. Baatz, 'Eine neue Inschrift vom Odenwald-Limes', *Bayerische Vorgeschichtsblätter,* 31 (1966), pp. 85–89

G. L. Cheesman, 'Numantia', *J.R.S.,* 1 (1911), pp. 180–6

E. Fabricius, 'Some notes on Polybius's description of Roman camps', *J.R.S.,* 22 (1932), pp. 78–87

R. W. Feachem, 'Six Roman Camps near the Antonine Wall', *P. Soc. Ant. Scot.,* 89 for 1955–6 (1958), pp. 329–39

A. Fox and W. L. D. Ravenhill, 'Early Roman outposts, on the north Devon coast, Old Burrow and Martinhoe', *P. Devon Arch. Expl. Soc.,* 24 for 1966 (1967), pp. 3–39

O. Germann and H. Isler, 'Der römische Grenzwachtturm von Rheinau', *Ur-Schweiz,* 18 (1954), pp. 5–14

G. D. B. Jones, 'Ystradfellte and Arosfa Gareg: Two Roman marching camps', *Bull. Board Celtic Studies,* 21 (1965), pp. 174–8

A. Oxé, 'Polybianische und Vorpolybianische Lagermasse und Lagertypen', *B.J.,* 143–4 (1939), pp. 47–74

I. A. Richmond, 'The four camps at Cawthorn in the North Riding of Yorkshire', *Arch. J.,* 89 for 1932 (1933), pp. 17–78

I. A. Richmond and J. McIntyre, 'The Roman camps at Reycross and Crackenthorpe', *T. Cumb. and West.,* 34 (1934), pp. 50–61

I. A. Richmond and G. S. Keeney, 'The Roman works at Chew Green, Coquetdalehead', *A.A.,* 4th ser., 14 (1937), pp. 129–50

I. A. Richmond, 'A Roman arterial signalling system in the Stainmore Pass', *Aspects of Archaeology in Britain and Beyond,* 1951, pp. 293–302

I. A. Richmond, 'The Roman siege-works of Masàda, Israel', *J.R.S.,* 52 (1962), pp. 142–55

H. Schönberger, 'Ein augusteisches Lager in Rödgen bei Bad Nauheim', *Saal.-J.,* 19 (1961), pp. 37–88, and *Germania,* 45 (1967), pp. 84–95

A. Schulten, *Numantia: die Ergebnisse der Ausgrabungen 1905–12,* vols 3 and 4, 1927–29

A. Schulten, 'Masàda', *Zeitschrift des Deutschen Palästina-Vereins, 56,* (1933)

ARMS AND EQUIPMENT

I. Alfs, 'Der bewegliche Metallpanzer im römischen Heer', *Zeitschrift für historische Waffen-und Kostumkunde,* Heft 3 and 4 (1941)

N. S. Angus, G. T. Brown and H. F. Cleere, 'The Iron Nails from the Roman Legionary fortress at Inchtuthil, Perthshire', *J. Iron and Steel Inst.*, 200 (1962), pp. 956–68

D. Baatz, 'Zur Geschützbewaffnung römischer Auxiliartruppen in der frühen und mittleren Kaiserzeit', *B.J.*, 166 (1966), pp. 194–207

J. E. Bogaers, 'Twee vondsten uit de Maas in midden-Limburg', *Berichten van de rijksdienst voor het oudheidkundig bodemonderzoek,* 9 (1959), pp. 85–97

J. E. Bogaers and J. Ypey, 'Ein neuer römischer Dolch mit silbertauschierter und emailverzierter Scheide aus dem Legionslager Nijmegen', *Berichten van de rijksdienst voor het oudheidkundig bodemonderzoek,* 12 and 13 for 1962–3, pp. 87–98

G. C. Boon, 'The Roman sword from Caernarvon—Segontium', *Bull. Board Celtic Studies,* 19 (1960), pp. 85–89

J. W. Brailsford, *Hod Hill, I: Antiquities from Hod Hill in the Durden Collection,* B.M., 1962

A. L. Busch, 'Die römerzeitlichen Schuh-und Lederfunde der Kastelle Saalburg, Zugmantel und Kleiner Feldberg', *Saal.-J.*, 22 (1965), pp. 158–210

P. Couissin, *Les Armes Romaines,* 1926

C. Daremberg and E. Saglio, *Dictionnaire des Antiquités Greques et Romaines*

A. von Domaszewski, Die Fahnen in römischen Heere', *Abh. d. Arch. Epigr. Seminars der Univers. Wien*

C. Dulière, 'Beschlagbleche aus Bronze mit dem Bild der römischen Wölfin', *G. pro Vindonissa,* (1965), pp. 5–14

A. Gansser-Burckhardt, 'Das Leder und seine Verarbeitung im römischen Legionslager Vindonissa', *Veröffentlichungen der Gesellschaft pro Vindonissa,* 1, Basel, 1942

A. Gansser-Burckhardt, 'Die Lederfunde aus dem Schutthügel von Vindonissa 1951', *G. pro Vindonissa* (1952), pp. 57–65

Victorine von Gonzenbach, 'Schwertscheidenbleche von Vindonissa aus der Zeit der 13. Legion', *G. pro Vindonissa* (1966), pp. 5–35

W. Groenman-van Waateringe, 'Een Romeins lederen schildfoedraal uit Valkenburg (Z.H.)', *Helinium,* 3 (1963), pp. 253–8

W. Groenman-van Waateringe, *Romeins lederwerk uit Valkenburg (Z.H.),* Groningen, 1967

von Groller, *Der Ursprung des Pilums,* 1911

V. Hoffiller, 'Oprema rimskoga vojnika ŭ prvo doba carstva', *Vjesnik Hrvatskoga Arheološkoga Društva (Zagreb),* N.S., 11 (1910–11), pp. 145 ff.; 12 (1912), pp. 16 ff. Deals with Roman arms

J. H. Holwerda, 'Een vondst uit den Rijn bij Doorwerth', *Oudheidkundige Mededeelingen,* supplement bij nieuwe reeks, 12 (1931), pp. 1–26

J. Keim and H. Klumbach, *Der römische Schatzfund von Straubing,* München, 1951

H. Klumbach, 'Bruchstücke eines römischen Helmes von Faurndau (Kr. Göppingen)', *Fundberichte aus Schwaben,* Neue Folge 14 (1957), pp. 107–12

H. Klumbach, 'Ein Paradeschildbuckel aus Miltenberg', *Bayerische Vorgeschichtsblätter,* 25 (1960), pp. 125–32

H. Klumbach, 'Ein römischer Legionarshelm aus Mainz', *Jahrbuch des Römisch-Germanischen Zentralmuseums Mainz,* 8 (1961), pp. 96–105

H. Klumbach, 'Römische Panzerbeschläge aus Manching', Festschrift Wagner, *Schriftenreihe zur Bayerischen Landesgeschichte,* 62 (1962), pp. 187 ff.

L. Lindenschmit, *Tracht und Bewaffnung des römischen Heeres während der Kaiserzeit,* 1882

F. Matz, *Die Lauerforter Phalerae,* in *Winckelmannsprogramm der archäologischen Gesellschaft zu Berlin,* 92, 1932

J. S. Milne, *Surgical Instruments in Greek and Roman Times,* 1907

A. Radnóti, 'Ein Legionarshelm aus Burlafingen, Landkreis Neu-Ulm', Festschrift Wagner, *Schriftenreihe zur Bayerischen Landesgeschichte,* 62 (1962), pp. 157 ff.

I. A. Richmond and J. McIntyre, 'Tents of the Roman Army and leather from Birdoswald', *T. Cumb. and West.,* 34 (1934), pp. 62–90

I. A. Richmond, 'Roman lead sealings from Brough-under-Stainmore', *T. Cumb. and West.,* 36 (1936), pp. 104–25

M. I. Rostovtzeff, ed., *The Excavation at Dura-Europos,* 6th season, 1936. Includes the military saps, horse armour and the painted shields

M. I. Rostovtzeff, '*Vexillum* and Victory', *J.R.S.,* 32 (1942), pp. 92–106

E. Sander, 'Die Kleidung des römischen Soldaten', *Historia,* 12 (1963), pp. 144–66

E. Schramm, *Die antiken Geschütze der Saalburg,* 1918

S. Piggott, 'Three metal-work hoards of the Roman period from Southern Scotland', *P. Soc. Ant. Scot.,* 87 for 1952–3 (1955), pp. 1–50

J. M. C. Toynbee and R. R. Clarke, 'A Roman decorated helmet and other objects from Norfolk', *J.R.S.,* 38 (1948), pp. 20–27

G. Ulbert, 'Silbertauschierte Dolchscheiden aus Vindonissa', *G. pro Vindonissa,* (1962), pp. 5–18

C. C. Vermeule, 'A Roman silver helmet in the Toledo (Ohio) Museum of Art', *J.R.S.,* 50 (1960), pp. 8–11

J. Ypey, 'Drei römische Dolche mit tauschierten Scheiden aus niederländischen Sammlungen', *Berichten van de rijksdienst voor het oudheidkundig bodemonderzoek*, 10–11 for 1960–1, pp. 347–62

J. Ypey, 'Twee Viziermaskerhelmen uit Nijmegen', *Numaga*, 13 (1966), pp. 187–99

SERVICE CONDITIONS, PAY AND RECRUITMENT

G. H. Allen, 'The Advancement of Officers in the Roman Army', *Supplementary Papers of American School of Classical Studies in Rome*, II, 1908

E. C. Baade, 'Two Yale Papyri dealing with the Roman Army in Egypt', *Akten des VIII, internationalen Kongresses für Papyrologie, Wien 1955*, 23 ff. = *Mitteilungen aus der Papyrussammlung der österreichischen Nationalbibliothek*, NS 5 (1956)

E. Birley, 'Beförderungen und Versetzungen im römischen Heere', *Carnuntum-J.* (1957), pp. 3–20

E. Birley, 'Promotions and Transfers in the Roman Army, II, the Centurionate', *Carnuntum-J.* (1965), pp. 21–33

P. A. Brunt, 'Pay and Superannuation in the Roman Army', *Pap. Brit. School at Rome*, 18 (1950), pp. 50–71

A. von Domaszewski, 'Der Truppensold der Kaiserzeit', *Neue Heidelberger J.*, 10

A. von Domaszewski, *Die Rangordnung des römischen Heeres*, 2nd ed. by Brian Dobson, Köln, 1967

H. Forni, *Il Reclutamento delle Legioni da Augusto a Diocleziano*, Rome, 1953

K. Kraft, *Zur Rekrutierung der Alen und Kohorten an Rhein und Donau*, Berne, 1951

K. Kraft, 'Zum Bürgerrecht der Soldatenkinder', *Historia*, 9 (1960), pp. 120 ff.

G. Lopuszanski, 'La transformation du corps des officiers supérieurs dans l'armée romaine du I^{e} au IIIe siècle après J.-C.', *Ecole française de Rome*, Mélange, d' archéologie et d'histoire, 55 (1938), pp. 131–83

J. C. Mann, '*Honesta Missio* and the Brigetio Table', *Hermes*, 81 (1953), pp. 496 ff.

R. Marichal, *L'Occupation romaine de la basse Égypte, le statut des Auxilia*, 1945. Deals merely with documents relating to pay

R. Marichal, 'Le Solde des armées romaines d'Auguste à Septime Sévère d'après les P. Gen. Lat. I et 4 et le P. Berlin 6866', *Mélanges Isidore Lévy*, Brussels, 1955, pp. 399 ff.

H. M. D. Parker, 'A note on the Promotion of the centurions', *J.R.S.*, 16 (1926), pp. 45–52

H. T. Rowell, *The* Honesta Missio *from the* Numeri *of the Roman Imperial Army*, 1939

E. Sander, 'Das Recht des römischen Soldaten', *Rheinisches Museum* 101 (1958), pp. 152 ff.

R. E. Smith, *Service in the Post-Marian Roman Army*, Manchester, 1958

G. R. Watson, 'The Birdoswald hoard: the pay and the purse', *T. Cumb. and West.*, 54 (1955), pp. 61–64

G. R. Watson, 'The Pay of the Roman Army: Suetonius, Dio and the *quartum stipendium*', *Historia*, 5 (1956), pp. 332–40

G. R. Watson, 'The Pay of the Roman Army: The Republic', *Historia*, 7 (1958), pp. 113–20

G. R. Watson, 'The Pay of the Roman Army: The Auxiliary Forces', *Historia*, 8 (1959), pp. 372–8

G. R. Watson, '*Immunis Librarius*', *Britain and Rome*, 1966, pp. 45–55

C. B. Welles, 'The *Immunitas* of the Roman legionaries in Egypt', *J.R.S.*, 28 (1938), pp. 41–49

C. B. Welles, R. O. Fink and J. F. Gilliam, *The Excavations at Dura-Europos*, 1959. The Parchments and Papyri include the military documents, The *Feriale*, daily reports and rosters

MILITARY UNITS, ORGANIZATION AND TACTICS

G. Alföldy, 'Die Auxiliartruppen der Provinz Dalmatien', *Acta Arch. Acad. Scient. Hungaricae*, 14, fac. 3–4 (1962), pp. 259 ff.

G. Alföldy, 'Die Legionslegaten der römischen Rheinarmeen', *Epigraphische Studien*, 3, 1967

D. Atkinson, '*Classis Britannica*', *Historical Essays in Honour of James Tait*, Manchester, 1933

D. Atkinson, *Report on Excavations at Wroxeter, 1923–27*, 1942. Includes a chapter on the diploma found in the Forum, pp. 184–93

W, Baehr, *De centurionibus legionariis*, 1900

A. Betz, 'Zur Dislokation der Legionen in der Zeit vom Tode des Augustus bis zum Ende der Prinzipatsepoche', *Carnuntina, Vorträge beim internationalen Kongress der Altertumsforscher, Carnuntum 1955*, 1956, pp. 17–24

E. Birley, 'A note on the title *Gemina*', *J.R.S.*, 18 (1928), pp. 56–60

H. Callies, 'Die fremden Truppen im römischen Heer des Prinzipats und die sogenannten nationalen Numeri', 45 *Bericht der Römisch-Germanischen Kommission 1964* (1965), pp. 130–227

G. L. Cheesman, *The* Auxilia *of the Roman Imperial Army,* Oxford, 1914

C.I.L., xvi, 1936 and Supplement 1945 for details of *diplomata* (army discharge certificates)

L. R. Dean, *A Study of the* Cognomina *of Soldiers in the Roman Legions,* Princeton, 1916

H. Dessau, 'Offiziere und Beamten des römischen Kaiserreichs', *Hermes,* 45 (1910), pp. 1–26

Brian Dobson, 'The *Praefectus Fabrum* in the Early Principate', *Britain and Rome,* 1966, pp. 61–84

A. von Domaszewski, *Die Phalangen Alexanders und Caesars Legionen,* 1926

M. Durry, *Les cohortes prétoriennes,* 1938

J. W. Eadie, 'The development of Roman mailed cavalry', *J.R.S.,* 57 (1967), pp. 161–73

J. F. Gilliam, 'The appointment of auxiliary centurions (P. Mich. 164)', *T. American Phil. Assoc.*, 88 (1958), pp. 155–68

A. Grenier, 'Recrutement des légionnaires en Narbonnaise', *Bull. de la Société nationale des antiquaires de France,* 1956, pp. 35 ff.

F. Haverfield and R. G. Collingwood, 'The provisioning of Roman forts', *T. Cumb. and West.,* 20 (1920), pp. 127–42

O. Hirschfeld, 'Die Sicherheitspolizei im römischen Kaiserreich', *Kleine Schriften,* 1913, pp. 576 ff.

M. G. Jarrett, 'Roman officers at Maryport', *T. Cumb. and West.*, 65 (1965), pp. 115–32

F. Kiechle, 'Die "Taktik" des Flavius Arrianus', 45. *Bericht der Römisch-Germanischen Kommission 1964* (1965), pp. 87–129

D. Kienast, *Untersuchungen zu den Kriegsflotten der römischen Kaiserzeit,* Bonn, 1966

W. Kubitschek and E. Ritterling, '*Legio*', in P-W., 1924

P. Lambrechts and H. van de Weerd, 'Note sur les corps d'archers au Haut-Empire', *Laureae Aquincenses* (1938), pp. 229–42

J. C Mann, 'A note on the *Numeri*', *Hermes,* 82 (1954), pp. 501–6

A. Passerini, *Le Coorti pretorie*, Rome, 1939

F. N. Pryce, 'A new diploma for Roman Britain', (from Brigetio), *J.R.S.,* 20 (1930), pp. 16–23

I. A. Richmond, 'The Sarmatae, *Bremetennacum veteranorum* and the *regio Bremetennacensis*', *J.R.S.*, 35 (1945), pp. 15–29

I. A. Richmond, 'The Roman Army medical service', *The University of Durham Medical Gazette*, June 1952

E. Sander, 'Zur Rangordnung des römischen Heeres: Die gradus ex Caliga', *Historia*, 3 (1954), pp. 87–105

E. Sander, 'Zur Rangordnung des römischen Heeres: Die Flotten', *Historia*, 6 (1957), pp. 347–57

E. Sander, 'Zur Rangordnung des römischen Heeres: Der Duplicarius', *Historia*, 8 (1959), pp. 239–47

E. Sander, 'Der praefectus fabrum und die Legionsfabriken', *B.J.*, 162 (1962), pp. 139–61

W. Schleiermacher, 'Zu Hadrians Heeresreform in Obergermanien', *Germania*, 35 (1957), pp. 117 ff.

M. Speidel, *Die Equites singulares Augusti: Begleittruppe der römischen Kaiser des zweiten und dritten Jahrhunderts*, Bonn, 1965

Th. Wegeleben, *Die Rangordnung der römischen Centurionen*, Berlin, 1913

RELIGION

F. Cumont, *The Mysteries of Mithras*, New York, repr. 1956

A. von Domaszewski, *Die Religion des römischen Heeres*, 1895

R. O. Fink, A. S. Hoey, and W. F. Snyder, 'The *Feriale Duranum*', *Yale Classical Studies*, 7 (1940), p. 1–222

R. O. Fink, 'Hunt's *Pridianum:* British Museum Papyrus 2851', *J.R.S.*, 48 (1958), pp. 102–16

E. and J. R. Harris, *The Oriental Cults in Roman Britain*, Leiden, 1965

H. Lehner, 'Orientalische Mysterienkulte im römischen Rheinland', *B.J.*, 129 (1924), pp. 36 ff.

I. A. Richmond, 'Roman legionaries at Corbridge, their supply-base, temples and religious cults', *A.A.*, 4th ser., 21 (1943), pp. 127–224

I. A. Richmond, *Archaeology and the After-Life in Pagan and Christian Imagery*, University of Durham, Riddell Memorial Lecture, 1950

I. A. Richmond and J. P. Gillam, 'The Temple of Mithras at Carrawburgh', *A.A.*, 4th ser., 29 (1951), pp. 1–92

M. J. Vermaseren, *Corpus Inscriptionum et Monumentorum Religionis Mithriacae*, i–ii, The Hague, 1956–60

CAMPAIGNS

A. Birley, *Marcus Aurelius*, London, 1966
G. A. T. Davies, 'Trajan's first Dacian war', *J.R.S.*, 7 (1917), pp. 74–97
B. W. Henderson, *Civil War and Rebellion in the Roman Empire,* A.D. *68–70*, 1908
R. P. Longden, 'Notes on the Parthian campaigns of Trajan', *J.R.S.*, 21 (1931), pp. 1–35
F. A. Lepper, *Trajan's Parthian War*, Oxford, 1948
S. N. Miller, 'The fifth campaign of Agricola', *J.R.S.*, 38 (1948), pp. 15–19
A. Mócsy, *Die Bevölkerung von Pannonien bis zu den Markomannenkriegen*, Budapest, 1959
C. Patsch, *Der Kampf um den Donauraum unter Domitian und Trajan*, Wien und Leipzig, 1937
E. Petersen, *Trajan's Dakische Kriege*, 1899 and 1903
T. Davies Pryce and E. Birley, 'The fate of Agricola's northern conquests', *J.R.S.*, 28 (1938), pp. 141–52
I. A. Richmond, 'Gnaeus Iulius Agricola', *J.R.S.*, 34 (1944), pp. 34–45

MONUMENTS DEPICTING ASPECTS OF THE ARMY

R. Amy, et al., *L'arc d'Orange*, supplement of *Gallia*, Paris, 1962
C. Caprino, A. M. Colini, G. Gatti, M. Pallottino and P. Romanelli, *La Colonna di Marco Aurelio*, Rome, 1955
C. Cichorius, *Die Reliefs der Traianssäule*, Berlin, 1896 and 1900
H. Daicoviciu, 'Osservazioni intorno alla colonna Traiana', *Dacia*, 3 (1959), pp. 311
G. A. T. Davies, 'Topography and the Trajan Column', *J.R.S.*, 10 (1920), pp. 1–28
A. von Domaszewski, *Die Marcus-Säule auf der Piazza Colonna in Roma*, with Petersen and Calderini, 1896
E. Doruçiu, 'Some observations on the military funeral altar of Adamclissi', *Dacia*, 5 (1961)
F. B. Florescu, *Monumentul de la Adamklissi Tropeaum Traiani*, 1959
A. Giuliano, *Arco di Costantino*, Milan, 1955
H. Hofmann, 'Römische Militärgrabsteine der Donauländer', *J. Öest.*, 5 (1905)

H. S. Jones, 'The historical interpretation of Trajan's Column', *Pap. Brit. School at Rome,* 5 (1910), pp. 435 ff.

K. Lehmann-Hartleben, *Die Trajanssaäule*, Berlin und Leipzig, 1926

I. A. Richmond, 'Trajan's Army on Trajan's Column', *Pap. Brit. School at Rome,* 13 (1935), pp. 1–40

I. A. Richmond, 'Adamklissi', *Pap. Brit. School at Rome*, 35 (1967), pp. 29–39

L. Rossi, *Trajan's Column and the Dacian Wars*, London, 1971

Addendum

E. W. Marsden, *Greek and Roman Artillery, Historical Development*, 1969; *Technical Treatises*, 1971, Oxford

Robert O Fink, *Roman Military Records on Papyrus*, 1971
The American Philological Association, Monogragh 26

R. W. Davies, 'The Roman Military Medical Service', *Saal. J.*, 27 (1970), pp. 84–104

David Breeze, 'Pay Grades and Ranks below the Centurionate', *J.R.S.*, 61 (1971), pp. 130–135

Michael G. Jarret and J. C. Mann, 'Britain from Agricola to Gallienus', *B.J.*, 170 (1970), pp. 178–210

R. W. Davies, 'The Medici of the Roman Armed Forces', *Epig. Stud.*, 8 (1969) pp. 83–99; 9 (1970), pp. 1–11

David Breeze, 'The Organisation of the Legion: the First Cohort and the *Equites Legionis*', *J.R.S.*, 59 (1969), pp. 50–55

H. Schönberger, 'The Roman Frontier in Germany: An Archaeological Survey', *J.R.S.*, 59 (1969), pp. 144–197

B. Dobson and D. J. Breeze, 'The Rome Cohorts and the Legionary Centurionate', *Epig. Stud.*, 8 (1969), pp. 100–124

Robert Saxer, 'Untersuchungen zu den Vexillationen des römischen Kaiserheeres von Augustus bis Diokletian', *Epig. Stud.*, 1 (1967)

Helmut Freis, 'Die Cohortes Urbanae', *Epig. Stud.*, 2, (1967)

Geza Alföldy, 'Die Legionslegaten der römischen Rheinarmeen', *Epig. Stud.*, 3 (1967)

Geza Alföldy, 'Die hilfstruppen der römischen provinz Germania Inferior', *Epig. Stud.*, 6 (1968)

R. W. Davies, 'The Roman Military Diet', *Britannia*, 2 (1971), pp. 122–142

G. R. Watson, *The Roman Soldier*, 1969, London

INDEX